T0295214

Substrate Integrated Suspended Line Circuits and Systems

For a complete listing of titles in the
Artech House Microwave Library,
turn to the back of this book.

Substrate Integrated Suspended Line Circuits and Systems

Kaixue Ma
Yongqiang Wang

ARTECH
HOUSE
BOSTON | LONDON
artechhouse.com

Library of Congress Cataloging-in-Publication Data
A catalog record for this book is available from the U.S. Library of Congress

British Library Cataloguing in Publication Data
A catalog record for this book is available from the British Library.

ISBN 13: 978-1-68569-029-8

Cover design by Andy Meaden

685 Canton St.
Norwood, MA

10 9 8 7 6 5 4 3 2 1

Contents

Preface

In the contemporary information society, radio technology has emerged as a key technology. It has significant and widespread uses in a variety of military and commercial domains, including radio astronomy, radar, and satellite-based wireless communication. The hardware support in the radio business is provided by the radio frequency (RF)/microwave/millimeter wave front-end, which is an essential part of radio technology. Also, with the advent of the 5G/6G wireless communication era, low-cost, high-performance, and miniaturized circuits have evolved into the fundamental requirements of new wireless communication technology. As a result, each module component needs to be both highly effective and compact. RF/microwave/millimeter wave circuits and systems place increased demands on cost, dimensions, integration, and other performance as the electromagnetic environment grows increasingly complicated.

The most fundamental part of the RF/microwave/millimeter wave front-end circuit is the transmission line. As the foundation of the RF/microwave/millimeter circuit, the structure and transmission characteristics of the transmission line have a direct or indirect impact on the performance of the entire front-end circuit. Metal waveguide, coaxial line, and other nonplanar transmission-line forms, as well as microstrip, stripline, coplanar waveguide, slotline, and other planar transmission line forms, are the most generally used transmission line types. The electromagnetic field of typical planar circuits is mostly spread in the medium as a result of the structure's restrictions, which leads to significant transmission loss and severe dispersion in the millimeter-wave frequency. Metal waveguides, coaxial cables, and suspension lines are examples of nonplanar transmission lines. They have the drawbacks of high volume and low integration.

A new style of transmission line with good qualities like low loss, low dispersion, low cost, and self-packaging is known as a substrate integrated suspended line (SISL). SISL is a type of quasi-planar transmission line with a multilayer printed circuit board (PCB) structure that has overcome the bulky structure and integration difficulties of traditional suspended lines. Many passive and active devices, including couplers, filters, multiplexers, baluns, power dividers, low-noise amplifiers, power amplifiers, switches, and VCOs, have been designed on the SISL platform recently. In conclusion, SISL is an excellent platform for self-packaging, high-performance, cost-effective circuits and systems in the RF, microwave, and millimeter-wave frequency bands.

This book introduces the principles and design methods of various passive and active circuits implemented on the SISL platform. This book addresses the technical challenges, limitations, and considerations faced by designers of RF/microwave/millimeter wave circuits and systems. This book can be used as a textbook for undergraduate and graduate students in integrated circuits, microelectronics, RF and microwave circuits, communication engineering, and other majors, and can also be used as a technical reference book for RF circuit design engineers.

CHAPTER 1

Introduction

1.1 Development of the Transmission Line

The possibility of electromagnetic waves propagating in a closed-air tube was considered by Heaviside as early as 1893 [1]. In 1897, Lord Rayleigh proved mathematically that wave propagation in a waveguide is possible [2]. Regardless of whether the cross-section of the waveguide is circular or rectangular, he also pointed out that there could be infinitely many transverse electric (TE) and transverse magnetic (TM) modes and that dielectric frequencies existed. After this, waveguides were largely forgotten until they were discovered again by researchers in 1936 [3], when George C. Southworth of American Telephone and Telegraph Company (AT&T) presented a paper on waveguides in 1937 [4], after conducting preliminary experiments in 1932. At the same conference, W. L. Barrow of the Massachusetts Institute of Technology (MIT) presented an article on circular waveguides and provided an experimental verification of wave propagation. The practical application of waveguides officially marked the microwave as a technical discipline. Since then, the first generation of three-dimensional (3D) microwave circuits represented by metal waveguides has been widely studied and applied. Due to the excellent performance of metal waveguide components with robust mechanical structure, high-quality factor, low insertion loss, good temperature characteristics, and high-power capacity, they are still widely used in special fields such as satellites, radio telescopes, and high-power microwave weapons [5–8].

Although metal waveguide-built components have many advantages, they also have large size, high production costs, high weight, are not convenient for system integration, and have other disadvantages. With the increasing demand for microwave systems and communications equipment miniaturization, microwave circuits from the bulky 3D structure to planar form, so a variety of microwave planar transmission lines came into being.

In the 1950s, Barrett proposed stripline [9], which was the first generation of printed-formed transmission lines in microwave technology. The stripline structure is shown in Figure 1.1(a). It is a transverse electromagnetic mode (TEM) transmission line consisting of two parallel grounding plates, a metal strip located in the middle of the grounding plates, and a dielectric between them. In the printed circuit board (PCB) process, the upper and lower ground plates are usually connected using vias. The via array keeps the potential of the upper and lower ground plates the same, avoiding the transmission of higher-order modes and suppressing radiation [10]. It has the advantages of wide operating bandwidth, high Q value, small size, good shielding, and is lightweight [11–14]. However, it is not convenient for external solid-state microwave devices, so it is less suitable for application in active circuits [15].

As the demand for transmission lines in various scenarios increased, microstrip lines, which are more widely used than striplines, emerged in 1952, developed by International Test Technologies Laboratories (ITT Labs) as a competitor to striplines. The first microstrip line used a thicker substrate that accentuated the characteristics of the non-TEM wave modes and the frequency dispersion on the line, a feature that made it less desirable compared to striplines, a situation that did not change until the 1960s when very thin substrates were applied, which reduced the frequency dependence of the transmission line, and now microstrip lines are often the best medium for microwave integrated circuits. The basic structure of a microstrip line is shown in Figure 1.1(b). Although microstrip lines appeared later than striplines, microstrip lines are compact, small, lightweight, and low-cost, and, more importantly, it is easy to connect microwave solid-state devices in series with the advantages of high integration, which makes it widely designed as coupler [16], divider/combiner [17], filter [18], feeding network [19], and antenna [20], and it is widely used in microwave integrated circuits and monolithic microwave integrated circuits.

In the 1960s, coplanar transmission lines entered the picture. Unlike the field structure of strip lines and microstrip lines, their electric field direction is parallel to the dielectric surface. In 1968, Cohn introduced the concept of a slotted line [21]. As shown in Figure 1.1(c), it is a balanced transmission line consisting of a medium and two large metal guide strips located on the same side of the medium. According to the characteristics of the slotted line, it is suitable for connecting the lumped elements in parallel between the two conduction bands, and due to its small edge coupling capacitance, it has a large impedance and is more suitable for applications with high characteristic impedance. The slotted line is widely used in power dividers [22], filters [23], and antennas [24]. However, it also has disadvantages such as high radiation, it cannot be connected directly to 50W systems, and it is unsuitable for connecting components in series. It is also unsuitable for application in monolithic microwave integrated circuits with high circuit area requirements due to the wide ground plane grounding electrode plates on both sides of the slot slit.

To meet the application needs of microwave integrated circuits and monolithic microwave integrated circuits, coplanar waveguide (CPW) has emerged. C. P. Wen proposed CPW in 1969 [25], and its characteristic impedance, phase velocity, and attenuation characteristics were calculated and measured. The coplanar waveguide, as shown in Figure 1.1(d), is composed of a central guide strip located on one side of the substrate and two conductor grounding plates on the same side. Compared with conventional microstrip lines, its main guide strip and grounding substrate are located on the same side of the dielectric plate, which can easily realize the series and parallel connection of the elements, with the advantages of convenient grounding, smaller dispersion effect, small radiation loss, and small crosstalk between adjacent coplanar waveguides [26, 27]. In addition, the characteristic impedance of coplanar waveguides is determined by the central signal trace and the slot widths on both sides, so their dimensions can be scaled when the characteristic impedance is determined. Therefore, circuits applying coplanar waveguide technology have the advantages of high integration density and high operating frequency, which makes coplanar waveguide very suitable for microwave integrated circuits and monolithic microwave integrated circuits [28, 29].

The suspended line, including the suspended microstrip line and the suspended stripline [30], is as shown in Figure 1.1(e). As early as the 1960s, scholars conducted in-depth research and analysis on suspended lines. Itoh et al. conducted a theoretical analysis of the transmission characteristics such as the characteristic impedance and wavelength of suspended-line multiconductor structures using the spectral domain method [31]. Simons investigated the effect of metal shielded enclosures on their dispersion, impedance, and relative permittivity for double dielectric layer suspended slotline structures, suspended multilayer slotline structures, and suspended broadside band symmetrically coupled slotline structures [32, 33]. The dispersion properties of suspended lines have been investigated by Chan et al. using methods such as full wave analysis [34, 35]. Yamashita et al. analyzed the effect of slotting the strip body at the edge of a suspended line on transmission characteristics such as characteristic impedance [36]. Wang et al. analyzed the dispersion characteristics of coupled suspended striplines and obtained closed analytical equations [37]. Cui et al. proposed a broadside coupled suspended microstrip line with smaller losses than conventional microstrip lines and suspended striplines operating in the even-mode state [38].

Among the nonplanar transmission lines, waveguide suspended line (including suspended stripline, suspended microstrip line) circuits are significantly superior transmission line systems, especially for applications in microwave devices such as various filters. Metal losses are significantly reduced compared to other planar transmission lines due to their larger cross-section and smaller current density. Thinner substrates are often used to keep the equivalent dielectric constant as low as possible so that the main electric field is distributed in the air cavity and the dispersion of the transmission line is reduced. The use of metal cavity packaging also makes the waveguide suspended line virtually radiation-free. However, the waveguide suspended line is typically smaller in size than the waveguide while combining some of the advantages of the conventional waveguide and is easily compatible with microwave transmission lines commonly used in microwave/millimeter-wave integrated circuits such as microstrip and CPW. The excellent properties of the waveguide suspended microstrip line and suspended stripline are unquestionable. However, there are inherent implementation drawbacks of this transmission line. It is similar to conventional waveguides in that it requires the machining of mechanical cassettes to form the necessary two or more air cavities while meeting the necessary mechanical support, impedance requirements, and electromagnetic shielding. The more complex requirement for waveguide suspended microstrip line and suspended striplines is that the necessary signal line needs to be mechanically machined or realized through circuit boards and needs to be mechanically assembled with metal conductor cavities. This assembly often requires high precision and many ancillary mechanical components, such as locating holes, pins, bolts, and nuts to complete the assembly, thus requiring additional machining and assembly work. Another disadvantage of the waveguide suspended line circuit is that, due to the presence of the electromagnetic field in the air cavity, the effective dielectric constant tends to be close to 1, and the size of the transmission line is usually large. As a result, waveguide suspended line circuits are large, relatively bulky, and expensive to process. Moreover, they require mechanical assembly later, making it difficult to form a large-scale production. These shortcomings seriously constrained the further development and application of waveguide suspended line, making excellent transmission line circuits, high-performance circuits, and systems based on it mainly limited to the military high-cost systems.

In the face of the development needs of modern communication electronic systems, both traditional metal waveguides and microstrip circuits have shown their limitations, namely, high Q value, and low loss and planarization of low cost are always challenging to consider. Especially in the Ku- to Ka-band, microstrip line loss rises sharply and crosstalk is serious, while metal waveguide has the defect of large size and inconvenient integration. In this situation, a combination of the metal waveguide and microstrip line advantages, especially for microwave millimeter-wave circuit waveguide structure, substrate integrated waveguide (SIW) came into being [39, 40].

In recent years, scholars have proposed the air-filled SIW structure and empty SIW [41, 42], replacing the consuming medium inside the SIW with air to reduce the dielectric loss further. SIW uses rows of metalized vias connecting the top and bottom of the PCB substrate, as shown in Figure 1.1(f), which are intended to implement the functions of conventional metallic waveguides on substrates, which are a type of rectangular metallic waveguide transmission line with the advantages of low cost, small size, light weight, easy processing, low radiation loss, and high Q value and are widely used in millimeter-wave systems [43, 44].

In 2006, a shielded, nondispersive substrate integrated coaxial line (SICL) was proposed by Gatti et al. [45]. As shown in Figure 1.1(g), SICL is longitudinally composed of two layers of the ground metal plane and the middle of the metal transmission line, and the transmission line has a row of metal holes on each side of the electrical wall. SICL allows TEM mode propagation and two rows of metal through-holes avoid unnecessary parallel plate mode propagation, thereby reducing leakage and crosstalk. Therefore, SICL has the advantages of single TEM mode transmission over a wideband, electromagnetic interference shielding, and being spurious-free. SICL meets the machining requirements of the multilayer PCB process, so it can be easily combined with other flat circuits that are also integrated on the PCB. In 2015, Belenguer et al. proposed empty SICL (ESICL) [46], which further reduced the dielectric loss based on traditional SICL, which can be applied to broadband communication systems.

Ridged waveguide (RGW) is a new millimeter-wave waveguide structure, which has the advantages of low loss. This technique consists of parallel flat waveguides and electrical tape gap (EBG) units, which can control waveguides to propagate in the desired path as much as possible [47], thus reducing loss. The earliest RGW used Computer Numerical Control (CNC) machine tools for processing, and the high processing difficulty resulted in an increase of cost. In 2014, Razavi et al. developed a thinner printed RGW (PRGW) based on PCB technology [48]. In the PRGW structure, the metal transmission line is surrounded by a mushroom-shaped structure located on the same substrate. The mushroom-shaped structure is the top circular patch connected by metalized via holes. The back of this substrate is covered by a whole ground plane.

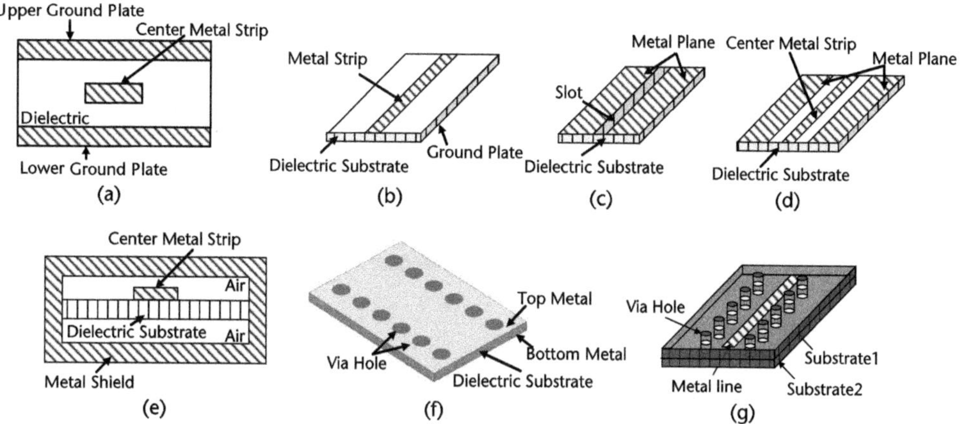

Figure 1.1 Structures of: (a) stripline, (b) microstrip line, (c) slotline, (d) coplanar waveguide, (e) traditional suspended line, (f) substrate integrated waveguide, and (g) substrate integrated coaxial line.

1.2 Substrate Integrated Suspended Line Platform

Ma proposed the substrate integrated suspended line platform (SISL) in 2007 [49]. Its basic structure has the following characteristics: (1) it is based on the multilayer board fabrication process, (2) it uses at least 3 layers of boards, (3) at least one substrate layer is locally hollowed out or grooved to form an air cavity or embedded cavity structure, (4) after the multilayer plates are compacted together in sequence, the air cavity structure required by the suspended circuit is formed, (5) the multilayer substrate board is provided with metalized holes through at least one layer of the board, which can realize the connection between the upper and lower metal layers of the single layer board and the vertical interconnection between different layers, and (6) the multilayer metalized holes around the air cavity form a better electromagnetic shielding environment and reduce the radiation loss of the internal circuit. Figure 1.2(a) shows the typical five-layer circuit structure of the SISL circuit, which comprises five layers of substrates. Substrate 2 and Substrate 4 are partially hollowed out, and the hollowed part forms the air cavity structure after the five-layer plate is compacted. The core circuit is mainly placed on Substrate 3.

Compared with the waveguide suspended line, SISL avoids the complicated process of fabricating and assembling the metal casing, presenting the advantages of small size, light weight, and low cost. Because SISL is based entirely on a multilayer substrate, it is easier to integrate with planar structures such as microstrip lines and striplines, giving it the advantage of high integration. Compared to the air-filled SIW, SISL is also based on the multilayer substrate structure and has the characteristics of self-packaging with almost identical advantages: low loss, compact, low cost, self-packaging, and high integration. However, the air-filled SIW uses the top and bottom substrate and via holes through the substrate to transmit signals through the construction of cavities of different shapes, such as the working mode of metal waveguides. In comparison, SISL can transmit signals by constructing cavities or designing circuits on the suspended substrate. The circuit on SISL can work like the signal transmission in planar transmission lines such as microstrip

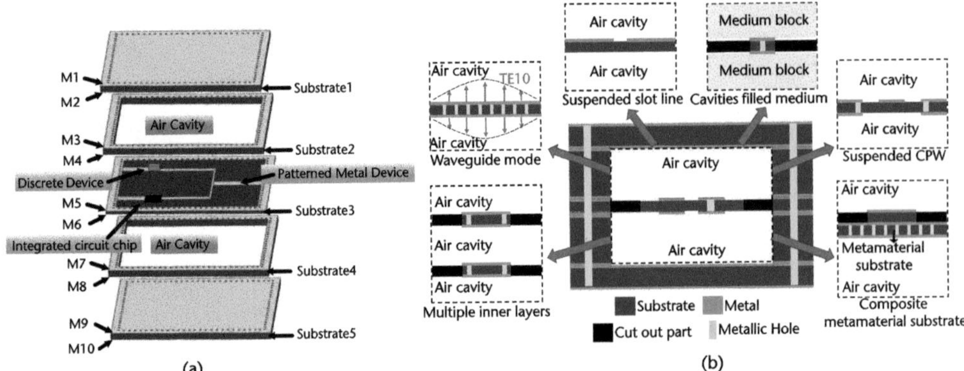

Figure 1.2 Multilayer SISL of: (a) the 3D view, and (b) the basic functions.

lines. For passive circuit design, SISL has comparable advantages to air-filled SIW. Active circuit designing requires soldered discrete components. SISL's mounted substrate is more accessible to insert and solder than air-filled SIW, thus facilitating active circuit design. In SISL, multilayer substrates and cavities are fully used for the medium embedding, metamaterial loading, design of suspended slotline and CPW, multiple inner layers coupling, and forming waveguide modes. Figure 1.2(b) shows the basic structure of these functions.

The remarkable structural and performance advantages of SISL have attracted scholars to carry out much research on passive and active circuits, including transitions [50, 51], couplers [52, 53], antennas [54, 55], filters [56, 57], power dividers [58, 59], oscillators [60, 61], low-noise amplifiers [62, 63], and power amplifiers (PAs) [64, 65]. Many SISL-based circuits are also included in [66].

This book can be used as a textbook for undergraduate and graduate students in integrated circuits, microelectronics, RF and microwave circuits, communication engineering, and other majors. It can also be used as a technical reference book for integrated circuit design engineers.

Acknowledgments

We would like to thank Jianteng Yang, Yuanjun Chai, and Haozhen Huang for their valuable support in this chapter.

References

[1] Heaviside, O., *Electromagnetic Theory*, Volume I, unabridged edition, Providence, RI: American Mathematical Society, 1971 (original, 1893, London: "The Electrician" Printing and Publishing Company).

[2] Lord Rayleigh, "On the Passage of Electric Waves Through Tubes," *Philosophical Magazine*, Vol. 43, 1897, pp. 125–132.

[3] Packard, K. S., "The Origin of Waveguides: A Case of Multiple Rediscovery," *IEEE Transactions on Microwave Theory and Techniques*, Vol. 32, No. 9, 1984, pp. 961–969.

[4] Southworth, G. C., "Some Fundamental Experiments with Wave Guides," *Proceedings of the Institute of Radio Engineers*, Vol. 25, No. 7, 1937, pp. 807–822.

[5] Riblet, H. J., and T. S. Saad, "A New Type of Waveguide Directional Coupler," *Proceedings of the IRE*, Vol. 36, No. 1, 1948, pp. 61–64.

[6] Mallach, M., and T. Musch, "Broadband Coaxial Line to Rectangular Waveguide Transition for a Microwave Tomography Sensor," *2017 11th European Conference on Antennas and Propagation (EUCAP)*, Paris, France, 2017, pp. 465–468.

[7] Ho, C. -H., and K. Chang, "A New Type of Annular Ring Waveguide Cavity for Resonator and Filter Applications," *1992 IEEE MTT-S Microwave Symposium Digest*, Albuquerque, NM, Vol. 3, 1992, pp. 1323–1326.

[8] Turkmen, C., and M. Secmen, "Circularly Polarized Hemispherical Antennas for Telemetry and Telecommand Applications in Satellite Communication," *2016 10th European Conference on Antennas and Propagation (EuCAP)*, Davos, Switzerland, 2016, pp. 1–5.

[9] Barrett, R. M., "Microwave Printed Circuits—The Early Years," *IEEE Transactions on Microwave Theory and Techniques*, Vol. 32, No. 9, 1984, pp. 983–990.

[10] Wadell, B. C., "Transmission Line Design Handbook," *Journal of Microwaves*, 1991.

[11] Fromm, W. E., "Characteristics and Some Applications of Stripline Components," *IRE Transactions on Microwave Theory and Techniques*, Vol. 3, No. 2, 1955, pp. 13–20.

[12] Rutz, E. M., "A Stripline Frequency Translator," *IRE Transactions on Microwave Theory and Techniques*, Vol. 9, No. 2, 1961, pp. 158–161.

[13] Davies, J. B., and P. Cohen, "Theoretical Design of Symmetrical Junction Stripline Circulators," *IEEE Transactions on Microwave Theory and Techniques*, Vol. 11, No. 6, 1963, pp. 506–512.

[14] Glance, B., and R. Trambarulo, "A Waveguide to Suspended Stripline Transition (Letters)," *IEEE Transactions on Microwave Theory and Techniques*, Vol. 21, No. 2, 1973, pp. 117–118.

[15] Vincent, B. T., "Ceramic Microstrip For Microwave Hybrid Integrated Circuitry," *G-MTT International Symposium Digest*, Palo Alto, CA, 1966, pp. 128–134.

[16] Arditi, M., "Characteristics and Applications of Microstrip for Microwave Wiring," *IEEE Transactions on Microwave Theory and Techniques*, Vol. 3, No. 2, 1955, pp. 31–56.

[17] Tang, C. -W., and J. -T. Chen, "A Design of 3-dB Wideband Microstrip Power Divider with an Ultra-Wide Isolated Frequency Band," *IEEE Transactions on Microwave Theory and Techniques*, Vol. 64, No. 6, 2016, pp. 1806–1811.

[18] Ma, K., et al., "A Compact Size Coupling Controllable Filter with Separate Electric and Magnetic Coupling Paths," *IEEE Transactions on Microwave Theory and Techniques*, Vol. 54, No. 3, 2006, pp. 1113–1119.

[19] Levine, E., et al., "A Study of Microstrip Array Antennas with the Feed Network," *IEEE Transactions on Antennas and Propagation*, Vol. 37, No. 4, 1989, pp. 426–434.

[20] Ramachandran, A., et al., "A Compact Triband Quad-Element MIMO Antenna Using SRR Ring for High Isolation," *IEEE Antennas and Wireless Propagation Letters*, Vol. 16, 2017, pp. 1409–1412.

[21] Cohn, S. B., "Slot Line—An Alternative Transmission Medium for Integrated Circuits," *1968 G-MTT International Microwave Symposium*, Detroit, MI, 1968, pp. 104–109.

[22] Ahmed, U. T., and A. M. Abbosh, "Wideband Out-of-Phase Power Divider Using Coupled Lines And Microstrip to Slotline Transitions," *2015 Asia-Pacific Microwave Conference (APMC)*, Nanjing, China, 2015, pp. 1–3.

[23] Chen, D., et al., "Differential Bandpass Filter on Dual-Mode Ring Resonator with Slotline Feeding Scheme," *Electronics Letters*, Vol. 51, No. 19, 2015, pp. 1512–1514.

[24] Lian, R., et al., "Design of a Low-Profile Dual-Polarized Stepped Slot Antenna Array for Base Station," *IEEE Antennas and Wireless Propagation Letters*, Vol. 15, 2016, pp. 362–365.

[25] Wen, C. P., "Coplanar Waveguide: A Surface Strip Transmission Line Suitable for Nonreciprocal Gyromagnetic Device Applications," *IEEE Transactions on Microwave Theory and Techniques*, Vol. 17, No. 12, 1969, pp. 1087–1090.

[26] Hasnain, G., A. Dienes, and J. R. Whinnery, "Dispersion of Picosecond Pulses in Coplanar Transmission Lines," *IEEE Transactions on Microwave Theory and Techniques*, Vol. 34, No. 6, 1986, pp. 738–741.

[27] Browne, J., "Coplanar Waveguide Supports Integrated Multiplier Systems," *Microwaves and RF*, Vol. 28, No. 3, 1989, pp. 137–138.

[28] Eblabla, A., et al., "Novel Shielded Coplanar Waveguides on GaN-on-Low Resistivity Si Substrates for MMIC Applications," *IEEE Microwave and Wireless Components Letters*, Vol. 25, No. 7, 2015, pp. 427–429.

[29] Zhang, Z., X. Liao, and K. Wang, "A Directional Inline-Type Millimeter-Wave MEMS Power Sensor for GaAs MMIC Applications," *Journal of Microelectromechanical Systems*, Vol. 24, No. 2, 2015, pp. 253–255.

[30] Yamashita, E., et al., "Effects of Side-Wall Grooves on Transmission Characteristics of Suspended Striplines," *IEEE Transactions on Microwave Theory and Techniques*, Vol. 33, No. 12, 1985, pp. 1323–1328.

[31] Itoh, T., "Generalized Spectral Domain Method for Multiconductor Printed Lines and Its Application to Turnable Suspended Microstrips," *IEEE Transactions on Microwave Theory and Techniques*, Vol. 26, No. 12, 1978, pp. 983–987.

[32] Simons, R., "Suspended Slot Line Using Double Layer Dielectric," *IEEE Transactions on Microwave Theory and Techniques*, Vol. 29, No. 10, 1981, pp. 1102–1107.

[33] Simons, R. N., "Suspended Broadside-Coupled Slot Line with Overlay," *IEEE Transactions on Microwave Theory and Techniques*, Vol. 30, No. 1, 1982, pp. 76–81.

[34] Chan, C. H., and Kouki, A. B., "Propagation Characteristic for a Suspended Substrate Microstrip Line with Pedestal," *Electronics Letters,* Vol. 24, No. 21, 1988, pp. 1342–1343.

[35] Polichronakis, I. P., and Kouris, S. S., "Higher order modes in suspended substrate microstrip lines," *IEEE Transactions on Microwave Theory and Techniques*, Vol. 41, No. 8, 1993, pp. 1449–1454.

[36] Yamashita, E., B. -Y. Wang, and K. Atsuki, "Effects of Side-Wall Grooves on Transmission Characteristics of Suspended Strip Lines," *IEEE MTT-S International Microwave Symposium Digest*, St. Louis, MO, 1985, pp. 145–148.

[37] Wang, Y. Y., G. L. Wang, and Y. H. Shu, "Analysis and Synthesis Equations for Edge-Coupled Suspended Substrate Microstrip Line," *IEEE MTT-S International Microwave Symposium Digest*, Long Beach, CA, Vol. 3, 1989, pp. 1123–1126.

[38] Cui, L., W. Wu, and D. G. Fang, "Low Loss Slow-Wave Unit Based on Even-Mode Bilateral Broadside-Coupled Suspended Microstrip Line," *Proceedings of 2011 Cross Strait Quad-Regional Radio Science and Wireless Technology Conference*, Harbin, 2011, pp. 463–467.

[39] Yoneyama, T., and S. Nishida, "Nonradiative Dielectric Waveguide for Millimeter-Wave Integrated Circuits," *IEEE Transactions on Microwave Theory and Techniques*, Vol. 29, No. 11, 1981, pp. 1188–1192.

[40] Deslandes, D., and K. Wu, "Integrated Microstrip and Rectangular Waveguide in Planar Form," *IEEE Microwave and Wireless Components Letters*, Vol. 11, No. 2, 2001, pp. 68–70.

[41] Parment, F., et al., "Air-Filled Substrate Integrated Waveguide for Low-Loss and High Power-Handling Millimeter-Wave Substrate Integrated Circuits," *IEEE Transactions on Microwave Theory and Techniques*, Vol. 63, No. 4, 2015, pp. 1228–1238.

[42] Belenguer, A., H. Esteban, and V. E. Boria, "Novel Empty Substrate Integrated Waveguide for High-Performance Microwave Integrated Circuits," *IEEE Transactions on Microwave Theory and Techniques*, Vol. 62, No. 4, 2014, pp. 832–839.

[43] Iqbal, A., et al., "Miniaturization Trends in Substrate Integrated Waveguide (SIW) Filters: A Review," *IEEE Access*, Vol. 8, 2020, pp. 223287–223305.

[44] Sun, X., et al., "Compact Substrate Integrated Waveguide Filtering Antennas: A Review," *IEEE Access*, Vol. 10, 2022, pp. 91906–91922.

[45] Gatti, F., et al., "A Novel Substrate Integrated Coaxial Line (SICL) for Wide-Band Applications," *2006 European Microwave Conference*, Manchester, U.K., 2006, pp. 1614–1617.

[46] Belenguer, A., et al., "High-Performance Coplanar Waveguide to Empty Substrate Integrated Coaxial Line Transition," *IEEE Transactions on Microwave Theory and Techniques*, Vol. 63, No. 12, 2015, pp. 4027–4034.

[47] Kildal, P. -S., et al., "Design and Experimental Verification of Ridge Gap Waveguide in Bed of Nails for Parallel-Plate Mode Suppression," *IET Microwaves, Antennas & Propagation*, Vol. 5, No. 3, 2011, pp. 262–270.

[48] Razavi, S. A., et al., "2x2-Slot Element for 60-GHz Planar Array Antenna Realized on Two Doubled-Sided PCBs Using SIW Cavity and EBG-Type Soft Surface Fed by Microstrip-Ridge Gap Waveguide," *IEEE Transactions on Antennas and Propagation*, Vol. 62, No. 9, 2014, pp. 4564–4573.

[49] Ma, K., and K. T. Chan, "Quasi-Planar Circuits with Air Cavities," *WO*, WO2007149046 A1. 2007.

[50] Li, L., et al., "A Novel Transition from Substrate Integrated Suspended Line to Conductor Backed CPW," *IEEE Microwave and Wireless Components Letters*, Vol. 26, No. 6, 2016, pp. 389–391.

[51] Wu, X., et al., "An FR4-Based DC to 60GHz Substrate Integrated Suspended Line to Suspended Coplanar Waveguide Transition," *Microwave and Optical Technology Letters*, Vol. 64, No. 2, 2022, pp. 203–207.

[52] Wang, Y., K. Ma, and S. Mou, "A Low Loss and Self-Packaged Patch Coupler Based on SISL Platform," *2017 IEEE MTT-S International Microwave Symposium (IMS)*, Honolulu, HI, 2017, pp. 192–195.

[53] Wang, Y., K. Ma, and S. Mou, "A Transformer-Based 3-dB Differential Coupler," *IEEE Transactions on Circuits and Systems I: Regular Papers*, Vol. 65, No. 7, 2018, pp. 2151–2160.

[54] He, Y., et al., "Dual-Band Monopole Antenna Using Substrate-Integrated Suspended Line Technology for WLAN Application," *IEEE Antennas and Wireless Propagation Letters*, Vol. 16, 2017, pp. 2776–2779.

[55] Newton, M. E., et al., "A Monopole Antenna Based on SISL with Dual Bands for WLAN Operations," *Microwave and Optical Technology Letters*, Vol. 60, No. 11, 2018, pp. 2784–2787.

[56] Ma, Z., et al., "Quasi-Lumped-Element Filter Based on Substrate-Integrated Suspended Line Technology," *IEEE Transactions on Microwave Theory and Techniques*, Vol. 65, No. 12, 2017, pp. 5154–5161.

[57] Wei, X., K. Ma, and Y. Guo, "Design of Miniaturized 5G SISL BPFs with Wide Stopband Using Differential Drive Inductor Resonators," *IEEE Transactions on Microwave Theory and Techniques*, Vol. 70, No. 6, 2022, pp. 3115–3124.

[58] Xiao, J. -K., X. -Y. Yang, and X.-F. Li, "A 3.9GHz/63.6% FBW Multi-Mode Filtering Power Divider Using Self-Packaged SISL," *IEEE Transactions on Circuits and Systems II: Express Briefs*, Vol. 68, No. 6, 2021, pp. 1842–1846.

[59] Qin, Z., et al., "An FR4-Based Self-Packaged Full Ka-Band Low-Loss 1:4 Power Divider Using SISL to Air-Filled SIW T-Junction," *IEEE Transactions on Components, Packaging and Manufacturing Technology*, Vol. 12, No. 3, 2022, pp. 587–590.

[60] Li, M., et al., "Design and Fabrication of Low Phase Noise Oscillator Using Q Enhancement of the SISL Cavity Resonator," *IEEE Transactions on Microwave Theory and Techniques*, Vol. 67, No. 10, 2019, pp. 4260–4268.

[61] Han, J., K. Ma, and N. Yan, "A Low Phase Noise Oscillator Employing Weakly Coupled Cavities Using SISL Technology," *IEEE Transactions on Circuits and Systems I: Regular Papers*, Vol. 70, No. 4, 2023, pp. 1503–1516.

[62] Zhang, K., et al., "A 0.7/1.1-dB Ultra-Low Noise Dual-Band LNA Based on SISL Platform," *IEEE Transactions on Microwave Theory and Techniques*, Vol. 66, No. 10, 2018, pp. 4576–4584.

[63] Zhang, J., et al., "An FR4-Based K-Band 1.0-dB Noise Figure LNA Using SISL Technology," *IEEE Microwave and Wireless Components Letters*, Vol. 32, No. 2, 2022, pp. 129–132.

[64] Feng, T., et al., "Band-Pass-Filtering Power Amplifier with Compact Size and Wideband Harmonic Suppression," *IEEE Transactions on Microwave Theory and Techniques*, Vol. 70, No. 2, 2022, pp. 1254–1268.

[65] Zhang, L., et al., "A Dual-Band and Dual-State Doherty Power Amplifier Using Metal-Integrated and Substrate-Integrated Suspended Line Technology," *IEEE Transactions on Microwave Theory and Techniques*, Vol. 70, No. 1, 2022, pp. 402–415.

[66] Ma, K., N. Yan, and Y. Wang, "Recent Progress in SISL Circuits and Systems: Review of Passive and Active Circuits Demonstrating SISL's Low Loss and Self-Packaging and Showcasing the Merits of Metallic, Shielded, Suspended Lines," *IEEE Microwave Magazine*, Vol. 22, No. 4, 2021, pp. 49–71.

CHAPTER 2

SISL Basics

This chapter describes the development history of the SISL platform, different types of SISL transmission lines, transitions, interfaces, capacitors, and inductors on the SISL platform.

2.1 The Development of SISL

2.1.1 Introduction

As modern communication and radar technologies rapidly advance, the miniaturization, planarization, and modularization of microwave and millimeter-wave systems are becoming increasingly crucial in the development of wireless systems and applied electromagnetics. The performance of the entire circuit and system can be directly or indirectly affected by the characteristics of transmission lines, such as loss, size, integration, cost, and weight.

SISL is a type of quasi-planar transmission line with a multilayer PCB structure that has overcome the bulky structure and integration difficulties of traditional suspended lines with metal boxes. SISL retains all the performance advantages of traditional suspended lines while enabling the packaging of passive and active circuits and systems. With its multilayer air cavity structure, the SISL platform ensures excellent integration flexibility and low dielectric loss as the electric field is primarily distributed in the air cavity. Therefore, SISL is an excellent platform for self-packaging, high-performance, cost-effective circuits and systems in the microwave and millimeter-wave frequency bands.

2.1.2 Historical Development

The SISL platform can be traced back to 2007 when Ma and Chan proposed a patent [1]. Since 2015, various circuits including passive circuits, active circuits, and antennas based on SISL have been reported.

A typical SISL structure includes a multilayer PCB, some of which need to be cut out to form air cavities that facilitate suspended line structures. Furthermore, these air cavities allow the design and integration of specific microwave devices, lumped elements, and chips. Figure 2.1 depicts a classic five-layer SISL structure consisting of five PCB layers and ten metal layers [2]. Substrate 2 and Substrate 4 are hollowed out to form cavities, while Substrate 1 and Substrate 5 serve as cover plates, eliminating the need for an additional packaging structure. SISL circuits can be fabricated by multilayer PCB or other multilayer processes.

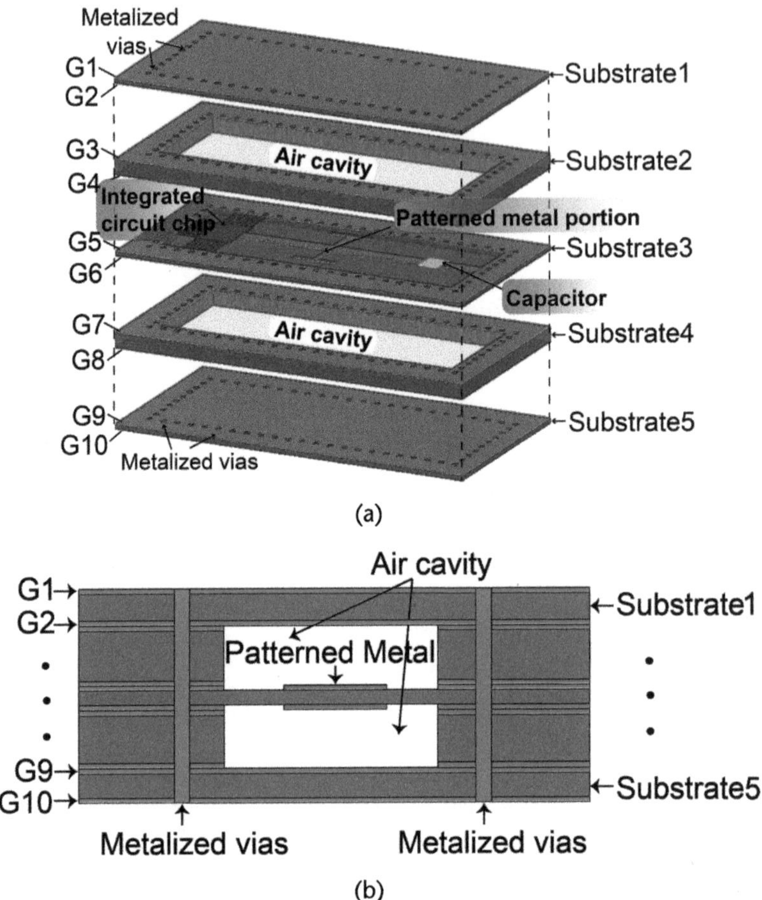

Figure 2.1 The typical structure of an SISL [2]: (a) 3D view and (b) side view.

As a result of the good performance of the SISL platform, researchers have proposed many passive components and circuits including transitions, filters, multiplexers, couplers, power dividers, Magic Tees, baluns, and many others. Moreover, a range of active circuits such as oscillators, voltage-controlled oscillators, low noise amplifiers (LNAs), mixers, and power amplifiers, as well as various SISL antennas and systems have also been developed.

2.2 Different Types of SISL Transmission Lines

In recent years, various transmission lines based on SISL platform have been developed, including single-ended transmission lines and differential transmission lines. Designers often choose between single-ended and differential circuit configurations for certain applications. For instance, opting for differential transmission lines proves beneficial when it is necessary to suppress outside noise or signal source. In addition to the inherent characteristics of single-ended and differential transmission lines, SISL transmission lines also have the advantages of low loss, self-packaging, low cost, and high integration.

Currently, SISL single-ended transmission lines mainly include suspended striplines and suspended coplanar waveguides, and SISL differential transmission lines include suspended slotlines, suspended parallel striplines, and suspended twisted differential lines. In this section, the structure and mechanism of these transmission lines based on SISL platform will be introduced in detail.

2.2.1 Single-Ended Transmission Line Based on SISL

2.2.1.1 Substrate Integrated Suspended Stripline

The typical substrate integrated suspended stripline structure is composed of five PCB layers with double-sided metal layers as shown in Figure 2.2(a). The second and fourth boards are hollowed to form the air cavities, which can allow electromagnetic fields to propagate in the air rather than in the substrate, thereby reducing dielectric loss. The top and bottom board layers, as well as the surrounding via holes, constrain the electromagnetic field inside the air cavity, and radiation loss can also be reduced [3].

The circuit losses generally encompass radiation, conductor, and dielectric loss. As described in the previous paragraph, the typical structure can effectively reduce dielectric and radiation losses. To reduce conductor loss, researchers designed a double-metal layer substrate integrated suspended line as shown in Figure 2.2(b),

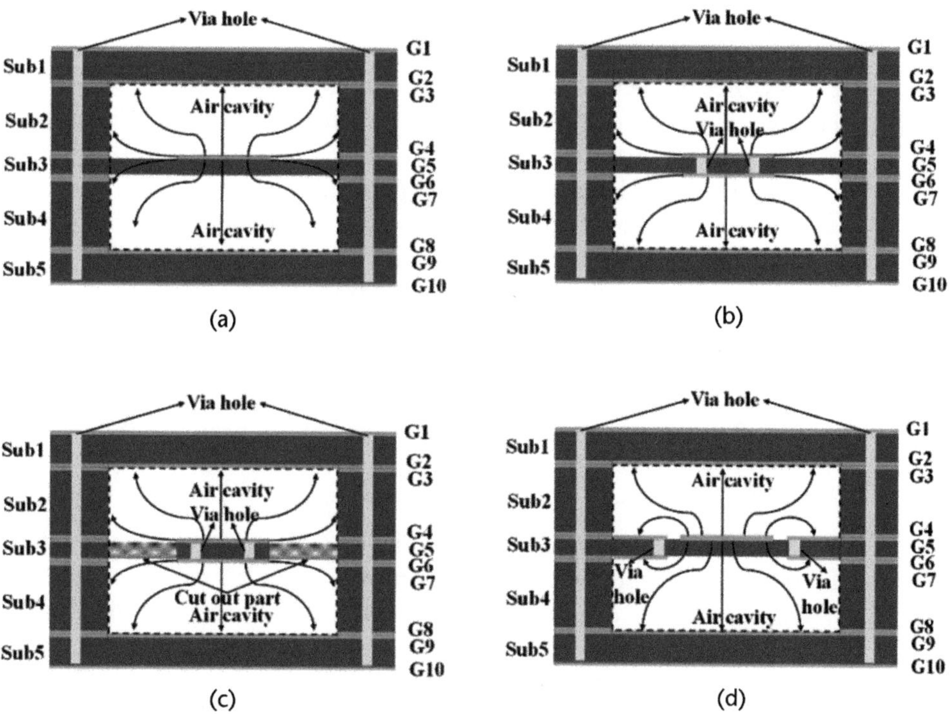

Figure 2.2 Cross-sectional view and electric field distribution of: (a) typical substrate integrated suspended stripline, (b) double-metal layer substrate integrated suspended line, (c) patterned substrate double-metal layer substrate integrated suspended line, and (d) coplanar waveguide based on SISL.

which features double metal traces and lots of via holes connecting metal traces. Generally, the double metal traces have the same patterns, which can be equivalent to a single metal trace with a larger cross-sectional area, so the double-layer substrate integrated suspended line will reduce the conductor loss. Besides, the electromagnetic shielding caused by the via-connected double metal trace makes less electric field distributed in the substrate layers where the core circuit is located.

Substrate excavation is an effective method to further reduce dielectric loss. As shown in Figure 2.2(c), the hollowed substrate around the transmission line is replaced by lossless air, and the dielectric loss will be reduced accordingly.

2.2.1.2 Substrate Integrated Suspended Coplanar Waveguide

The cross-sectional view and electric field distribution of the structure of the substrate integrated suspended coplanar waveguide (SISCPW) are shown in Figure 2.2(d). Compared to the general coplanar waveguide, SISCPW has added two extra air cavities on both sides of the transmission line, which results in more electromagnetic fields distributed in the air rather than in the substrate, so the substrate loss is reduced. Additionally, G2 and G9 act as shielding layers to confine the electromagnetic field within the air cavities, which reduces radiation loss.

2.2.2 Differential Transmission Line Based on SISL

In the early days of electronic technology, a single transmission line was sufficient for signal transmission due to the low operating frequencies. However, with the continuous increase in electronic circuit density and operating frequency, crosstalk has become an inevitable problem in electronic design. In many designs, engineering experience has been relied upon to avoid crosstalk to ensure that normal circuit operation is unaffected. However, as circuit miniaturization and operating frequency continue to improve, the practicality of relying solely on engineering experience has become increasingly limited by actual application conditions. This is where the use of differential transmission lines comes into play. Differential transmission lines are often employed to transmit high-speed signals as a single transmission line is easily affected in a complex electromagnetic environment.

Due to the low loss and self-packaging characteristics of the SISL platform, various differential transmission lines are developed based on SISL, such as substrate integrated suspended slotline (SISSL), substrate integrated suspended parallel strip line (SISPSL), and substrate-integrated suspended twisted line (SISTL).

2.2.2.1 SISSL

Slotline is a type of transmission line designed with a planar geometry that supports a TE dominant mode, much like the dominant mode in a rectangular waveguide (RWG). The slotline can be embedded inside SISL to form an SISSL platform [4], which can minimize the radiation loss with the self-packaging characteristics of SISL. The cross-sectional view of SISSL is shown in Figure 2.3(a). The SISSL platform consists of five double-sided substrates, where the top and bottom metal layers of G2 and G9 will constitute the electromagnetic shielding environment, which is equivalent to the effect of the metal shielding box in Figure 2.3(b), thereby minimizing the radiation loss of the internal slotline.

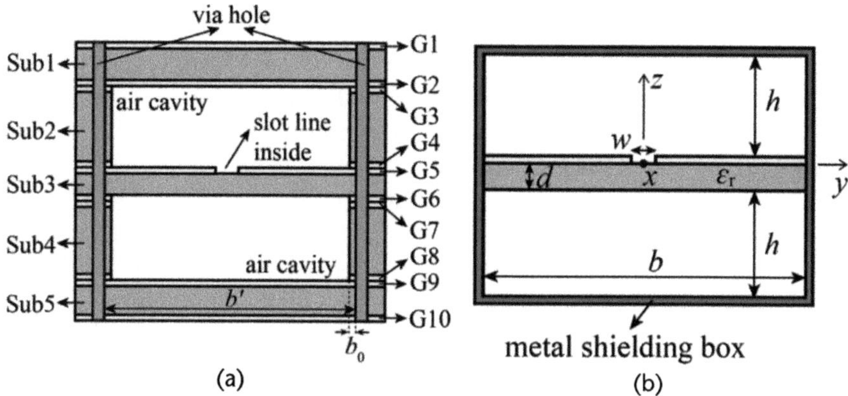

Figure 2.3 Self-packaged SISSL platform: (a) cross-sectional view, and (b) its equivalent structure [4].

2.2.2.2 SISPSL

SISPSL is a differential transmission line that embeds the parallel stripline inside SISL [5]. Compared to conventional parallel striplines, SISPSL helps to reduce radiation loss and solve the packaging problem. The cross-section of SISPSL is shown in Figure 2.4. The parallel striplines embedded in SISPSL include a substrate layer and two signal conduction strips on both sides. Both sides of the signal conduction strips are air cavities, which reduce the dielectric loss. Moreover, the presence of metalized via holes through five substrate layers and metal layers creates an effective electromagnetic shielding environment, which helps to minimize the radiation loss of the internal circuit.

2.2.2.3 SISTL

The increasing electronic circuit density and frequency have made crosstalk [6] a significant issue in electronic design, leading to differential signals. The twisted differential line (TDL) is effective in reducing crosstalk. To further improve signal transmission quality, SISTL, which embeds twisted lines within the self-packaged

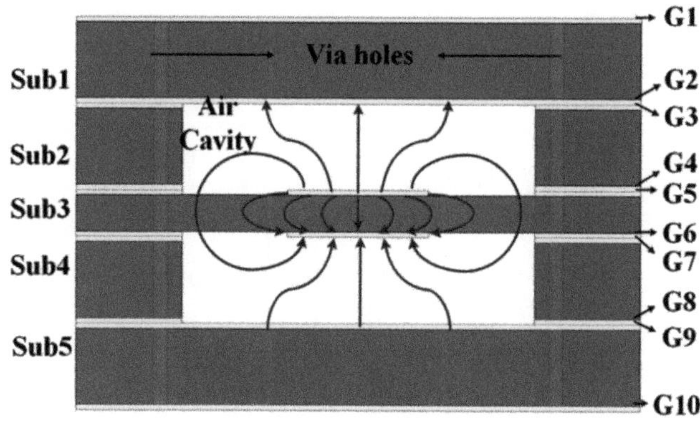

Figure 2.4 Cross-sectional view of SISPSL [5].

SISL platform, has been proposed [7]. The SISTL structure employs a circular arc transition mechanism to compensate for the misalignment of via holes on various twisted line traces, thus ensuring impedance matching [8–11]. The mixed-mode scattering parameters and eye diagrams are employed to showcase the viability and signal quality of transmission.

As illustrated in Figure 2.5, the twisted differential line (TDL) comprises numerous disjointed line segments that interconnect via holes, utilizing the centerline as the axis. The two transmission lines alternate continuously between different layers, interconnecting via holes and maintaining an uninterrupted conductive path between them. To mitigate interference between the intended transmission lines and their surroundings, metallic via holes are intentionally incorporated into the design to mitigate crosstalk and electromagnetic interference. The top and bottom layers, and metallic vias form an analogous shielding enclosure [12, 13], which safeguards against electromagnetic radiation from the transmission line.

2.3 Transitions

2.3.1 Introduction

Numerous high-performance transmission lines have been developed on the SISL platform. To test self-packaged SISL circuits and integrate them with other transmission line types (such as microstrip lines, coplanar waveguides, coaxial lines, and waveguides), a high-performance transition is necessary.

2.3.2 Transitions on SISL Platform

2.3.2.1 SISL to CBCPW Transition

Coplanar waveguides are suitable for high-frequency and broadband applications due to their lower losses compared to microstrip lines, as well as their weaker sensitivity to substrate thickness. During practical testing, the backside of the coplanar waveguide's dielectric material is often in contact with the ground plane or coated

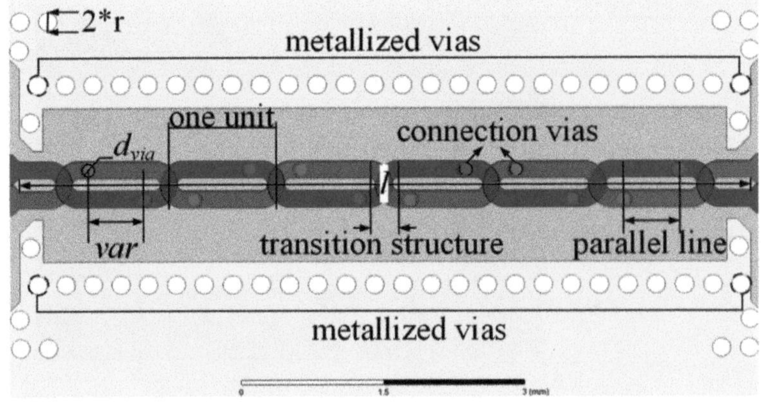

Figure 2.5 Top view of the substrate integrated suspended twisted line [7].

with a metal layer, resulting in a conductor-backed coplanar waveguide (CBCPW). A high-performance transition from SISL to CBCPW is crucial for facilitating the testing of SISL circuits [14].

Figure 2.6(a) illustrates the 3D view of an SISL-to-CBCPW transition and its transition between metal layers M1 to M10. The transition utilizes a 5-layer substrate, with Substrates 1 and 5 serving as top and bottom cover plates for the SISL structure. Substrates 2 and 4 are hollowed out in the middle to form cavities using low-cost FR4 materials with a thickness of 20 mil. The middle circuit layer, Substrate 3, uses Rogers 4003c with a thickness of 8 mil. The metal layers M2 and M9 are connected to the ground through metalized vias, ensuring good anti-electromagnetic interference capability and electromagnetic compatibility. Figure 2.6(b) shows the transition consisting of three parts: SISL, stripline, and CBCPW. The main circuit is distributed on the M5 metal layer, and the M9 metal layer is used as the ground of the CBCPW. Impedance matching is ensured by controlling the line width of each part of the transmission line and the metalized vias 1, 2, and 3. Metalized vias along the circuit can also suppress the excitation of slab modes. To avoid discontinuities, tapered transitions are used at the junction of different

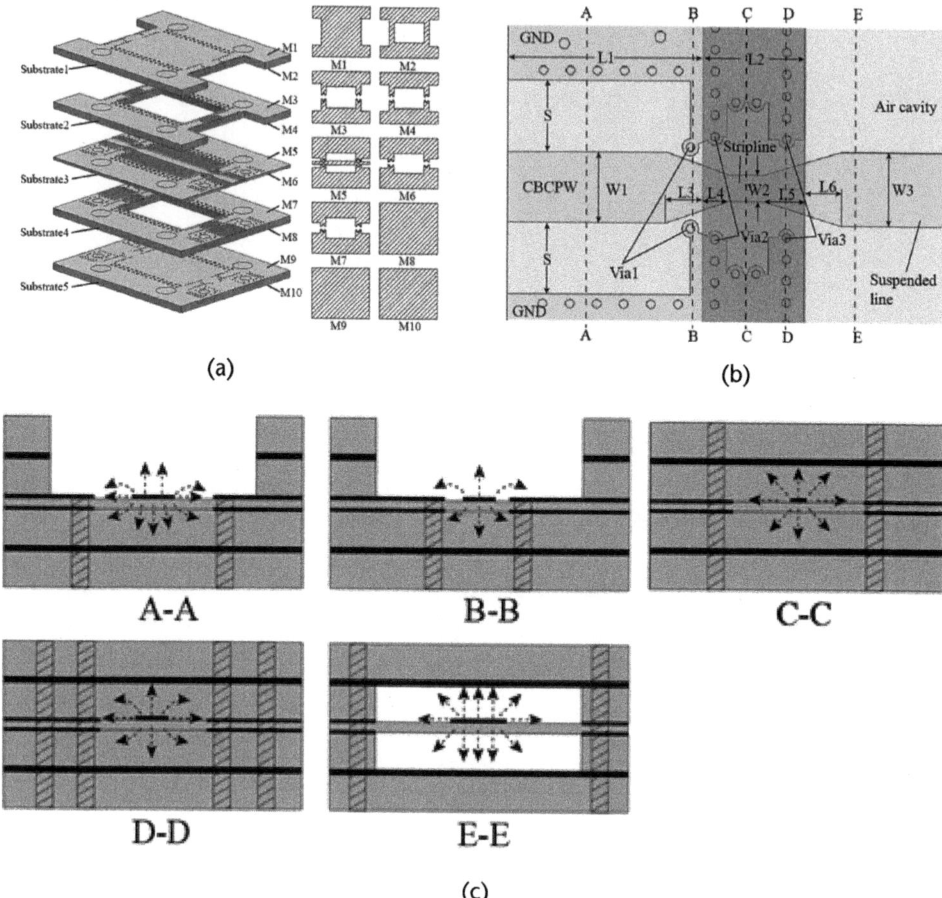

Figure 2.6 SISL-to-CBCPW transition: (a) the 3D view of the model, (b) the top view of the main circuit, and (c) the electric field at each cross-section [14].

transmission lines. The cross-sectional electric field of the transition is shown in Figure 2.6(c). At the same time, to avoid the propagation of the waveguide modes, the cavity size needs to be considered.

The test structure of the back-to-back SISL to CBCPW transition is shown in Figure 2.7(a). After the actual testing, the obtained S-parameters in Figure 2.7(b) exhibit excellent agreement with the simulated S-parameters. Within the frequency range of dc to 8 GHz, a return loss exceeding 15 dB and an insertion loss below 0.6 dB were achieved. However, in the actual test, it was found that S21 has an offset phenomenon, which is caused by the discontinuity between the connector and the transmission line.

2.3.2.2 SISL to SCPW Transition

The coplanar waveguides are well-suited for high frequency and, as mentioned earlier, while they have lower loss compared to microstrip lines, the impact of

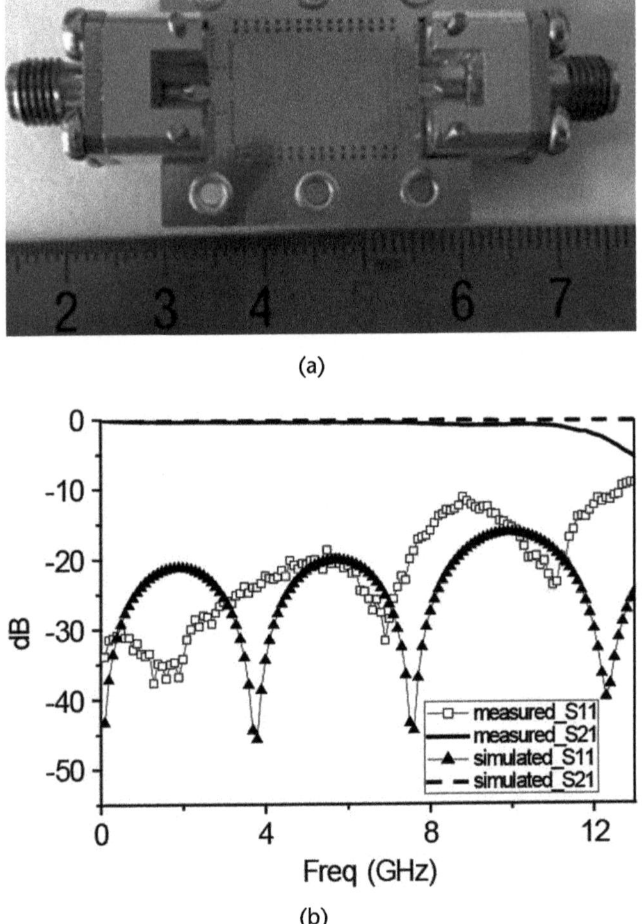

Figure 2.7 (a) Photograph of the test fixture, and (b) simulated and measured S-parameters of SISL-to-CBCPW transition [14].

dielectric loss cannot be disregarded when multiple substrate plates are stacked to form a circuit. Therefore, a suspended coplanar waveguide (SCPW) structure is employed, with some scholars investigating the transition structure from the SISL transmission line to SCPW [15].

The proposed broadband dc-to-60-GHz SISL to suspended CPW transition structure is illustrated in Figure 2.8(a), showcasing its dimensional structure. The entire circuit consists of five cost-effective FR4 boards, referred to as Sub1–Sub5 for the substrate layers and G1–G10 for the metal layers. Each board has a specific thickness of 0.3, 0.6, 0.127, 0.6, and 0.3 mm, respectively. Figure 2.8(b) provides a top view of Sub2 and Sub4, which have been hollowed out with a length (a) of 8 mm and width (b) of 16 mm to create two air cavities required for the suspended line circuit.

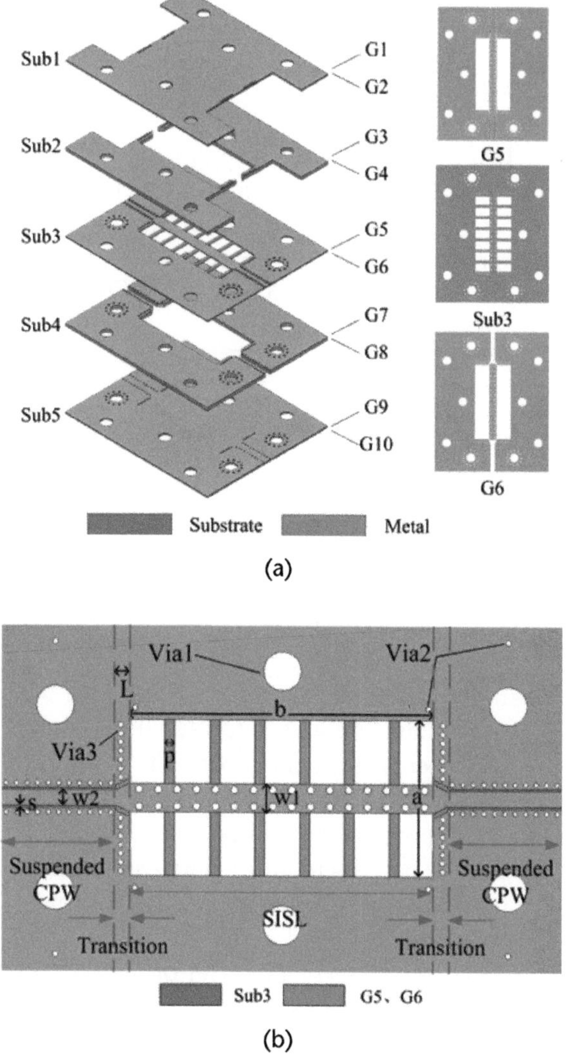

Figure 2.8 The structure of back-to-back SISL to suspended CPW transition [15]: (a) a 3D view of the whole transition, and (b) the top view on Sub3.

2.3.2.3 SISL to RWG Transition

RWGs are extensively utilized in a variety of components owing to their minimal loss, elevated quality factor, and robust power-handling capabilities. Therefore, it is also necessary to design a high-performance transition from SISL to RWGs. Based on the SISL platform, SISL-to-RWG transition designs in Ka-band and V-band have been proposed [16, 17].

The researchers successfully implemented a transition from SISL to RWG by introducing a T-patch.

The depicted SISL-to-RWG transition with a back-to-back configuration is illustrated in Figure 2.9(a), providing a 3D perspective. This transition consists of 6 substrates, named substrates 1 to 6 in sequence. Considering performance and cost, the research team opted for Rogers 5880 as Substrate 3, featuring dimensions of 0.254 mm in thickness, a dielectric constant of 2.2, and a loss tangent of 0.0009. Substrates 1 to 6 are uniformly composed of FR-4 material with a dielectric constant of 2.2 and a loss tangent of 0.02. The thickness of the metal layers M1 to M12 is 0.035 mm. Substrate 2 and Substrate 4 of the SISL structure are hollowed out to create 2 embedded air cavities. To allow for sufficient operating space, the length of the SISL air cavity L_1 has been set to 28 mm, while the width of the 50Ω SISL is $W_2 = 1.75$ mm.

To achieve broadband performance, it is necessary to ensure single-mode propagation. Therefore, the first higher-order mode, TE10 mode, in the SISL should be suppressed to produce quasi-TEM mode only. The TE10 mode arises due to the SIW channel present in the SISL, and its cutoff frequency is f_c.

$$f_c = \frac{c}{2W_{\text{SISL}}} \times \sqrt{1 - \frac{h_3}{h_2 + h_3 + h_4} \times \left(\frac{\varepsilon_r - 1}{\varepsilon_r}\right)} \tag{2.1}$$

The effective channel width, W_{SISL}, is defined as the vertical distance between the centers of the through vias on both sides of the SISL. It can be expressed as $W_{\text{SISL}} = W_1 + 2 \times W_6$, where W_1 is the width of the SISL air cavity, $W_6 = 0.25$ mm is the metal edge, and the through via the distance between the center. The channel height H_2 represents the distance between metal layers in SISL, that is, $H_2 = h_2 + h_3 + h_4$, where h_3 is 0.254 mm, ε_r is 2.2, and c is the velocity of the electromagnetic wave in vacuum. Through calculation, choose W_1 to be 2.9 mm, that is, W_{SISL} to be 3.4 mm. The calculated cutoff frequency f_c is 42 GHz, which is higher than the Ka-band.

AF-SIW components are created through the process of excavating the interiors of substrates 1, 2, 4, and 5 and introducing passageways around the cavities. Substrate 3 provides support for the probes. The metal layer M11 is assigned as the AF-SIW short plan. The spatial separation between the matching probe (located on the upper surface of substrate 3) and the waveguide's short plane (situated on the upper surface of substrate 6) is denoted as H_1, calculated as the sum of h_3, h_4, and h_5. Its theoretical value corresponds to a quarter of the waveguide's wavelength, equating to $H_1 = 2.9$ mm, at a frequency of 33 GHz. According to the actual situation, it is more practical to select the thickness of the substrate as $h_1 = h_5 = 1.8$ mm and $h_2 = h_4 = h_6 = 0.6$m. As shown in Figure 2.9(b), the distance between the edge of the

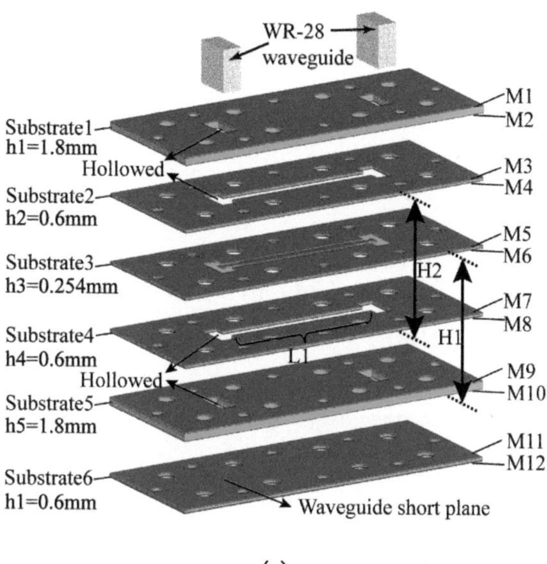

(a)

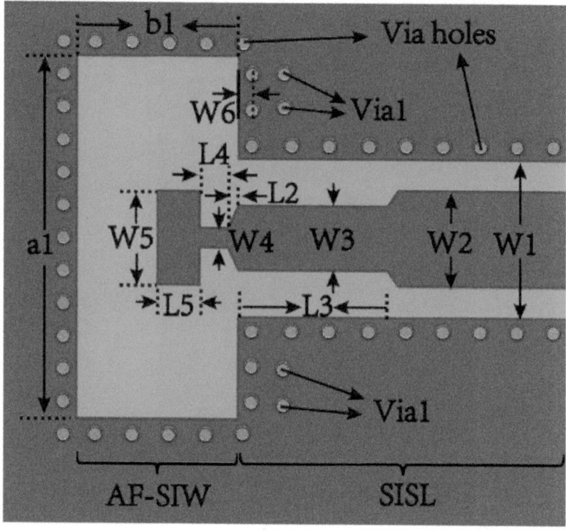

(b)

Figure 2.9 SISL-to-RWG transition [16]: (a) a 3D view of the whole transition, and (b) the top view on Sub3.

metal and the center of the via is $W_6 = 0.25$ mm. The size of the AF-SIW air cavity is determined by a_1 and b_1, where $a_1 = 6.72$ mm and $b_1 = 3.056$ mm.

Figure 2.9(b) demonstrates the improved impedance matching using quarter-wave matching lines (W_3 = 1.18 mm, L_3 = 2.8 mm). To weaken the effect of the discontinuity between each part of the circuit, the connection is chamfered with a length of $L_2 = 0.2$ mm. A T-shaped patch (consisting of W_5, L_5, W_4, and L_4) formed by two orthogonal rectangles was optimized to enable electromagnetic field conversion between SISL and AF-SIW.

Considering the balance between bandwidth and performance, W_5 has been selected as 1.72 mm, and the other dimensions of the T-shaped patch are $W_4 = 0.35$ mm, $L_4 = 0.53$ mm, and $L_5 = 0.8$ mm.

Figure 2.10(a–c) illustrates the SISL to RWG transition achieved using a standard PCB fabrication process. The S-parameters corresponding to the back-to-back transition were both simulated and measured, and these results are presented in Figure 2.10(d).

The measured return loss consistently surpasses 12 dB and nearly reaches 15 dB within the frequency range of 24.6 to 38.5 GHz, which accounts for approximately 44% of the total bandwidth. The insertion loss for the back-to-back transition

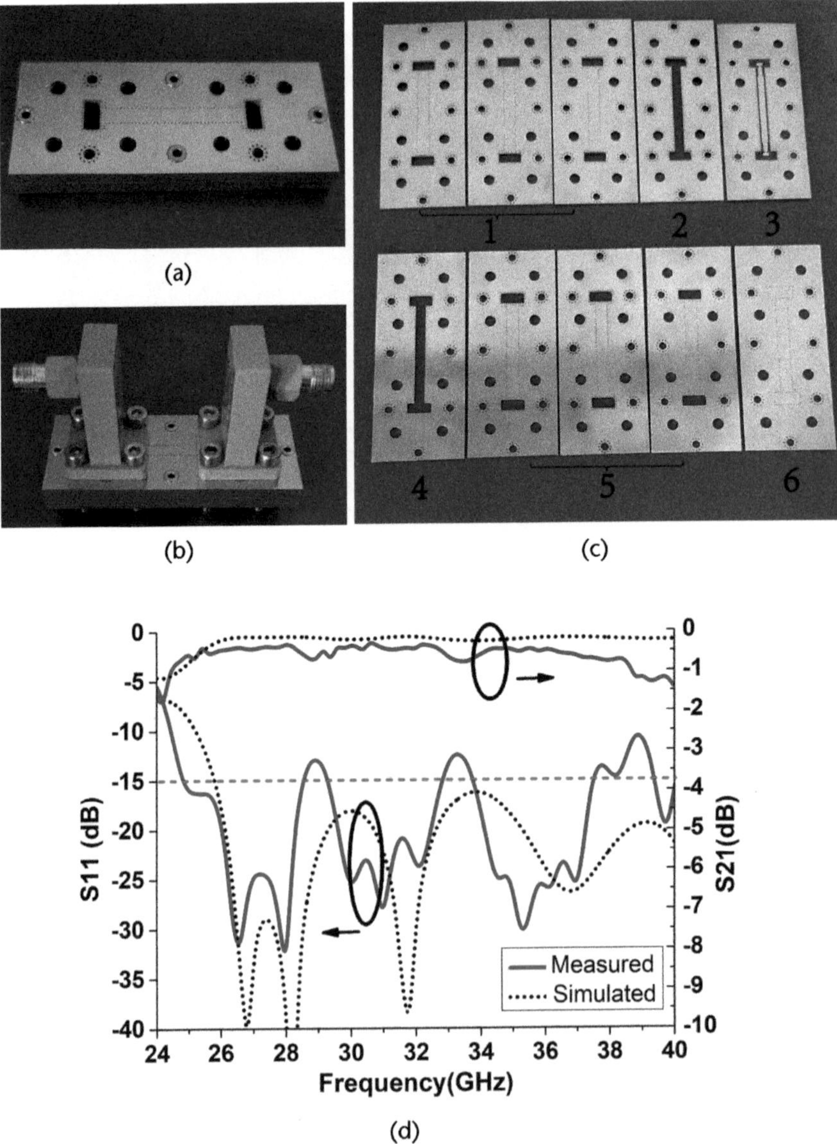

Figure 2.10 (a–c) Photograph of the test fixture, and (d) simulated and measured S-parameters of SISL-to-RWG transition [16].

remains under 0.8 dB, attaining a minimum of 0.34 dB. The transition itself comprises a 22-mm SISL transmission line, demonstrating an estimated loss of 0.12 dB. Consequently, for a single transition from SISL to RWG, the maximum insertion loss is 0.34 dB, while the minimum insertion loss is recorded at 0.11 dB within the bandwidth. Despite the limitations imposed by the adapter's performance, the conversion measurements demonstrate a broad operational bandwidth of 44% and an impressive level of insertion loss.

2.3.2.4 SISL to AF-SIW Transition

Substrate integrated waveguide (SIW) is a planar form of the traditional rectangular waveguide. SIW is favored due to its combination of the advantages of planar printed circuits and metal waveguides. In recent years, there has been extensive attention on air-filled SIW (AF-SIW), which uses air as the medium, replacing the intermediate substrate in the traditional SIW structure. This substitution has reduced the loss to a certain extent.

In 2019, some researchers proposed a compact transition structure, as shown in Figure 2.11. The transition structure consists of six base layers made of FR4 material. These layers are aptly named Sub1-6, and they are associated with 12 metal layers named M1–12. Sub1 and Sub5 form a closed structure that provides metal shielding and reduces the propagation of higher-order mode waves, while Sub2 and Sub4 are hollowed out to create air cavities for potential packaging of AF-SIW cavities and active devices. Sub3 forms the suspended line in SISL and the bottom cap in AF-SIW. Finally, Sub6 acts as the short end for the I/O port, acting as a WR-28 RWG. There are two air cavities in the SISL, and the transition from SISL to single-layer AF-SIW is first introduced. The lower air cavity on the Sub4 is partially blocked in the AF-SIW, allowing it to be used for the assembly of active devices/circuits and as a cavity resonator.

The transition from RWG to AF-SIW has a total length of only $L_2 = 9.45$ mm ($1.04\lambda g$) [18]. A detailed illustration of the transition on Sub3 (left half) is provided in Figure 2.11(b), while the other layers have the same dimensions as Sub3 at their respective positions.

The equation is used to estimate the initial physical dimensions of the unilateral width of AF-SIW sections.

$$a = W_0 + p\left(0.766e^{\frac{0.482d}{p}} - 1.176e^{-\frac{1.214d}{p}}\right) \tag{2.2}$$

$$W_0 = \frac{c}{2f_c\sqrt{\varepsilon_r}} \tag{2.3}$$

a represents the width of the Sub3 cavity, while p and d refer to the pitch and diameter of the vias shown in Figure 2.11, respectively, and the symbol ε_r represents the relative permittivity. The TEM mode propagation in SISL is transformed to TE10 mode propagation in AF-SIW with a double-sided taper line. The prototype

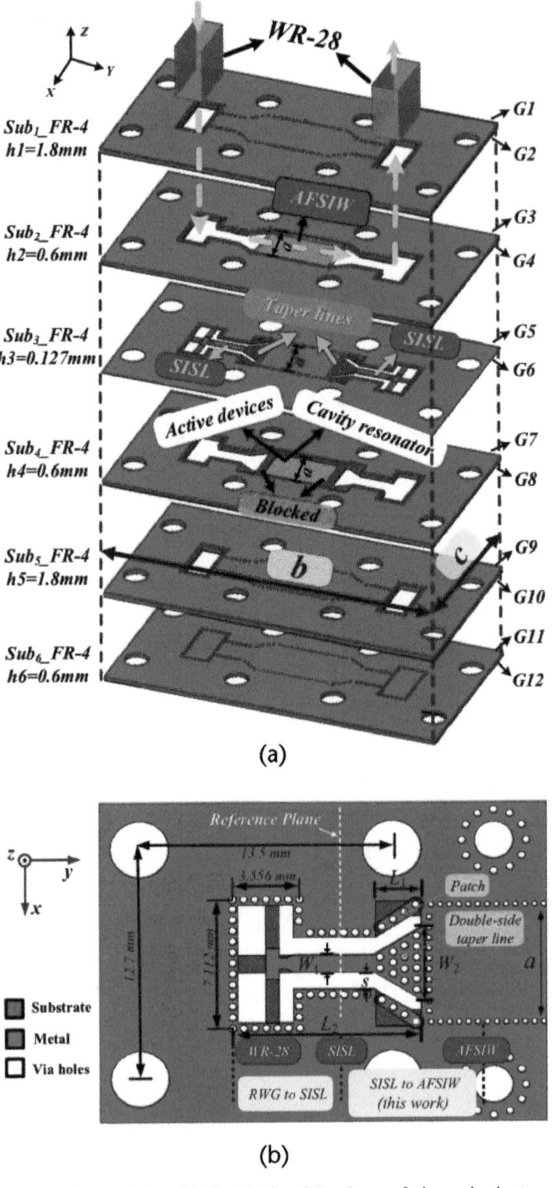

Figure 2.11 SISL-to-AF-SIW transition [18]: (a) the 3D view of the whole transition, and (b) the top view on Sub3 (left half).

exhibits a return loss of over 13 dB and an insertion loss ranging from 0.4 to 1.03 dB (0.2–0.52 dB for a single transition) in the frequency range of 26.5 to 40 GHz.

2.4 Interfaces

2.4.1 Introduction

To test the performance of designed components based on SISL, all kinds of interfaces are needed to connect the SISL circuit and test instruments (such as vector network analyzer (VNA)). According to developed transitions based on SISL recently, RF coaxial connectors and coaxial-to-waveguide adapters are often used in the measurement of SISL designs.

RF coaxial connectors are designed to connect PCBs and cables to an external device. SISL is compatible with PCB technology, which makes it easy to connect to coaxial cables through RF coaxial connectors. Thus, RF coaxial connectors are widely applied for the measurement of SISL circuits.

Coaxial-to-waveguide adapters are devices used to transition signals from coaxial transmission lines to waveguide interfaces. The waveguide is fed using a coaxial probe pierced into the wall of the waveguide. The coaxial probe is connected to a source of EM waves. SISL has a multi-layer structure with air cavities inside, which allows it to form waveguide structures. Therefore, SISL can utilize coaxial-to-waveguide adapters as its interfaces.

In this section, to clarify the position of the interface in the measurement system, the connection types among SISL, CPW, coaxial line, and VNC will be demonstrated. To provide a guide for SISL designers to conveniently select the appropriate interface for their own SISL designs, the methods to select interfaces for SISL will be introduced.

2.4.2 The Connection Types

An example of a connection for measurement is shown in Figure 2.12. The signal from SISL is usually not directly measured, and one of the best ways is to use a transition from SISL to CPW. Using VNA, the coaxial line, and the interface, the signal from SISL can be measured with VNA.

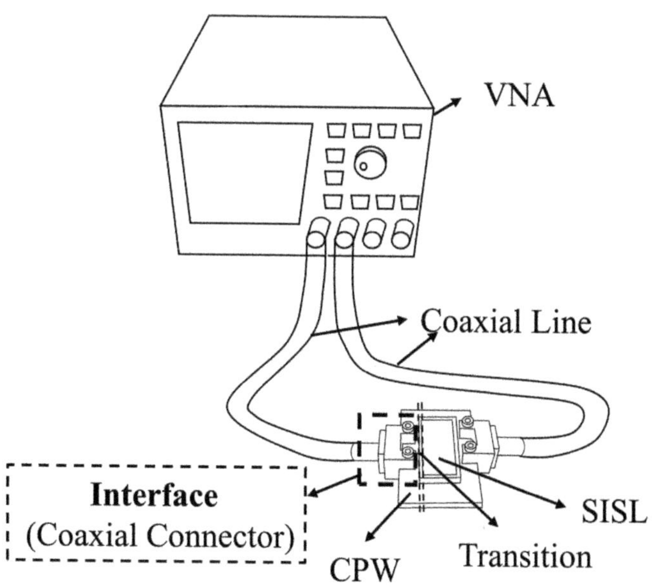

Figure 2.12 An example of a connection for measurement.

2.4.3 Selection of Interfaces for SISL

The selection of interfaces for SISL mainly depends on the type of transition and the frequency range of the designed SISL circuit. Some transitions enable the interconnection between SISL and planar transition lines, including conductor-backed CPW (CBCPW) [14], suspended coplanar waveguide [15], and microstrip [19]. Some coaxial connectors can be selected as interfaces for CPW. Some transitions realize the interconnection between SISL and rectangular waveguide transition [16], and the coaxial-to-waveguide adapters can be utilized to measure the SISL circuit with this transition.

When selecting an RF connector, it is necessary to choose a connector series that will support the frequency requirements of the application. The coaxial connector can be classified into several types, such as SMA connectors, 2.92-mm connectors, and 1.85 mm connectors, which typically support a wide frequency range of dc to 18 GHz, dc to 40 GHz, and dc to 67 GHz.

The polarity of coaxial connectors and cables is a crucial factor that requires careful consideration. Standard RF plugs are typically male, with threads located on the inside of the shell, while standard RF jacks are female, with threads located on the outside of the shell. Designers need to ensure that the coaxial connectors and coaxial cables have the same polarity. Furthermore, other factors such as size, mechanical dimension, voltage standing wave ratio (VSWR), and insertion loss also need to be considered.

2.5 Capacitors

2.5.1 Introduction

Lumped-element and quasi-lumped-element circuits, different from distributed circuits, can achieve more compact sizes at low frequencies. Due to its advantage of miniaturizing circuits, the lumped element technique is widely used in the design of miniaturized RF/microwave circuits. Lumped element devices, such as capacitors and inductors, are important components of lumped element circuits. The performance of lumped element devices directly affects the overall performance of lumped element circuits. Therefore, the research of lumped element devices is very essential for the design of lumped element circuits.

Lumped-element capacitors are important devices in the lumped-element circuit, and there are some issues with the practical application of capacitors based on SISL platform. One of the most important issues is that the SISL-based capacitors usually occupy a large area, relative to inductors.

Based on the SISL platform, researchers have conducted extensive research on lumped element capacitors with high capacitance density. Lots of lumped element capacitors with high capacitance density are researched, analyzed, and applied.

Capacitance density is an important parameter of the capacitor. For a three-dimensional capacitor, its capacitance density C_d can be expressed with the overall capacitance C_T and the volume V as

$$C_d = \frac{C_T}{V} \tag{2.4}$$

However, the capacitors often have complex structures, and it is difficult to calculate the overall capacitance according to its structures and dimensions. Therefore, it is necessary to provide a convenient formula for calculating capacitance [20].

Fortunately, with the help of some equivalent circuit methods and electromagnetic simulation software, capacitors can be extracted according to the following methods. If the capacitance to the ground plays a major role, the designed capacitor can be equivalent to a T network. Z_{C11} is the input impedance at Port 1 when Port 2 is an open circuit, Z_{C12} is the transmission impedance between Port 1 and Port 2, and Z_{C22} is the input impedance at Port 2 when Port 1 is an open circuit. Through the T network, the capacitance to the ground can be calculated by

$$C_T = -\frac{1}{2\pi f \times im\left(Z_{C12}\right)} \tag{2.5}$$

where Z_{C12} can be obtained with ease by using electromagnetic simulation software.

If the capacitance between the two ports of designed capacitors plays a major role, the designed capacitor can be equivalent to a π network. Y_{C11} is the input admittance at Port 1 when Port 2 is a short circuit, Y_{C12} is the transmission admittance between Port 1 and Port 2, and Y_{C22} is the input admittance at Port 2 when Port 1 is a short circuit.

Through the π network, the capacitance to the ground can be calculated by

$$C_T = -\frac{im\left(Y_{C12}\right)}{2\pi f} \tag{2.6}$$

where Y_{C12} can be easily obtained through electromagnetic simulation software.

In this section, we will introduce several types of capacitors with high capacitance density based on the SISL platform and illustrate the principle of capacitors.

2.5.2 Interdigitated Capacitors

A good way to increase the capacitance density of the capacitor on the SISL platform is to use interdigitated capacitors. In [21–24], double interdigital capacitors (DIDCs) are utilized to realize lumped circuits. DIDC is usually designed on the two metal layers of the SISL core circuit. The 3D model of DIDC is shown in Figure 2.13(a) and the electric field distribution of DIDC is shown in Figure 2.13(b). The DIDC utilizes both the vertical and lateral electric fields, whereas conventional parallel-plate capacitors only employ the vertical electric field, and interdigital capacitors only utilize the lateral electric field. This characteristic of DIDCs allows for a higher capacitance density to be achieved. Additionally, because of the low dielectric loss characteristics inherent to the SISL platform, the DIDC can achieve a high-quality factor.

In Figure 2.13(a), the width of the gap between each finger is represented with $s1$ and it is closely associated with the lateral capacitor. A comparison between the capacitance of DIDC and a conventional parallel-plate capacitor of identical

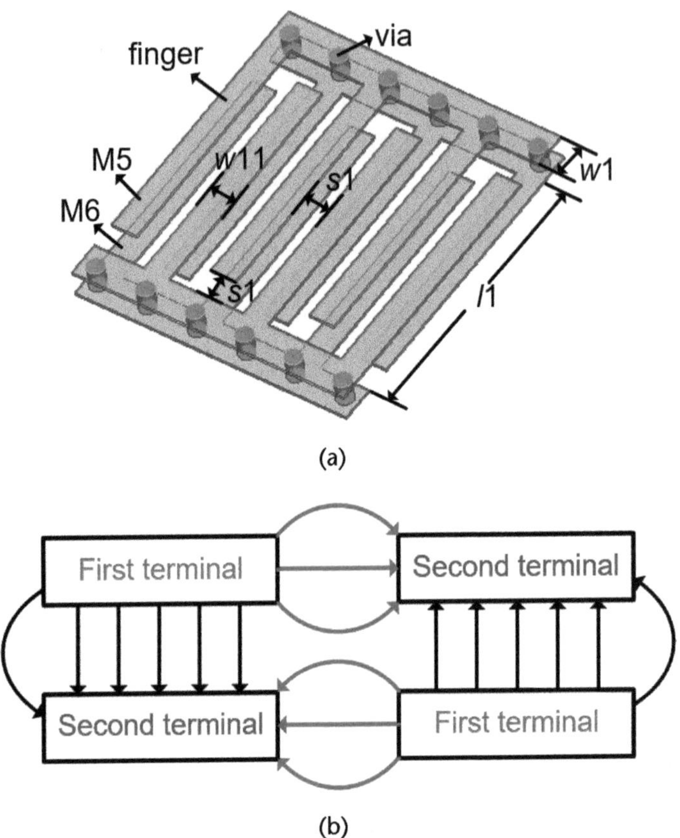

Figure 2.13 (a) The 3D model of DIDC, and (b) the distribution of electric field in DIDC [21].

dimensions on the SISL platform is presented, and the capacitance variation across $s1$ is also plotted in [21]. As demonstrated, a decrease in $s1$ results in an increase in the capacitance of DIDC, while the capacitance of the parallel-plate capacitor changes little due to the small, adjusted value of $s1$. For values of $s1$ below 0.20 mm, the capacitance of the DIDC exceeds that of the parallel-plate capacitor. However, it is also not feasible to decrease $s1$ indefinitely to obtain a larger capacitance because the processing accuracy also needs to be taken into account. It is evident that for the same capacitance, DIDC has a smaller physical size compared to traditional parallel-plate capacitors, thus possessing a higher capacitance density.

For the DIDC, smaller physical dimensions lead to a decrease in series resistance, thereby resulting in an improved quality factor. Reference [21] studied the variation in capacitance as the number of fingers changes from 4 to 14. The DIDC consistently demonstrates higher capacitance than the parallel-plate capacitor. With an increasing number of fingers (N), this distinction becomes more prominent, leading to the DIDC possessing a superior quality factor to the parallel-plate capacitor.

2.5.3 Metal-Insulator-Metal Capacitors

The metal-insulator-metal (MIM) capacitor is a classical capacitor that is formed by two or more parallel plates. It is difficult to increase the capacitor density based on the traditional transmission line. However, due to the multilayer structure and flexible design based on the SISL platform, we can adopt various methods to improve capacitance density.

One of the solutions to design a dielectric-filled capacitor is shown in Figure 2.14 [25], in which a stack of five substrate layers, labeled Sub1 through Sub5, is employed. The circular plate located in the middle of the stack employs a pair of metal layers, namely G5 and G6, which are connected through a hole. G2 functions as the upper ground plane, while G9 serves as the lower ground plane. The capacitor maintains its structure as a parallel plate capacitor; however, the introduction of high dielectric constant blocks between the circular plate and the ground plate serves to enhance the density of the capacitor. The center sections of Sub2 and Sub4 are hollowed to form a cylindrical shape for the placement of the dielectric blocks, which are fixed in place using the dielectric rings. All five substrates are fabricated with economical FR4 material (dielectric constant: 4.4). The dielectric block with a higher relative dielectric constant of 38.3 is chosen. Sub2 and Sub4 possess a thickness of 2 mm, with Sub3 being 0.34 mm thick. Sub1 and Sub5 serve as cover boards, each having a 0.6-mm thickness.

The equivalent circuit of the dielectric-filled capacitor is shown in Figure 2.14(c). The capacitance of the parallel plate capacitors shunted to the top or bottom ground planes is denoted by C_1, which can be calculated by

$$C_1 = \frac{\varepsilon_0 \varepsilon_r \pi d^2}{4h} \tag{2.7}$$

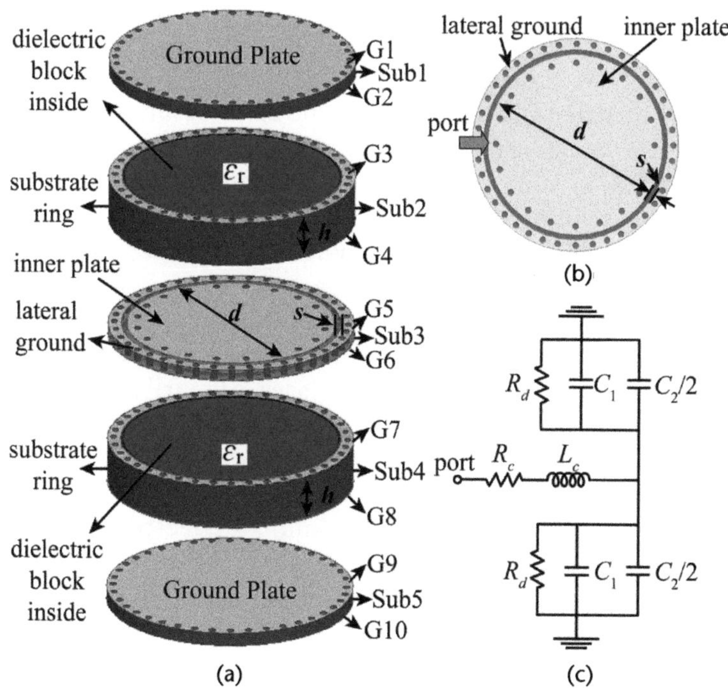

Figure 2.14 Self-packaged SISL capacitor with filled dielectric block: (a) a 3D view, (b) a planar view of Substrate 3, and (c) an equivalent circuit [25].

where ε_0 represents the vacuum permittivity, ε_r represents the relative dielectric constant of the dielectric block, d denotes the diameter of the circular plate, and h corresponds the thickness of the Sub2 or Sub4. The parasitic capacitance C_2 is formed between the circular plate and the lateral ground boundary, and the gap between them is denoted by s in Figure 2.14. The results indicate an approximately inverse relationship between C_2 and the gap s, which can be expressed as

$$C_2 = \frac{k}{s} \tag{2.8}$$

where the coefficient k primarily correlates with the diameter, thickness, and dielectric constant of the dielectric block. Then the capacitance of the overall capacitor C_T can be given by

$$C_T = 2C_1 + C_2 = \frac{\varepsilon_0 \varepsilon_r \pi d^2}{2h} + \frac{k}{s} \tag{2.9}$$

Generally, C_1 is the main part of C_T. The parasitic series inductance L_c can be obtained by

$$L_c = \frac{1}{4C_T \pi^2 \mathrm{SRF}^2} \tag{2.10}$$

where SRF represents the self-resonant frequency. R_c refers to the effective resistance of the metal plate, and R_d corresponds to the effective resistance arising from the substrate. When lossless conditions prevail, R_c is equal to zero and R_d tends towards positive infinity.

The calculated C_T amounts to around 29 pF, resulting in a significantly increased capacitance density of 0.067 pF/mm^3. This is notably larger than the values reported in [16] (0.028 pF/mm^3) and [26] (0.019 pF/mm^3).

Another solution is to utilize the multilayer characteristics of the SISL. The miniaturized multiplate capacitor proposed in [20] achieves higher capacitance density and is compatible with the conventional multilayer PCB process.

The SISL multiplate miniaturized capacitor consists of six substrate layers, represented as S1 to S6, along with 12 metal layers identified as G1 to G12. S1 and S6, with metal layers on both surfaces, primarily serve the purpose of confining the electromagnetic field within the cavity. S2 and S5 are primarily utilized to provide support to the air cavity, and the miniaturized capacitor is mainly designed with S3 and S4, as well as the metal layers on both sides of them. By utilizing two substrates, S3 and S4, the miniaturized capacitor can achieve a higher capacitance density.

The SISL miniaturized capacitor primarily consists of three parts: capacitor C_1, which is formed between G7 and G8; capacitor C_2, established between G6 and G5; and capacitor C_3, formed between the square patch on G7 and the surrounding ground metal. It is important to note that G6 and G7 are connected when substrate layers are stacked together. The primary contributors to the overall capacitance are capacitors C_1 and C_2, with a lesser role played by C_3. These three capacitors are connected in parallel to yield the overall capacitance C_T. Similar to (2.9), the overall capacitance value of this miniaturized capacitor can also be expressed as

$$C_T = C_1 + C_2 + C_3 \tag{2.11}$$

where C_1 and C_2 are considered parallel plate capacitors, and the formulas for calculating them are given as

$$C_i = \frac{\varepsilon_0 \varepsilon_{ri} ab}{h_i} \quad i = 1,2 \tag{2.12}$$

where a and b symbolize the length and width of the square patch of the capacitor, ε_0 stands for the vacuum permittivity, ε_{ri} represents the relative dielectric constant of the dielectrics S3 and S4, and h_i signifies the thickness of board. Similar to (2.8), C_3 can also be given as

$$C_3 = \frac{k}{s} \tag{2.13}$$

where s denotes the gap width, and constant k is primarily influenced by the thickness and dielectric constant of the substrate. The overall capacitance C_T can be easily given as

$$C_T = \frac{\varepsilon_0 \varepsilon_{r1} ab}{h_1} + \frac{\varepsilon_0 \varepsilon_{r2} ab}{h_2} + \frac{k}{s} \tag{2.14}$$

When $h_1 = h_2 = h$, (2.14) can be rendered in a simplified form as

$$C_T = \frac{\varepsilon_0 \left(\varepsilon_{r1} + \varepsilon_{r2}\right) ab}{h} + \frac{k}{s} \tag{2.15}$$

Based on (2.15), the dielectric constants of the two substrates are summed to enhance the capacitance density and reduce the capacitor area correspondingly. The presented capacitor in this study utilizes the metal layers on both sides of the two substrates (S3 and S4), which results in a significant increase in capacitance density.

Due to the complexity in directly calculating C_3, the overall capacitance C_T can be obtained using (2.5). The miniaturized capacitor achieves a higher capacitance density, reaching 0.1 pF/mm^3.

2.6 Inductors

2.6.1 Introduction

Inductors, like capacitors, are critical components in lumped-element circuit designs. High-performance miniaturized lumped-element circuits can be easily designed using small-sized and low-loss inductors. The SISL platform has advantages in designing high-performance inductors due to its low-loss, multilayered, low-cost, and self-packaging characteristics.

Similar to capacitors, it is also a challenging task to calculate the inductance directly from the structure of the inductors. By utilizing electromagnetic simulations and equivalent p network, the extract formula for calculating the inductance with two ports is given as:

$$L_T = \frac{1}{2\pi f \times im(Y_{L12})} \tag{2.16}$$

where Y_{L12} denotes the transmission impedance between Port 1 and Port 2.

Besides, the extraction of the inductance for a single-ended inductor (where one side of the inductor is grounded) L_{11} can be performed as follows

$$L_{11} = \frac{im(1/Y_{L11})}{2\pi f} \tag{2.17}$$

where Y_{L11} represents the input impedance at Port 1 when Port 2 is an open circuit.

The quality factor Q_{11} of the SISL inductor can be expressed by

$$Q_{11} = -\frac{im(Y_{L11})}{re(Y_{L11})} \tag{2.18}$$

In this section, we will introduce two types of inductors based on SISL: spiral inductors and meander inductors.

2.6.2 Spiral Inductors

The SISL spiral inductor proposed in [27] realizes a high Q factor. The spiral inductor is designed with a classical SISL structure with five substrate layers and ten metal layers. Figure 2.15 displays the SISL inductor from a 3D view, along with the top view of Substrate 3 both without and with the patterned substrate. The SISL inductor is located on the metal layers M5 and M6, and the five substrates are interconnected to form a self-packaging system. Plated via holes surrounding the air cavity serve the purpose of shielding, effectively minimizing radiation loss and the propagation of leaky waves from the sides of the board. Therefore, with the use of surrounding vias and the grounding of the SISL inductors (M2 and M9), the SISL circuit and system exhibit good EMI and EMC performance. For further enhancement of the inductor's quality factor, a double-sided SISL (DSISL) inductor with the patterned substrate is presented in Figure 2.15(c), where the suspended Substrate 3 is hollowed to further decrease the substrate loss. The substrate corresponding to the SISL inductor strip is not etched, and several narrow strips of the suspended substrate are intentionally reserved to provide crucial mechanical support for the centrally suspended inductor. Meanwhile, the vias are employed to connect M5 and M6 of the SISL inductor strip, establishing a dual-sided interconnected stripline layout that effectively increase the metal thickness of the SISL inductor. According to (2.17) and (2.18), the inductance of the single-ended inductor and quality factor can be extracted.

To facilitate analysis, the inductor is equivalent to the π model. In the SISL inductor, L_s represents the series inductance while R_s denotes the effective resistance of the spiral strips, which accounts for the skin effect associated with the metal trace. In the shunt branch, R_p and C_p denote the resistance and capacitance between the SISL inductor and the grounds on its both sides (M2 and M9), which also include the resistance and capacitance arising from substrates. The parameters of the frequency-dependent circuit model can be calculated by

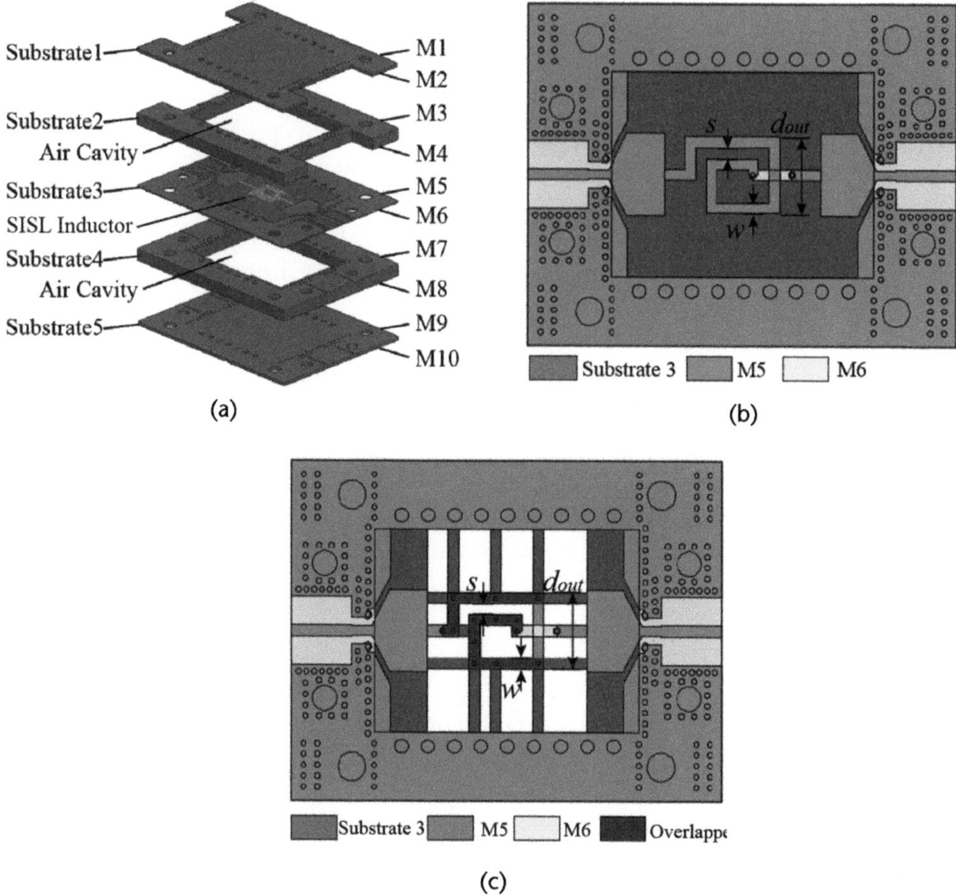

Figure 2.15 SISL inductor [19]: (a) a 3D view of the SISL inductor, (b) the top view of the SISL inductor without patterned substrate, and (c) the top view of the SISL inductor with patterned substrate.

$$L_s(\omega) = \frac{im\left(1/Y_{L11}(\omega)\right)}{2\pi f} \tag{2.19}$$

$$R_s(\omega) = \frac{1}{re\left(-Y_{L12}(\omega)\right)} \tag{2.20}$$

$$C_p(\omega) = \frac{im\left(Y_{L11}(\omega) + Y_{L21}(\omega)\right)}{2\pi f} \tag{2.21}$$

$$R_p(\omega) = \frac{1}{re\left(Y_{L11}(\omega) + Y_{L21}(\omega)\right)} \tag{2.22}$$

where Y_{L21} denotes the transmission impedance between Port 1 and Port 2.

By employing the π network, the design efficiency of SISL inductors and circuits is elevated.

2.6.3 Meander Inductors

Generally, meander inductors have lower inductors per unit area compared to spiral inductors. However, in some cases, spiral inductors are not convenient for layout in the circuit design process due to their coil shape. In addition, the spiral inductors generally occupy multiple metal layers, which may also affect the layout of other components. The meander inductors only require a single metal layer and allow for a more flexible layout. Therefore, the meander inductors can be used as a substitute for the spiral inductors.

In [28], an SISL lumped coupler using meander inductors is proposed. The lumped coupler is also designed with a classical SISL structure with five substrates and ten metal layers, and the inductor is implemented with a narrow line in a meander line shape, which is designed on the single layer of G5 or G6. With the flexible layout of the SISL meander inductors, as well as the SISL capacitors, the lumped-element coupler achieves a compact size.

Acknowledgments

We would like to thank Hanyong Wang, Yongyun Wang, Haiyang Zhang, and Haozhen Huang for their valuable support in this chapter.

References

[1] Ma, K., and K. T. Chan, Quasi-Planar Circuits with Air Cavities, PCT Patent, WO/2007/149046, 2007.

[2] Ma, K., N. Yan, and Y. Wang, "Recent Progress in SISL Circuits and Systems: Review of Passive and Active Circuits Demonstrating SISL's Low Loss and Self-Packaging and Showcasing the Merits of Metallic, Shielded, Suspended Lines," *IEEE Microwave Magazine*, Vol. 22, No. 4, April 2021, pp. 49–71.

[3] Wang, Y., and K. Ma, "Loss Mechanism of the SISL and the Experimental Verifications," *IET Microw. Antennas Propag.*, Vol. 13, No. 11, 2019, pp. 1768–1772.

[4] Wang, Y., M. Yu, and K. Ma, "Substrate Integrated Suspended Slot Line and Its Application to Differential Coupler," *IEEE Transactions on Microwave Theory and Techniques*, Vol. 68, No. 12, December 2020, pp. 5178–5189.

[5] Liu, J., et al., "A Differential SISPSL Branch-Line Coupler with Common-Mode Suppression Using Compensated Stub," *IEEE Transactions on Circuits and Systems II: Express Briefs*, Vol. 70, No. 2, February 2023, pp. 511–515.

[6] Hatirnaz, I., and Y. Leblebici, "Modelling and Implementation of Twisted Differential On-Chip Interconnects for Crosstalk Noise Reduction," *2004 IEEE International Symposium on Circuits and Systems (IEEE Cat. No.04CH37512)*, May 2004.

[7] Wang, Y., et al., "Wideband Millimeter-Wave Substrate Integrated Suspended Twisted Line for High-Speed Transmission," *IEEE Transactions on Components, Packaging and Manufacturing Technology*, Vol. 12, No. 12, December 2022, pp. 1959–1968.

[8] Kam, D. G., et al., "A New Twisted Differential Line Structure on High-Speed Printed Circuit Boards to Enhance Immunity to Crosstalk and External Noise," *IEEE Microw. Wirel. Compon. Lett.*, Vol. 13, No. 9, September 2003, pp. 411–413.

[9] Kam, D. G., and J. Kim, "A Novel Twisted Differential Line on PCB: Crosstalk Model and Its Application to High-Speed Interconnect Circuit Design," *2002 IEEE 11th Topical Meeting on Electrical Performance of Electronic Packaging*, October 2002, pp. 153–156.

[10] Kam, D. G., H. Lee, and J. Kim, "Twisted Differential Line Structure on High-Speed Printed Circuit Boards to Reduce Crosstalk and Radiated Emission," *IEEE Transactions on Advanced Packaging*, Vol. 27, No. 4, November 2004, pp. 590–596.

[11] Kam, D. G., et al., "GHz Twisted Differential Line Structure on Printed Circuit Board to Minimize EMI and Crosstalk Noises," *52nd Electronic Components and Technology Conference 2002. (Cat. No.02CH37345)*, May 2002, pp. 1058–1065.

[12] Khan, Z. A., "A Novel Transmission Line Structure for High-Speed High-Density Copper Interconnects," *IEEE Transactions on Components, Packaging and Manufacturing Technology*, Vol. 6, No. 7, July 2016, pp. 1077–1086.

[13] Liao, C., et al., "Wideband Electromagnetic Model and Analysis of Shielded-Pair Through-Silicon Vias," *IEEE Transactions on Components, Packaging and Manufacturing Technology*, Vol. 8, No. 3, 2018, pp. 473–481.

[14] Li, L., et al., "A Novel Transition from Substrate Integrated Suspended Line to Conductor Backed CPW," *IEEE Microw. Wirel. Compon. Lett.*, Vol. 26, No. 6, June 2016, pp. 389–391.

[15] Wu, X., et al., "An FR4-Based dc to 60 GHz Substrate Integrated Suspended Line to Suspended Coplanar Waveguide Transition," *Microw. Opt. Technol. Lett.*, Vol. 64, No. 2, 2022, pp. 203–207.

[16] Chen, Y., K. Ma, and Y. Wang, "A Ka-Band Substrate Integrated Suspended Line to Rectangular Waveguide Transition," *IEEE Microw. Wirel. Compon. Lett.*, Vol. 28, No. 9, September 2018, pp. 744–746.

[17] Chen, Y., K. Ma, and Y. Wang, "A Novel V-Band Substrate Integrated Suspended Line to Rectangular Waveguide Transition," *2018 IEEE/MTT-S International Microwave Symposium—IMS*, June 2018, pp. 186–189.

[18] Xu, W., et al., "Ka-Band SISL-to-AFSIW Transitions with Fabrication Tolerance Characteristics," *IEEE Transactions on Components, Packaging and Manufacturing Technology*, Vol. 9, No. 10, October 2019, pp. 2097–2103.

[19] Li, L., K. Ma, and S. Mou, "Modeling of New Spiral Inductor Based on Substrate Integrated Suspended Line Technology," *IEEE Transactions on Microwave Theory and Techniques*, Vol. 65, No. 8, August 2017, pp. 2672–2680.

[20] Zhang, H., Y. Wang, and K. Ma, "A Compact Low-Pass Filter Using a Miniaturized Capacitor and Honeycomb Cavity on Substrate Integrated Suspended Line Platform," *Microw. Opt. Technol. Lett.*, Vol. 65, No. 7, 2023, pp. 1843–1849.

[21] Ma, Z., et al., "Quasi-Lumped-Element Filter Based on Substrate-Integrated Suspended Line Technology," *IEEE Transactions on Microwave Theory and Techniques*, Vol. 65, No. 12, December 2017, pp. 5154–5161.

[22] Ma, Z., K. Ma, and S. Mou, "An Ultra-Wide Stopband Self-Packaged Quasi-Lumped-Element Low Pass Filter Based on Substrate Integrated Suspended Line Technology," *2017 IEEE MTT-S International Microwave Symposium (IMS)*, June 2017, pp. 1084–1087.

[23] Feng, T., K. Ma, and Y. Wang, "A Miniaturized Bandpass Filtering Power Divider Using Quasi-Lumped Elements," *IEEE Transactions on Circuits and Systems II: Express Briefs*, Vol. 69, No. 1, January 2022, pp. 70–74.

[24] Feng, T., and K. Ma, "A 0.9 GHz Self-Packaged Power Amplifier Based on SISL Platform," *2017 IEEE MTT-S International Microwave Workshop Series on Advanced Materials and Processes for RF and THz Applications (IMWS-AMP)*, September 2017, pp. 1–3.

[25] Wang, Y., M. Yu, and K. Ma, "A Compact Low-Pass Filter Using Dielectric-Filled Capacitor on SISL Platform," *IEEE Microw. Wirel. Compon. Lett.*, Vol. 31, No. 1, January 2021, pp. 21–24.

[26] McDaniel, J. W., et al., "A Low-Loss Fully Board-Integrated Low-Pass Filter Using Suspended Integrated Strip-Line Technology," *IEEE Transactions on Components, Packaging and Manufacturing Technology*, Vol. 8, No. 11, November 2018, pp. 1948–1955.

[27] Li, L., K. Ma, and S. Mou, "Modeling of New Spiral Inductor Based on Substrate Integrated Suspended Line Technology," *IEEE Transactions on Microwave Theory and Techniques*, Vol. 65, No. 8, August 2017, pp. 2672–2680.

[28] Wang, Y., K. Ma, and S. Mou, "A Compact Self-Packaged Lumped-Element Coupler Using Substrate Integrated Suspended Line Technology," *2016 IEEE MTT-S International Microwave Symposium (IMS)*, May 2016, pp. 1–3.

CHAPTER 3

SISL Filters and Multiplexers

The rapid development of modern communication systems has led to increasingly high-performance requirements for RF transceiver systems. Filters, which filter signals at the front end of RF transceiver systems, are devices that can affect the quality of the system. Therefore, filter devices with advantages such as miniaturization, low insertion loss, high out-of-band rejection, and ease of integration have become the focus of developers' attention. This chapter uses SISL as the design platform, which has advantages over filters such as double-sided circuits, low loss, high Q value, self-packaging, and ease of integration.

3.1 Introduction

RF/microwave filters are essential components in modern microwave relay communications, satellite communications, wireless communications, and electronic countermeasure systems. They are also the most important and technologically advanced microwave passive devices. With the rapid development of modern communication demands, available spectrum resources are becoming increasingly scarce; thus, the requirements for the frequency selectivity characteristics of filters are becoming higher. In order to improve communication capacity and avoid interference between adjacent channels, filters are required to have steep out-of-band attenuation; to improve the signal-to-noise ratio (SNR), low insertion loss is required in the passband; and to reduce signal distortion, a flat amplitude-frequency characteristic and group delay characteristic is required in the passband. In order to meet the trend of miniaturization of modern communication terminals, filters are required to have a smaller size and weight. Traditional Butterworth filters and Chebyshev filters have difficulty in meeting these requirements. Filters with cross-coupling structures with finite transmission zeros are the most commonly used and best choice currently. Compared with traditional filters, this type of filter not only can meet the high selectivity requirements of the passband but also can reduce the number of resonant cavities, lowering design costs and filter size.

3.2 Compact Filters on the SISL Platform

3.2.1 Bandpass Filters

3.2.1.1 Design of Miniaturized 5G SISL BPFs

The differential inductor, which is frequently employed as an inductive component, is transformed into a resonator in [1]. This differential drive inductor resonator

(DDIR) may grow a second-order mode to a small size based on the varied electric field distributions on distinct modes. The spiral line resonator is frequently employed to create hybrid structures, as seen in Figure 3.1(a). Naturally, this hybrid structure's turns provide mutual capacitances. Additionally, mutual magnetism results from turns with the same current directions. Figure 3.1(b) depicts the surface current's direction. With a larger equivalent inductor, there is consequent reciprocal magnetism. As a result, the spiral line resonator's resonant frequency can be lowered.

It is suggested that the differential inductor, which is widely used as an inductive component, can be transformed into a DDIR, as shown in Figure 3.1(c). The turns of the DDIR still contain mutual capacitances and magnetism, which lower the resonate frequency to that of a spiral line resonator. Additionally, the crossover components will significantly affect the resonating mechanism. The capacitors cause the electrical length along the lines to be less than when DDIR is operating in basic resonance mode. Field symmetry exists. Edges are where charges are most prone to buildup, and this is also where the electric field's amplitude is strongest.

An electric dominating coupling structure that is frequently utilized is the interdigital capacitor. The layout of the second-order bandpass filter (BPF) is shown in the top view in Figure 3.2(a). The whole 3D view and the top view of this double-layer interdigital capacitor are also shown. This double-layer interdigital capacitor can achieve the desired value with a smaller footprint than single-layer interdigital capacitors or parallel-plate capacitors. The interdigital capacitor can be calculated using six variables, as illustrated in Figure 3.2(a).

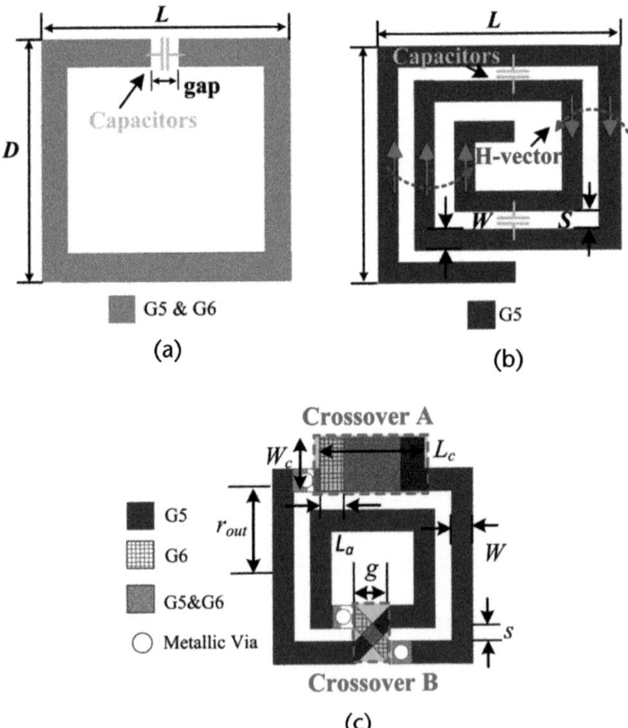

Figure 3.1 Layout of: (a) spiral line resonator in basic mode with generated capacitors, current direction, and magnetic field line and (b) DSRR in basic mode with generated capacitors; and (c) structure of DDIR on Sub3 [1].

The design and construction of the second-order BPF1 with electric dominant coupling have been completed. Figure 3.2(b) displays the outcomes of BPF simulation and measurement. It has a 770-MHz frequency center and an 8.4% fractional bandwidth (FBW). The insertion loss is 1.14 dB. The magnitude of the second-order mode is 5.7 times that of the fundamental band.

The symmetrical fourth-order BPF containing four poles and four zeros is likewise produced under the electric coupling structure. Figure 3.2(c) displays the simulation and measurement findings. BPF2 has an FBW of 10.5% and is centered

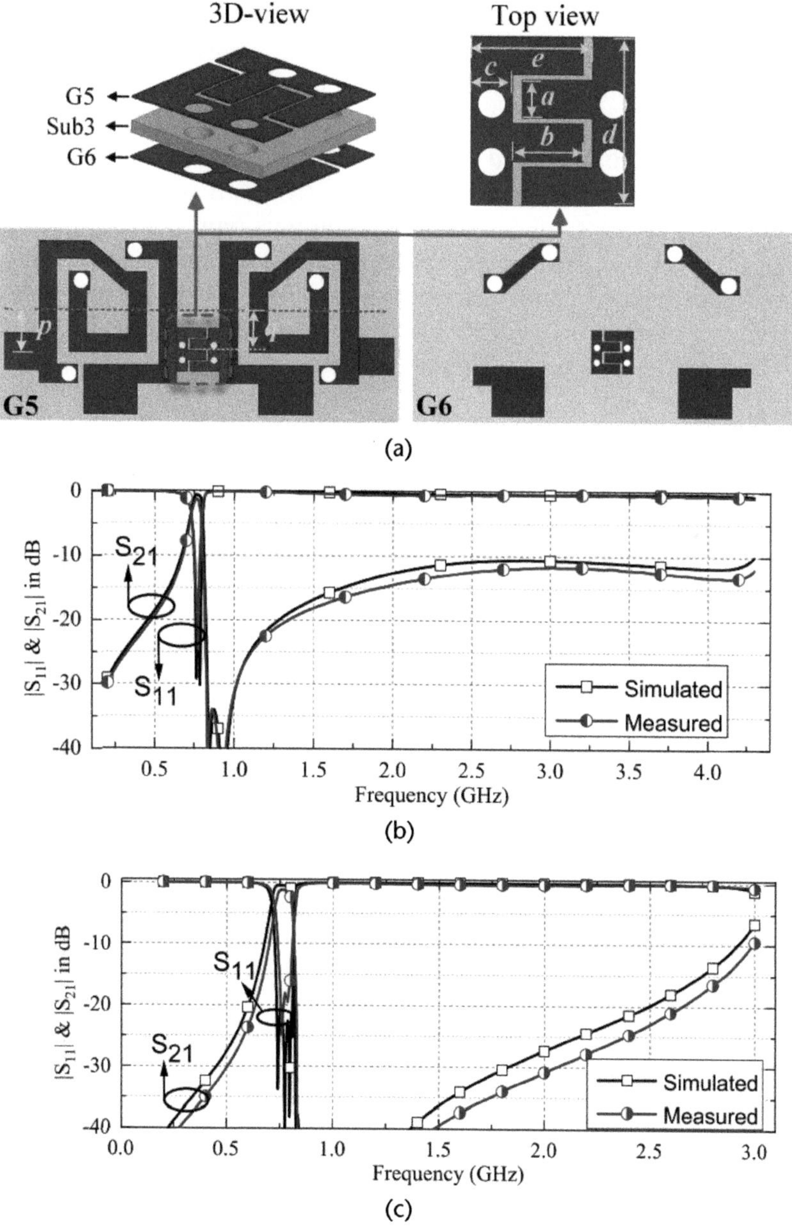

Figure 3.2 (a) Top view of the second-order bandpass filter on Sub3. Simulated and measured results of (b) BPF1 and (c) BPF2 [1].

at 760 MHz, and 1.01 dB is the insertion loss. The miniature size is obtained with 0.0052 λ_g^2. The second-order mode is 3.9 times the fundamental band in magnitude. Both the very small BPF1 and BPF2 perform well without the aid of lumped components or substrates with a high permittivity. The second-order modes are widened without adding more circuits in the cascade.

For the aforementioned designed BPFs, the lower stopband rejection is lower than the higher stopband rejection. The mathematical analysis indicates that TZ1 is generated by the DDIR itself. TZ1 always appears above the frequency of the DDIR's self-resonance. When magnetic coupling dominates mixed coupling, TZ2 can be relocated to the lower stopband to enhance the lower stopband's roll-off.

A T-shaped coupling structure is initially suggested on SISL to achieve the ideal placement of zeros by strengthening the magnetic coupling and reducing the electric coupling, hence improving the rejection of the stopband.

Edge coupling is used to produce magnetic dominant coupling, as seen in Figure 3.3(a). Narrower gap g_s and thinner W_s can strengthen the connection. The frequency response of S_{21} versus W_s and g_s is represented by black lines in Figure 3.3(b). TZ1 is in the lowest stopband in this magnetic dominant coupling. As W_s and g_s drop, TZ1 can lower its frequency and widen its bandwidth. However, the coupling coefficient is still not enough to provide the required 100-MHz bandwidth. There are limitations on the manufactured circuit board's production capability.

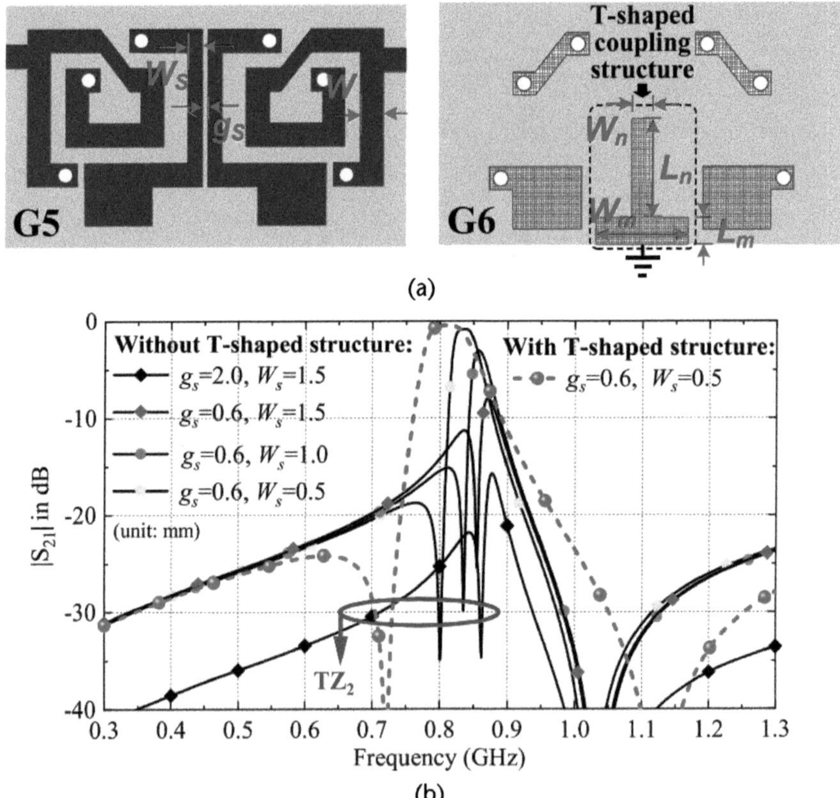

Figure 3.3 (a) Construction of a second-order filter with a T-shaped structure using magnetic dominating coupling, and (b) frequency response of S_{21} versus g_s and W_s [1].

However, loss is greater in transmission lines that are narrower and have a higher characteristic impedance.

In the context of magnetic-dominant coupling, the total coupling, denoted as $K = K_M - K_E$, is obtained by subtracting electric coupling from magnetic coupling. Consequently, elevating the overall coupling can be achieved by both diminishing the electric coupling and augmenting the magnetic coupling.

The results for the second-order filter with magnetic-dominant coupling (BPF3) combine simulations and measurements. Simulated and measured outcomes closely match. BPF3 had an 11.9% FBW centered at 798 MHz. Compared to BPF1, the 20-dB roll-off rate dropped significantly by 58.8% from 12.3 to 5.07. The second-order mode was about 4.6 times higher than the fundamental mode. The physical size was 13.5 mm × 23.4 mm, corresponding to only 0.034 $\lambda_g \times 0.062\ \lambda_g$.

The fourth-order filter is created and constructed employing both magnetic and electric dominating coupling. The results from simulation and measurement agree well. The BPF had a 14.1% FBW and operated at 778 MHz. The fundamental mode was 4.48 times the second-order mode frequency. Its physical size was 15 mm × 52.7 mm, equivalent to just 0.039 $\lambda_g \times 0.137\ \lambda_g$.

3.2.1.2 Design of a Self-Packaged BPF with Controllable TZs

The construction of a high characteristic BPF with controlled TZs is presented in [2] using a unique dual-mode circular patch resonator (DMCPR).

Figures 3.4(a, b) illustrate the structure of the conventional circular patch resonator and the proposed DMCPR, respectively, under conditions of weak coupling. In Figure 3.4(c), the frequency responses of two resonators are displayed under weak coupling conditions. Figure 3.4(c) emphasizes the significance of the two inserted notches in effectively distinguishing the degenerate modes of the circular resonator and achieving a lower center frequency to minimize the size of this DMCPR.

In Figure 3.4(d), two modes are excited using high-impedance lines on the M5 layer of double circuits. The metal layers M5 and M6 are connected via plated-through-holes. Additionally, two feeders contribute to cross-coupling from source to load, introducing multiple transmission zeros (TZs) across the passband. L1 adjusts magnetic coupling by altering the width of the square patch, while T1 fine-tunes electric coupling by varying the separation between the two feeders. These adjustments are made to distinctly separate electric and magnetic coupling in the source-to-load connection.

The observed data shows the BPF has a center frequency of 5.48 GHz and an FBW of 14.8%, while the simulated data indicates a center frequency of 5.54 GHz and an FBW of 11.4%.

This 5.5-GHz BPF, which is self-packaged and cost-effective, not only allows for convenient control of TZs but also simplifies degenerate mode operation. Additionally, the steep rejection in the stopbands can be attributed to multiple TZs present in the upper and lower sidebands of the passband.

3.2.1.3 Design of Ka-Band BPFs Using Folded Half-Wavelength Resonators

In [3], we introduce a novel folded half-wavelength ($\lambda/2$) resonator based on the SISL platform. This innovative resonator design capitalizes on the equipotential property

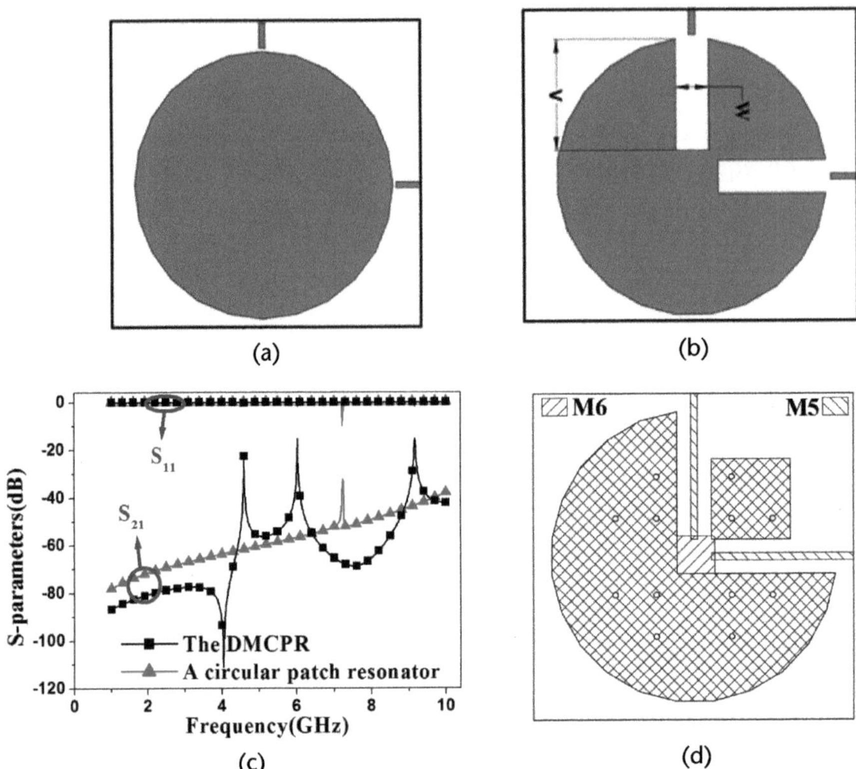

Figure 3.4 (a) A circular patch resonator, (b) the DMCPR, (c) the frequency response of two resonators under weak coupling, and (d) the top-side view of this double-layer BPF [2].

of folded resonators and employs substrate excavation technology to effectively disperse electromagnetic fields in the air. This approach substantially minimizes dielectric substrate losses and reduces the required substrate thickness.

The conventional $\lambda/2$ resonator with short-circuited ends on both sides, as depicted in Figure 3.5(a), serves as the basis for our new foldable structure, as illustrated in Figure 3.5(b). By folding the resonant unit and connecting it via holes, we achieve an equal upper and lower electric potential along the transmission line, as evident in Figure 3.5(b). Consequently, there is no potential difference and minimal substrate loss. This approach offers a dual benefit: it reduces the size by folding in half and significantly mitigates losses.

We used SISL technology to model the field distribution of these two resonators, confirming reduced losses in the folded resonator design. The air cavity serves as the primary location for storing electric-field energy. Figure 3.5(c) shows significant energy dissipation on Substrate 3, while Figure 3.5(d) demonstrates minimal loss on Substrate 3 in the folded resonator. Consequently, folded resonators exhibit lower dielectric losses compared to conventional designs.

To validate our theoretical findings, we created and processed second-order and fourth-order BPFs based on stepped-impedance resonator (SIR). In Figure 3.5(e), the electromagnetic simulations and measurements for the second-order filter closely match. We observed a TZ in the upper passband, likely due to electromagnetic

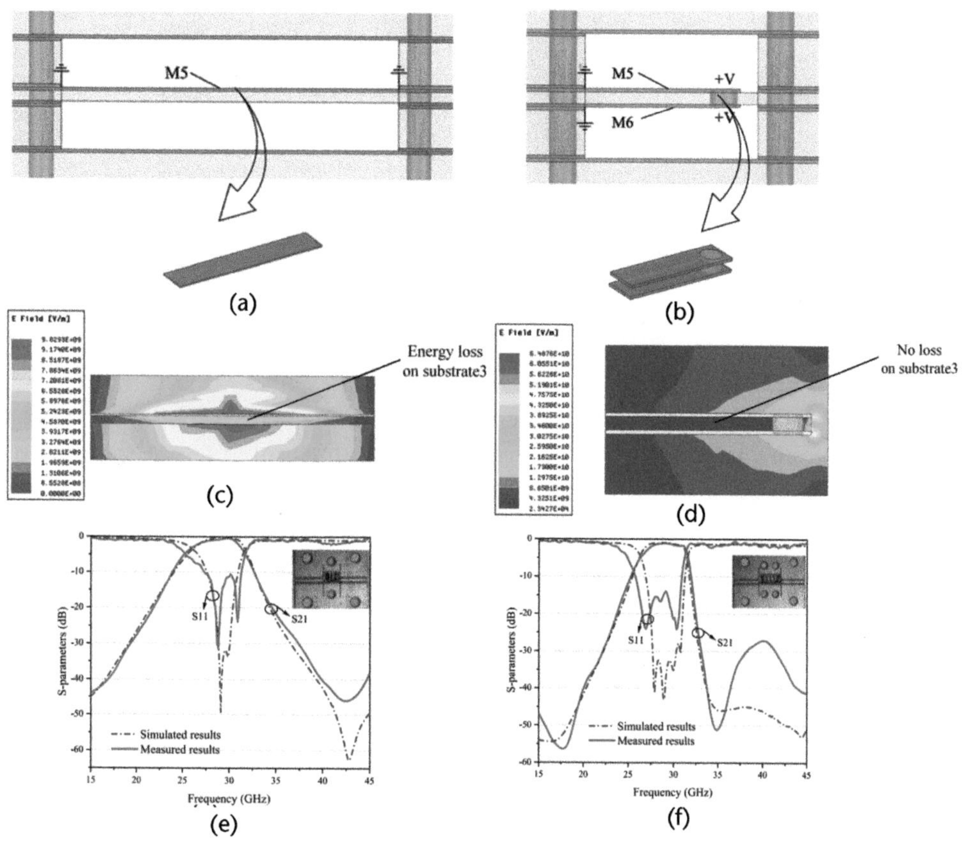

Figure 3.5 (a) Cross-sectional view of the conventional unfolded short-circuited $\lambda/2$ resonator structure. (b) Cross-sectional view of the proposed folded short-circuited $\lambda/2$ resonator structure. EM field distribution of (c) conventional unfolded short-circuited $\lambda/2$ resonator. (d) Folded short-circuited $\lambda/2$ resonator. Measured and simulated S-parameters of the designs of (e) proposed folded second-order BPF, and (f) proposed folded fourth-order BPF [3].

coupling between the resonators, as described in Hong et al.'s classical theory. The measured insertion loss (IL) is approximately 0.78 dB at the passband center frequency of 29.2 GHz.

The design process for the fourth-order filter closely resembles that of the second-order filter. Figure 3.5(f) displays the simulation findings in comparison to the measurement. At 29.2 GHz, its observed IL is 1.19 dB. Fabrication tolerances are what caused the measurement to slightly change.

Furthermore, it is evident that each passband, both upper and lower, exhibits two TZs. We postulate that these dual TZs are likely a result of cross-coupling between the resonators. This arises because each resonator is not isolated, given that their ground terminals are interconnected. Nevertheless, the filter maintains a reasonably compact and small size, a characteristic supported by electromagnetic simulations. The coupling coefficient between resonators i and j is denoted as Mij. The filter's symmetrical structure implies a symmetric coupling matrix, with $M12 = M34$, as depicted in the following:

$$M = \begin{pmatrix} 0 & 1.9031 & 0.1799 & -0.0898 \\ 1.9031 & 0 & 1.5990 & 0.1699 \\ 0.1799 & 1.5990 & 0 & 1.9031 \\ -0.0898 & 0.1699 & 1.9031 & 0 \end{pmatrix} \tag{3.1}$$

The enormous amount of information and the urgent need for portability in modern society have driven the rapid development of communication technology towards broadband, high capacity, and miniaturization. Single-band filters are increasingly unable to meet the requirements of fast-paced information transmission [3], and dual-frequency and multifrequency filters have gradually taken the stage of the era. The biggest feature of dual-frequency filters is that they can simultaneously select two different frequency bands, saving system volume and cost, enabling communication systems to operate in two frequency bands at the same time, better utilizing spectrum resources, and reducing the size of filters required to achieve dual-band filtering. This is a big step in the development process of wireless communication systems. The following are some dual-band filters based on the SISL structure.

3.2.2 Dual-Band Bandpass Filters

3.2.2.1 Design of Multiple-Zeros Dual-Band Bandpass Filter

On the SISL platform, you can design not only single-passband filters but also dual-passband filters [4–9]. Figure 3.6(a) illustrates the dual-band cascaded quadruple filter structure that was suggested in [5]. Resonators R1 and R1′ together form the first band, known as band 1, while R2 and R2′ together form the second band, known as band 2. Magnetic couplings M1, M2, and M3 as well as electronic couplings E1, E2, and E3 are employed. According to the separate electric and magnetic coupling path (SEMCP) theory from [10], E1 and M1 are used between synchronously tuned resonators R1 and R1′ to produce one TZ1 in band 1. Similar to this, due to the distinct coupling path, E2 and M2 produce another TZ near band 2. A TZ is produced by applying E3 and M3 to the synchronously tuned resonators R1 and R2, in accordance with [10]. As a result, this DBBPF's out-of-band property produces at least three independently controllable TZs, which makes it attractive.

An equivalent circuit model, depicted in Figure 3.6(b), clarifies the operation of various resonators under various coupling routes. In contrast to the initial study in [10] on SEMCP between synchronously tuned resonators, this architecture uses SEMCP between asynchronously tuned resonators. The first two resonators, R1 and R2, might resonate at $\omega_{01} = (L_1C_1) - 1/2$ and $\omega_{02} = (L_2C_2) - 1/2$, where C_m and L_m stand for magnetic and electric coupling, respectively. With regard to both electric and magnetic coupling, [11] discussed the transmission poles of asynchronously tuned resonators, shedding light on the resonant frequencies.

$$\omega_1 = \sqrt{\frac{\Re_B - \Re_C}{\Re_A}}, \ \omega_2 = \sqrt{\frac{\Re_B + \Re_C}{\Re_A}} \tag{3.2}$$

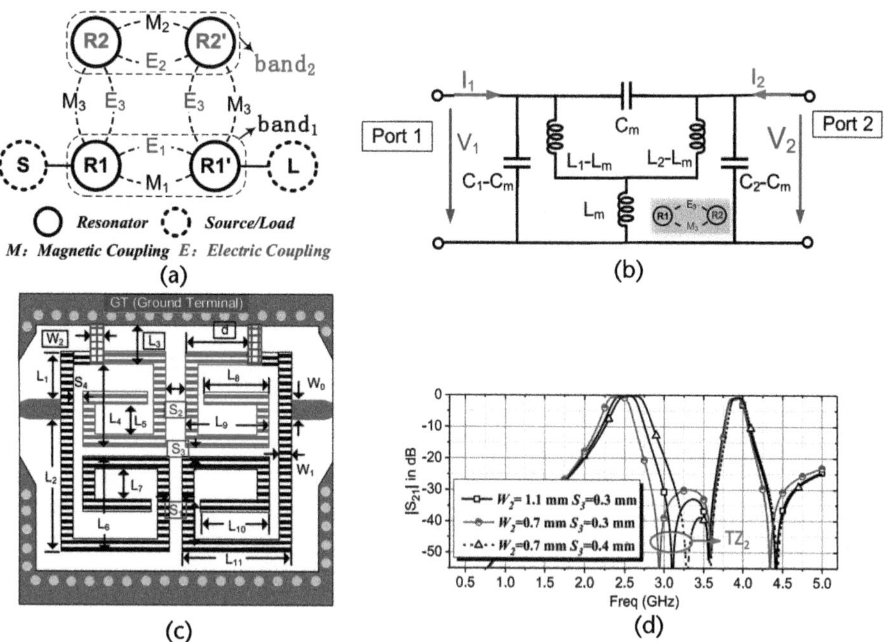

Figure 3.6 (a) The topology structure of the proposed DBBPF, (b) circuit model of asynchronously tuned resonators under mixed electric and magnetic coupling, (c) top view of the DBBPF layout on G5, and (d) the S_{21} of DBBPF under different width of W2 and width of S3 [5].

in which

$$\Re_A = 2\left(L_1C_1L_2C_2 - L_m^2C_1C_2 - L_1L_2C_m^2 + L_m^2C_m^2\right)$$
$$\Re_B = \left(L_1C_1 + L_2C_2 - 2L_mC_m\right)$$
$$\Re_c = \sqrt{\Re_B^2 - 2\Re_A}$$

The electric coupling coefficient k_e and magnetic coupling coefficient k_m are defined as

$$k_e = \frac{C_m}{\sqrt{C_1C_2}},\quad k_m = \frac{L_m}{\sqrt{L_1L_2}} \tag{3.3}$$

and $C_m < C_1$ and C_2, $L_m < L_1$ and L_2. The overall coupling coefficient is

$$k_x = \left|k_e - k_m\right| \tag{3.4}$$

The discussion has covered the coupling coefficient and resonant frequencies; however, there are limited studies on the first-order response of asynchronously tuned resonators using SEMCP. The admittance matrix for this equivalent circuit can be expressed as:

$$\begin{bmatrix} I_1 \\ I_2 \end{bmatrix} = \begin{bmatrix} Y_{11} & Y_{12} \\ Y_{21} & Y_{22} \end{bmatrix} \cdot \begin{bmatrix} V_1 \\ V_2 \end{bmatrix} \tag{3.5}$$

With

$$\begin{aligned} Y_{12} &= \left.\frac{I_1}{V_2}\right|_{V_1=0} = j\omega C_m - \frac{L_m}{j\omega\left(L_1L_2 - L_m^2\right)} \\ &= j\omega C_m - \frac{1}{j\omega L_m\left(\dfrac{L_1L_2}{L_m^2} - 1\right)} = Y_{21} \end{aligned} \tag{3.6}$$

When

$$\omega = \omega_m = \frac{1}{\sqrt{L_mC_m\left(\dfrac{L_1L_2}{L_m^2} - 1\right)}} \tag{3.7}$$

the Y_{12} (Y_{21}) equals zero. Since

$$S_{21} = \frac{-2Y_{21}Y_0}{\left(Y_{11} + Y_0\right)\left(Y_{22} + Y_0\right) - Y_{12}Y_{21}} \tag{3.8}$$

The transmission zero of S_{21} is also located at ω_m. Additionally, synchronously tuned resonators with SEMCP can use the equation for ω_m in (3.7).

According to (3.7), under electric and magnetic coupling, synchronously tuned resonators will experience a transmission zero. The following guidelines can be used to pinpoint where the zero is:

- ω_m changes to a higher frequency as K_m rises. In addition, the frequency of ω_m also rises when mutual inductance L_m rises or starting inductances L_1 and L_2 fall.
- However, as capacitance C_m rises, ω_m shifts to a lower frequency. K_e is altered when C_1 and C_2's values are altered without affecting C_m, but ω_m's position is unaffected.
- By comparing the values of ω_1, ω_2, and ω_m in (3.2) and (3.7), it is possible to determine the relationship between ω_m, ω_1, and ω_2 as $\omega_1 > \omega_m$, $\omega_1 < \omega_m < \omega_2$, or $\omega_m > \omega_2$.

The configuration of the planned DBBPF, which uses spiraling $\lambda/4$ wavelength resonators, is depicted in Figure 3.6(c). The ground terminals (GT) on the blue stubs are shared by R1 (R1′) and R2 (R2′). The red stubs and GT (stub2) in Figure 3.6(c) correspond to R2 and R2′ in Figure 3.6(a), while the black stubs and GT (stub1) in Figure 3.6(a) represent R1 and R1′. The hybrid layout allows for numerous electric and magnetic couplings as needed, while the spiraling stubs have the benefit of being small in size. Each $\lambda/4$ resonator's input impedance can be calculated as follows:

$$Z_{\text{in}} = jZ_0 \tan(\beta z) \tag{3.9}$$

The basic transmission pole is produced when $z = \pi/4$ at resonance $Z_{\text{in}} = \infty$. The lengths of stub 1 and stub 2 are used to calculate the center frequencies. Consequently, the frequency ratio can range from 1/3 to 3. The gaps' widths, namely S_1, S_2, and S_3, encircled in Figure 3.6(c), independently control the edge coupling of E_1, E_2, and E_3. The dimensions (W_2, L_3, and d) of GT dictate the magnetic coupling strength of M_1, M_2, and M_3. The picture is split into two halves, synchronously tuned coupling and asynchronously tuned coupling, to clarify the zero mechanism of the DBBPF.

A. Asynchronously Tuned SEMCP Resonators (R1 and R2)

According to Figure 3.6(c), the coupling between R1 (R1′) and R2 (R2′) is what gives band 2 its energy. Between R1 and R2, there are two distinct coupling pathways, E_3 and M_3. The distinct E_3 and M_3 coupling paths result in a transmission zero (TZ2), which is based on the mathematical computation in Section II. While the width of W_2 and the length of L_3 control M_3, the length of L_{10} and the width of S_3 control E_3. Figure 3.6(d) shows the impact of TZ2 produced by the SEMCP between R1 and R2 on the breadth of W_2 (M_3) and S_3 (E_3). E_3 predominates the connection, and TZ2 appears between two passbands. Stronger M_3 results in TZ2 shifting to a higher frequency, while decreasing E_3 causes TZ2 to move toward a higher frequency.

B. Synchronously Tuned SEMCP Resonators

The following are synchronously tuned SEMCP resonators:

1. Figure 3.7(a) shows a weak coupling circuit made up of S-R1-R1′-L that alters the response of band 1 in order to illustrate the coupling mechanism between synchronously tuned SEMCP resonators R1 and R1′. Between R1 and R1′, there is an electric connection (E_1) and a magnetic coupling (M_1). The length of L_6 and the breadth of S_2 can be changed to alter E_1, which mostly determines the coupling coefficient. As a result, S_2 and L_6 can be used to directly alter the bandwidth of band 1. The strength of M_1 is weak and is controlled by the separation of two GTs (d). The mathematical analysis in the discussion in the previous section and [10] both indicated that a TZ can be produced from SEMCP at the lower stopband. M_1 has a bigger impact on where the TZ is located than E_1, which predominates in terms of coupling coefficient. While Figure 3.7(e) displays the frequency response of the

entire circuit as a function of d, Figure 3.7(c) shows the frequency response of the S-R1-R1′-L circuit for various values of d. Notably, the TZband1 of the S-R1-R1′-L circuit and the TZ1 of the full circuit clearly overlap in frequency. Both TZband1 and TZ1 are located in the same general area, moving to higher frequencies as d decreases, indicating a stronger M_1. As a result, it can be said that the TZ1 can be controlled and is produced by SEMCP between R1 and R1′.

2. R2 and R2′: Figure 3.7(b) illustrates the S-R2-R2′-L circuit used to investigate the operating mechanism of band 2. The frequency response of the S-R2-R2′-L circuit is shown in Figure 3.7(d). The dominant coupling E_1, which is defined by the size of S_2 and the length of L_4, can affect the bandwidth of band 1 by using the same operating principle as band 2. Furthermore, the creation of TZband2 at lower bands will be aided by the weak M_1 coupling from GT. The simulation findings shown in Figure 3.7(d, e) show that the DBBPF's TZ3 comes from the SEMCP between R2 and R2′, and that as M_2 rises, its location shifts to a higher frequency. Three TZs are produced by the SEMCP theory in this second-order DBBPF, while a fourth zero, TZ4, is produced via harmonic effects, which improves the rejection of the higher stopband. With these multiple TZs, excellent out-of-band rejection is guaranteed. The position of the feeder lines can change the Qe of this DBBPF; a shorter L1 will have a weaker coupling and a higher Qe [11].

For 5G sub-6-GHz applications, this section [6] suggests a cross-coupled dual BPF. The filter utilizes SEMCPs and introduces source-to-load couplings for the flexibility of heterogeneous feed lines. The dual-band filter produces a total of five controllable TZs and achieves good overall performance by employing two pairs of quarter-wavelength stepped-impedance resonators (SIRs). To reduce the filter size and increase the controllable range of the second passband, the SIRs forming the second passband are folded onto two different layers. The BPF arrangement is shown in Figure 3.8(a) with four resonators labeled R1, R3, R4, and R2 in order from left to right. In Figure 3.8(b), the BPF architecture is shown.

Case 1: Analysis of Single-Band Performance

Figure 3.8(c) depicts the setup where SIR resonators R1 and R2 provide the first passband in the frequency range of 3.3–3.6 GHz along with the feed lines (SIR-I). With an additional degree of freedom and more convenience for modifying capacitance and coupling coefficient, the load stubs on top of R1 and R2 can affect electric coupling. The topology of Figure 3.8(d) shows the coupling architecture of SIR-I. The electric (E) coupling in SIR with SEMCPs [10] is expressed by the equivalent gap capacitance, C_m, which also represents the E coupling path. The magnetic (M) coupling path is also formed by the common shorted transmission line, and it is symbolized by the analogous inductance, L_m. The coupling coefficient can then be expressed as follows:

$$C = \frac{\omega_{\text{odd}}^2 - \omega_{\text{even}}^2}{\omega_{\text{odd}}^2 + \omega_{\text{even}}^2} = F\left(Y_c L_m - C_m Z_c\right) = M - E \tag{3.10}$$

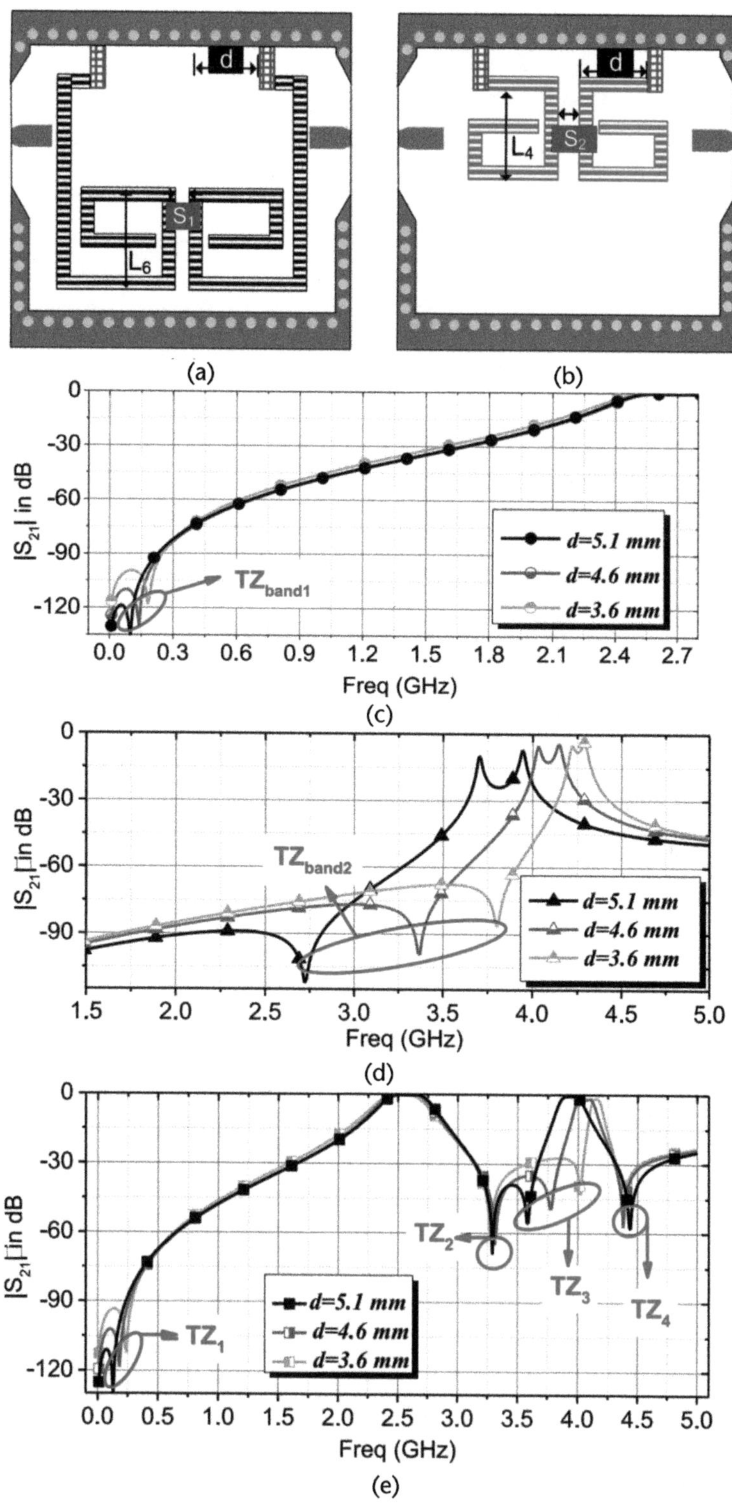

Figure 3.7 Circuit structure of (a) *S*-R1-R1′-*L* and (b) *S*-R2-R2′-*L*. The simulation results of: (c) circuit *S*-R2-R2′-*L* in (a), (d) circuit *S*-R1-R1′-*L* in (b), and (e) complete circuits [5].

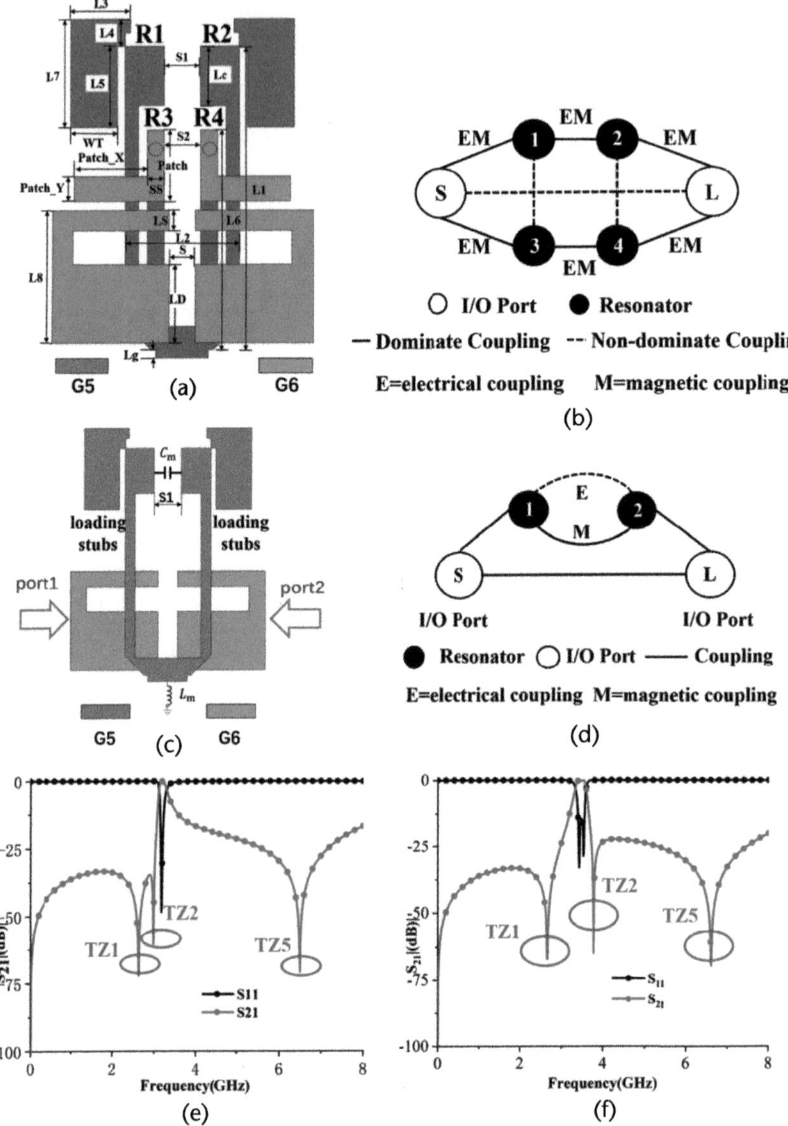

Figure 3.8 (a) Configuration of the dual-band BPF, (b) filter topology, (c) main coupling pattern of SIR-I, (d) topology of SIR-I, (e) *M* dominates the performance of SIR-I, and (f) *E* dominates the performance of SIR-I [6].

Here, ω_{odd} and ω_{even} are the resonant angular frequencies for odd and even modes. Z_c is $1/Y_c$ and stands for the characteristic impedance of SIR. Additionally,

$$F = \frac{4\left[\sqrt{\varepsilon_{\text{re}}}l + \left(Y_c L_m - C_m Z_c\right)c\right]}{\left(2Y_c L_m + \sqrt{\varepsilon_{\text{re}}}l\right)^2 + \left(2C_m Z_c + \sqrt{\varepsilon_{\text{re}}}l\right)^2} \tag{3.11}$$

For SIR-I, the length of the resonators is equal to l_1, and the variable l denotes their length. The symbol for the speed of light in empty space is c. The link between the center frequency during resonance and the TZ frequency of S_{21} is determined by the relationship between M and E in [12]. Equation (3.10) demonstrates that the coupling coefficient C, which, in turn, determines the FBW, is also determined by the difference between M and E. As a result, the TZ point is closer to center frequency, the values of C and FBW are less, and the coupling strength between E and M is stronger. The cancellation of the E and M coupling results in a TZ (TZ2) in the upper stopband when M ($M > E$) dominates SIR-I, as depicted in Figure 3.8(e). The E coupling (C_m) rapidly grows and approaches the intensity of the M coupling as the gap S_1 shrinks, causing TZ2 to approach center frequency. As E eventually surpasses M in SIR-I, TZ2 enters the lower stopband, as depicted in Figure 3.8(f).

R3 and R4 are added to the original structure to add an extra passband inside the 4.8 to 5 GHz frequency range. By using the R3, R4, and U-type feed lines (SIR-II), the second passband is produced. The ends of R3 and R4 are folded and extended onto G5 to make best use of the available space and broaden the range of controlled center frequencies. The topology of SIR-I and SIR-II is identical, as seen in Figure 3.8(d). M coupling also predominates in SIR-II, leading to the production of three TZs. SEMCPs produce TZ4, much like TZ2 does. The E couplings of SIR-II are enhanced when the value of S_2 is decreased, shifting TZ4 closer to the center frequency.

Case 2: Analysis of Dual-Band Performance

Combining the parallel single-band responses between the source and load will yield the dual-band filter features [13]. Thus, a dual-band BPF can be achieved by connecting the SIR-I and SIR-II in parallel.

Let's examine the first passband's performance first. TZ2 is seen in the upper stopband when SIR-I is M dominant. The TZ2 gradually shifts to the left as the E coupling (C_m) rises by decreasing the gap S_1. In addition, by lowering the coupling coefficient C, the enhanced E coupling (C_m) can reduce the FBW of the first passband.

TZ2 appears in the lower stopband as SIR-I shifts from M to E dominance and slowly advances to the left as the E coupling (C_m) or the M coupling (L_m) changes. The FBW of the second passband is similarly decreased by the higher E coupling (C_m).

Moreover, [7] proposed a dual-band bandpass filter (DBBPF) designed using a short-stub-loaded resonator and a pair of modified feed lines that excite the circuit on G6 and simultaneously form the second passband and introduce TZs through input-output coupling. The use of a patterned suspended substrate structure leads to loss reduction of 0.22 and 0.24 dB in the two passbands. Figure 3.9 illustrates the structure of the SISL DBBPF.

Two modified feed lines on G_5 and an SSLR on G_6 make up the proposed double-balanced bandpass filter. The SSLR, which is shown in Figure 3.9(a), creates the first passband, which may be examined using both even-mode and odd-mode analysis. Figures 3.9(b, c) show examples of the even and odd modes, respectively. In this study, C_m stands for the equivalent gap capacitance, or the electric coupling path, and L_m for the equivalent inductance, or magnetic coupling path, of the shorted transmission line.

Reference [8] provided the frequencies of the resonator for both even and odd modes:

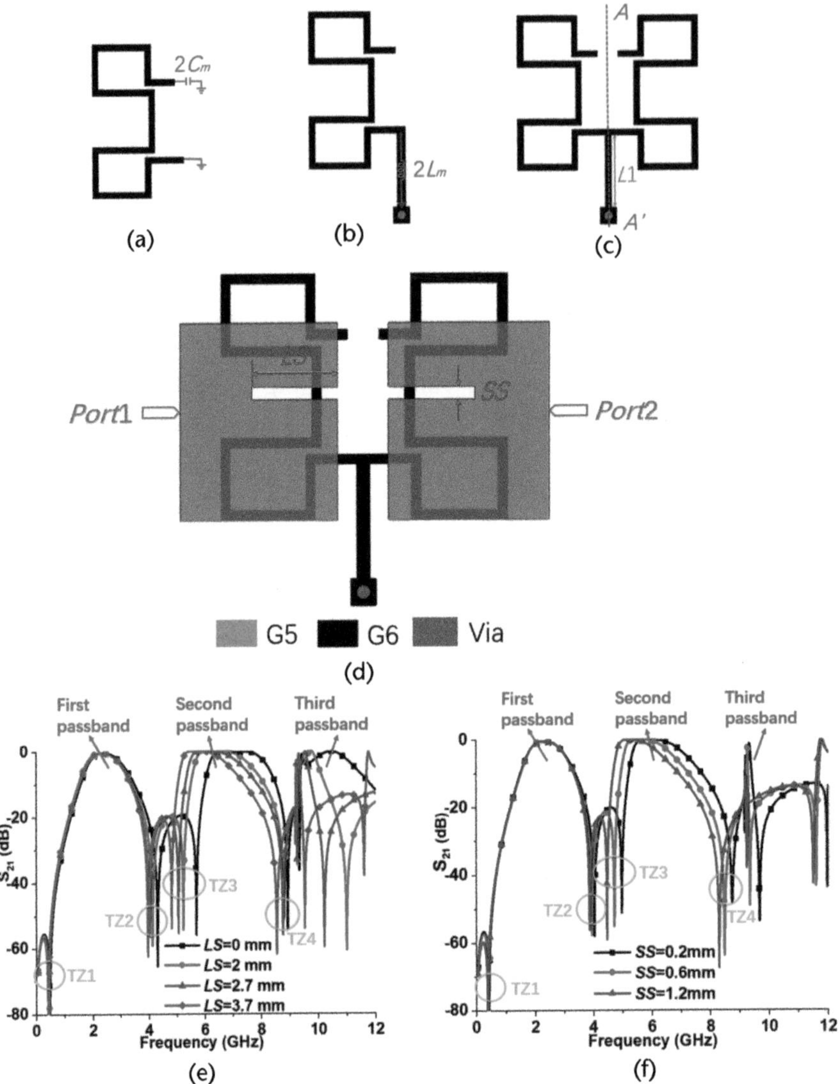

Figure 3.9 (a) SSLR, (b) odd mode, (c) even mode, (d) SSLR with strong coupling, (e) operating frequencies against slot length L_S, and (f) operating frequencies against slot width S_S [7].

$$f_e = \frac{\omega_e}{2\pi} = \frac{1}{4\left(2Y_cL_m + A\right)} \tag{3.12}$$

$$f_o = \frac{\omega_o}{2\pi} = \frac{1}{4\left(Z_cC_m + A\right)} \tag{3.13}$$

Let $A = \sqrt{\varepsilon_{\text{re}}}l/c$ where l represents the length of the transmission line. The working frequency f can be expressed as $f = (f_e + f_o)/2$. As the value of $L_1(L_m)$ increases in Figure 3.9(a), the even-mode frequency decreases, while the odd-mode frequency remains relatively unchanged. This results in a variation of the fractional bandwidth of the first passband. SSLR's interstage coupling (C) comprises two components: electrical coupling (E) and magnetic coupling (M).

The SSLR design under discussion is an M-dominant filter because of its poor electric coupling and substantial magnetic coupling. According to [7], the value of C_m determines whether an extra TZ exists. In particular, a single TZ is produced between the two passbands when $C_m = 0$, whereas two TZs are produced when $C_m > 0$. As a result, the frequency response of the filter exhibits two TZs, designated TZ2 and TZ3, between the two passbands, as shown in Figure 3.9(b). Source-to-load coupling is added in Figure 3.9(d) to improve the filter's selectivity and make it easier to create more TZs. The extra TZs (TZ1 and TZ4) are situated in the lower and higher stopbands, respectively, as shown in Figure 3.9(e). The second passband of the filter is formed by the chamfered feed lines in Figure 3.9(d), which can be seen as pairs of U-type resonators. Together, the SSLR on the G_6 and the redesigned feed lines on the G_5 produce the third passband. While electromagnetic coupling forms at the short ends of the circuits on G_6 with G_5, the open ends of the resonators on G_6 couple electrically with the G_5 circuit. We pick the slot position in the center of the resonator to ensure appropriate coupling with the G_6 circuit. The second passband shifts to a lower frequency when the slot size is increased, creating a longer current route. L_m affects the electromagnetic coupling between the circuits on G_5 and G_6. As seen in Figure 3.9(e), increasing L_1 extends the second passband. In particular, the frequency of the second passband is maximum when $L_S = 0$ (i.e., the feed lines are uncut). The frequency response of the filter is also influenced by the slot width S_S, as shown in Figure 3.9(f). The second passband shifts to a lower frequency as S_S increases and the third passband is suppressed.

The computed and measured results, with measured bandwidths of 1.95–3.04 GHz and 4.92–6.05 GHz. As a result, at 2.4/5.2 GHz, the two passbands' fractional bandwidths and insertion losses are 45.43%/21.64% and 0.59 dB/0.55 dB, respectively. This suggested filter has advantages for self-packaging, numerous TZs, and low insertion loss.

3.2.2.2 Design of SISL Filters Based on Patch Resonators

This section introduces two compact filters that employ patch resonators based on the SISL platform [8]. The first filter utilizes multiple coupling schemes and a shorted-stub-loaded patch resonator. The BPF achieves third-order filtering, resulting in a size reduction of 98.18%. Furthermore, the BPF produces five TZs through mixed electric and magnetic coupling, cross-coupling, and source-to-load coupling. The second filter is an SISL DBBPF that employs an improved slotted patch resonator, resulting in six TZs.

Figure 3.10(a) demonstrates a new configuration that employs three resonators to achieve separate E/M coupling of the source to load. Additionally, resonators R1 and R3 are cross-coupled, enabling multiple and controllable TZs in filter designs by utilizing various coupling schemes. The self-packaging SISL platform is used to implement the suggested topology.

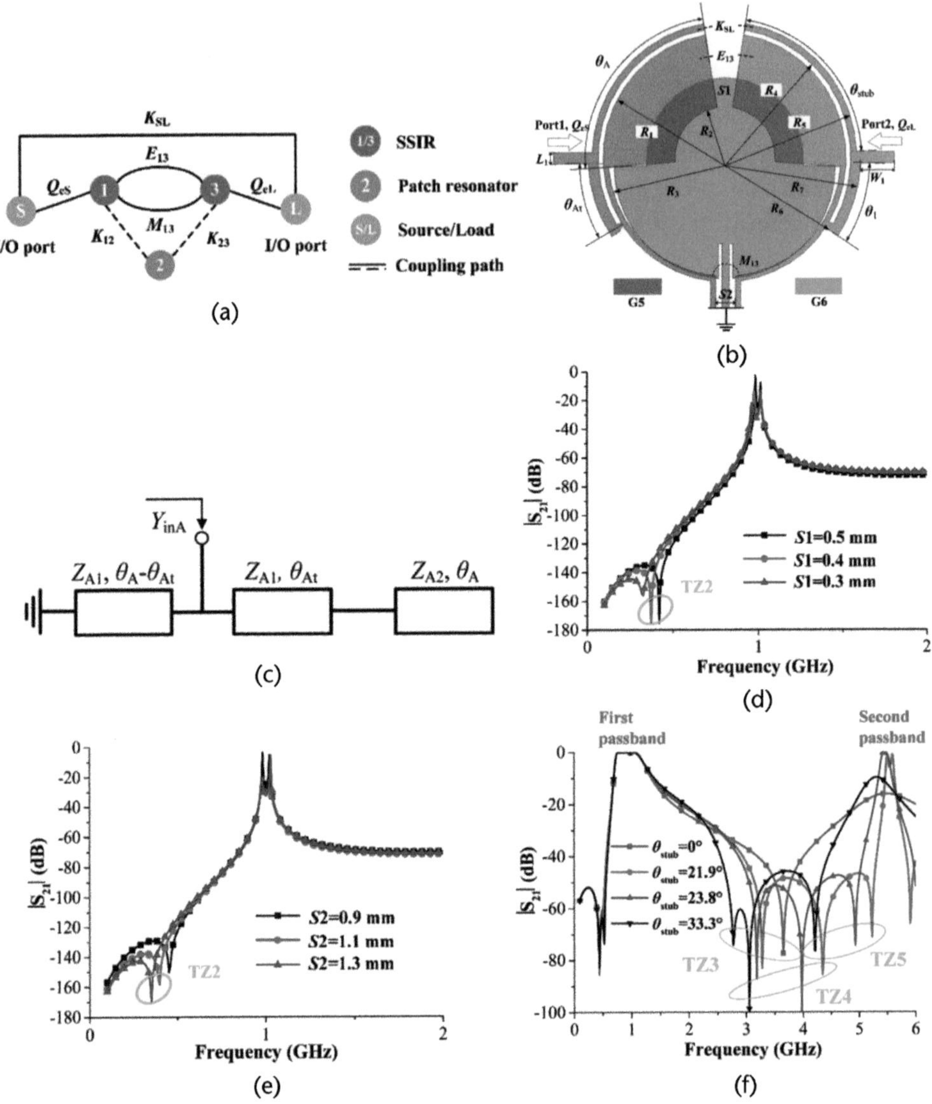

Figure 3.10 (a) The topology of the proposed filters, (b) flowchart for the proposed patch resonator's design process, (c) an SSIR's circuit model, (d) the proposed SSIRs' frequency response to various gap S1s, (e) the proposed SSIRs' frequency response to various gap S_2 values, and (f) simulation of the single-band BPF's $|S_{21}|$ against various stub [8].

Patch I, a patch resonator with a rectangle and two semicircles of various radii, is the basic resonator that is used. To downsize Patch I, a shortened stub is attached to its internal, with a length of L_3. At a resonance frequency of 1 GHz, Patch I and Patch II exhibit resonant behavior, with R_{patch1} and R_{patch2} values of 60 mm and 80 mm for Patch I, and 7.8 mm and 11 mm for Patch II, respectively.

A shorter stub allows Patch II to significantly reduce its size compared to Patch I by 98.18%. A clearly defined skirt is produced by Patch II's inherent TZ (TZ1) in the lower stopband. A high degree of isolation between the patch resonator's first two resonant modes may be seen in the first harmonic mode of Patch II, which occurs at 7.5 GHz.

Reference [11] determined the external quality factors Q_{eS} and Q_{eL}, as shown in Figure 3.10(a).

$$Q_e = \frac{\text{RBW}}{f_c \times k_{S1}^2} = \frac{\text{RBW}}{f_c \times k_{L3}^2} \tag{3.14}$$

The passband's center frequency and ripple bandwidth (RBW) are denoted as f_c and RBW, respectively. The approach described in [11] can be used to compute the coupling coefficient (k_{mn}) and external quality factors (Q_e) represented in Figure 3.10(a).

$$\begin{aligned} k_{mn} &= \pm\frac{\text{RBW}}{2f_c}\left(\frac{f_{0m}}{f_{0n}} + \frac{f_{0n}}{f_{0m}}\right) \\ &\times\sqrt{\left(\frac{f_m^2 - f_n^2}{f_m^2 + f_n^2}\right)^2 - \left(\frac{f_{0m}^2 - f_{0n}^2}{f_{0m}^2 + f_{0n}^2}\right)^2}, \quad m \neq n \end{aligned} \tag{3.15}$$

$$Q_e = Z_0\pi f \left.\frac{\partial(Y \text{ in } A)}{\partial f}\right|_{f-f_c} \tag{3.16}$$

As shown in Figure 3.10(b), Patch II is coupled to two ring shorted stepped-impedance resonators (SSIRs) to produce single-band performance. This setup's SSIR transmission line model is displayed, and its input admittance is denoted by a one-port:

$$Y_{\text{in}A} = Y_{\text{in}A1} + Y_{\text{in}A2} \tag{3.17}$$

where

$$Y_{\text{in}A1} = -j\frac{\cot\left(\theta_A - \theta_{At}\right)}{Z_{A1}} \tag{3.18}$$

$$Y_{\text{in}A2} = j\frac{Z_{A1} + Z_{A2}\cot\theta_A\tan\theta_{At}}{Z_{A1}Z_{A2}\cot\theta_A - Z_{A1}^2\tan\theta_{At}} \tag{3.19}$$

SSIR exhibits resonance when Im $Y_{\text{in}A} = 0$. To achieve greater suppression in the stopband, a smaller impedance ratio Z_{A1}/Z_{A2} is selected. The proposed SSIR, which resonates at 1.007 GHz, can be satisfied by optimizing θ_A to 33.7 at 1 GHz after setting Z_{A1} to 107.14 and Z_{A2} to 16.91. The suggested SSIR's first harmonic frequency occurs at 6.101 GHz. The mixed E/M coupling between the SSIRs, which results in overall coupling throughout them, is shown in Figure 3.10(a).

$$K_{13} = \frac{(E_{13} - M_{13})\text{RBW}}{(1 - M_{13}E_{13})} \quad (3.20)$$

The coupling coefficient between the first and third resonators, k_{13}, is 0.0352, while the coupling coefficient between the first and second resonators and between the second and third resonators, k_{12} and k_{23}, is 0.3179. The ideal answer for Q_e can be obtained by maximizing θ_{At} to 13.6. The overlapping area between the SSIR and the patch resonator can be modified to control k_{12} and k_{23}, while the gap between the two SSIRs can be adjusted to regulate k_{13}.

According to [12], the *E*/*M* coupling relationship is a key factor in defining the TZ frequency and center frequency of the filter. Gap S_1 is used to control the *E* coupling while gap S_2 is used to control the *M* coupling to produce separate *E*/*M* coupling. The SSIRs exhibit a strong electric coupling and a weak magnetic coupling, which makes them *E*-dominant filters. The cancellation of E_{13} and M_{13} coupling results in a TZ (TZ2) located in the lower stopband and its frequency f_{EMC} is determined by:

$$f_{EMC} = f_{01}\sqrt{\frac{M_{13}}{E_{13}}} = f_{03}\sqrt{\frac{M_{13}}{E_{13}}} \quad (3.21)$$

Figures 3.10(d, e) illustrate the variations of TZ2 with respect to gap S_1 and gap S_2. The E_{13} coupling steadily decreases as the gap S_1 widens, leading TZ2 to travel in the direction of center frequency. However, when the gap S_2 narrows, the M_{13} coupling steadily increases, causing TZ1 to shift in the direction of center frequency. A pair of ring stubs attached to the feed lines in Figure 3.10(b) introduce source to load coupling and have capacitive coupling with the SSIRs. The proposed BPF is simulated with other parameters held constant to examine how the electrical length θ_{stub} of the ring stubs influences the TZs. Figure 3.10(f) shows the simulated $|S_{21}|$ of the suggested filter against θ_{stub}. Only one TZ is produced at the upper stopband when θ_{stub} is 0. The second passband is formed and two additional TZs appear at the upper stopband when θ_{stub} is equal to 21.9. The first passband is approached by TZ4 and TZ5 as θ_{stub} rises, while the second passband is repressed. A TZ (TZ3) is produced through a cross-coupling path between resonators R1 and R3 (*S*-R1-R3-*L* and *S*-R1-R2-R3-*L*), in addition to the TZs previously examined. The overlapping region between the circuit on G_5 and G_6 is what essentially produces the interstage coupling K_{12} and K_{23}. Figure 3.10(f) shows the simulated $|S_{21}|$ of the suggested filter when θ_{stub} is set to 0. The interstage coupling K_{12} and K_{23} gradually decline as the parameter W_{patch} rises, causing TZ3 to drop to center frequency.

When $\theta_{\text{stub}} = 23.8°$, the proposed single-band BPF generates a secondary passband at 5.5 GHz, as shown in Figure 3.10(f). The secondary passband is controlled by the slots to produce the DBBPF.

3.2.2.3 Design of Dual Passband Filter Based on Stubs Loading

The research [9] has reported a DBBPF with a controllable multimode resonator based on SISL technology, as shown in Figure 3.11. In [14], a $\lambda/2$ SIR was provided.

References [14, 15] provided the first spurious resonance frequency f_{s1} and resonance condition.

$$\tan\theta_1 \tan\theta_2 = K \tag{3.22}$$

$$f_{s1} = \frac{\pi}{2\tan^{-1}\sqrt{K}} f_0 \tag{3.23}$$

By compromising between stopband suppression and the compact configuration layout, the electrical length ratio $u = \theta_2/(\theta_1 + \theta_2) = 0.25$ is used. A high impedance restricted SISL PCB method is necessary for low-impedance ratio designs. The impedance ratio $K = Z_2/Z_1 = Y_1/Y_2 = 0.3$ is taken into account because of the high conductor loss [16], causing the frequency ratio to be more than 2.5.

The resonator in Figure 3.11(a) must fulfill $K = Z_2/Z_1 = Z_4/Z_3 = 0.3$ and $u = 0.25$, [17, 18] began by selecting the following parameters: $Z_1 = 213\Omega$ (0.2 mm, lowering the filter size, while the SISL 50Ω transmission line width is around 6 mm), $Z_2 = 65\Omega$ (4 mm), $Z_3 = 180\Omega$ (0.8 mm), $Z_4 = 55\Omega$ (12 mm), and $\theta_1 = 4$, $\theta_2 = 4$, and θ ($\theta = 0.04\ \lambda_g$).

Advanced design system (ADS) or the method in [19] are used to calculate the characteristic impedance. The equivalent even-mode/odd-mode circuits are shown in Figures 3.11(b, c) based on the even-mode/odd-mode method. The admittance of the transmission-line mode from the open plane is obtained as

$$Y_{\text{odd}} = jY_2 \frac{Y_2 \tan\theta_1 \tan\theta_2 - Y_1}{Y_2 \tan\theta_1 + Y_1 \tan\theta_2} \tag{3.24}$$

$$Y_{\text{even}} = jY_2 \frac{Y_2 \tan\theta_1 \tan\theta_2 - Y_1}{Y_2 \tan\theta_1 + Y_1 \tan\theta_2} + jY_4 \frac{Y_4 \tan\theta_1 \tan\theta_2 - Y_3}{Y_4 \tan\theta_1 + Y_3 \tan\theta_2} \tag{3.25}$$

The resonance condition can be described as follows

$$\tan\theta_1 \tan\theta_2 = \frac{Y_1}{Y_2} = \frac{Z_1}{Z_2} \tag{3.26}$$

$$\tan\theta_1 \tan\theta_2 = \frac{Y_3}{Y_4} = \frac{Z_4}{Z_3} \tag{3.27}$$

The two fundamental modes in the lower passband and two first spurious frequencies in the spurious passband are introduced by the resonator in Figure 3.11(a). The additional degree of freedom is introduced to control the operation of the spurious passband to generate a dual-band response. By loading the stubs in Figure 3.11(b), it is possible to change two resonant modes of the spurious passband. Finding suitable feeders to satisfy the coupling requirements without introducing

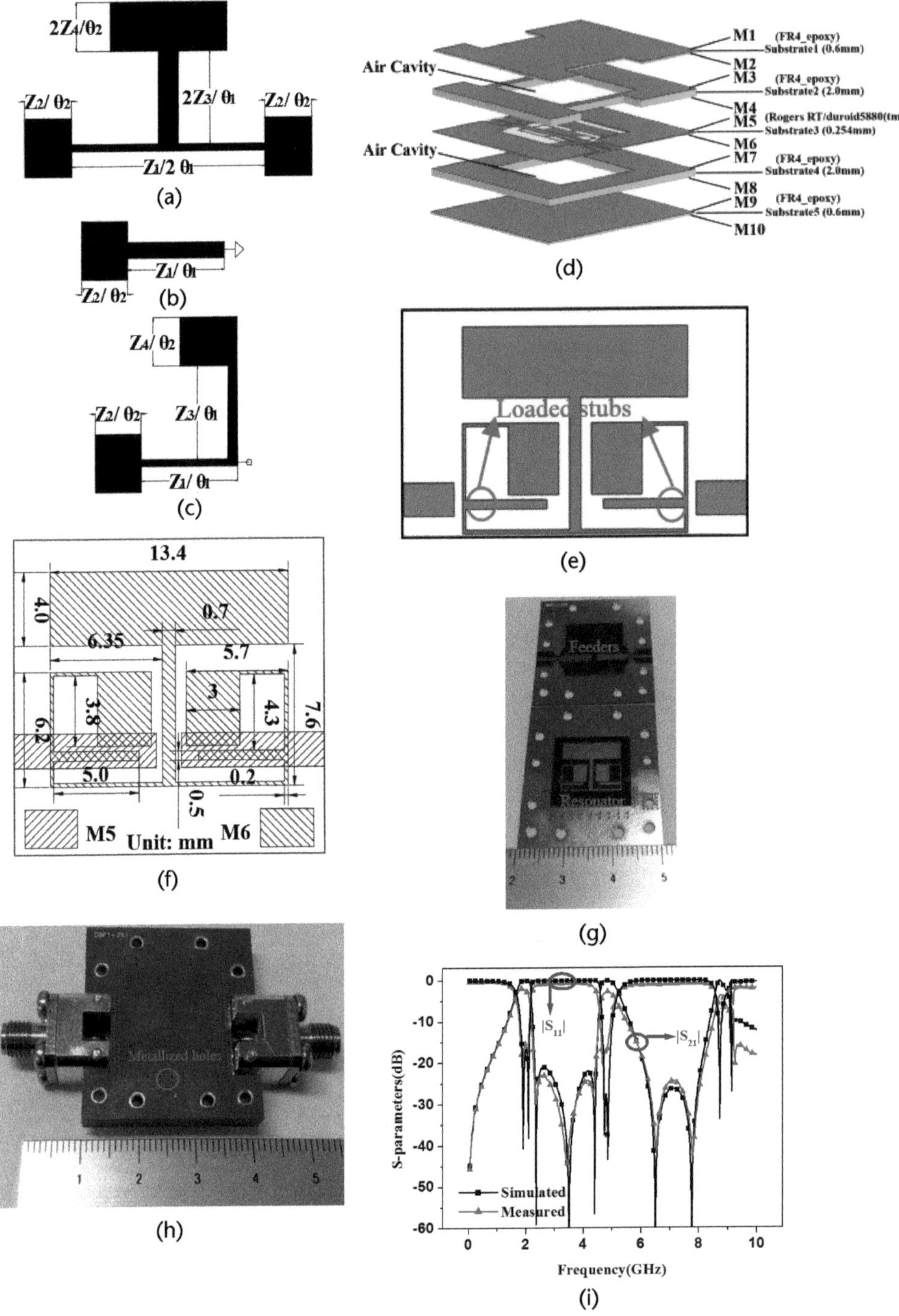

Figure 3.11 (a) Transmission-line model, (b) odd-mode circuits, (c) even-mode circuits, (d) a 3D view of the SISL DBBPF and M_1–M_{10} are metal layers, (e) multimode resonators with loaded stubs, (f) top-side view of this DBBPF with geometric dimensions, (g) photographs of inner circuits, (h) photographs of self-packaged DBBPF, and (i) simulated and measured results of the DBBPF [13].

any external dual-band impedance-transformer feeds is the next challenge. The two high-impedance feeders are implemented with M_5 and the resonator is designed on M_6, as seen in Figure 3.11(d). The chosen feeders primarily perform two tasks: simultaneously stimulate four modes and, by introducing source to load coupling, obtain multiple TZs. The measured insertion losses of 1.96/2.86 dB of the two passbands, which contain 0.27/0.42-dB insertion losses of the SMA connectors and the transition structure, result in actual insertion losses of only 1.69/2.44 dB at 1.90 and 4.85 GHz. The SISL DBBPF shows a very compact size of 0.12 $\lambda_g \times 0.10\ \lambda_g$ and two independently controlled passbands at 1.90 and 4.85 GHz.

A dual-band BPF is designed with controllable characteristics and low loss by adopting a simple ring resonator loading open stubs based on SISL platform [20]. By adjusting the distance of two feeders and the length of the loaded open stub, the TZs and FBWs of two passbands can be controlled. The dual-band BPF shows self-packaging and low losses with 0.45/0.83 dB of two passbands. In Figure 3.12(a), the designed SISL structure is composed of two pieces of FR4_epoxy with air cavity, two pieces of FR4_epoxy, and one piece of Rogers RT/Duroid 5880, which is conducive to the manufacturing process. Four modes can be obtained by the simple resonator in Figure 3.12(b). The four modes are divided into two groups naturally. According to [21], four modes can be generated by the circuit topology. Figure 3.12(c) shows the overall layout of the self-packaged dual-band BPF. The filter shows

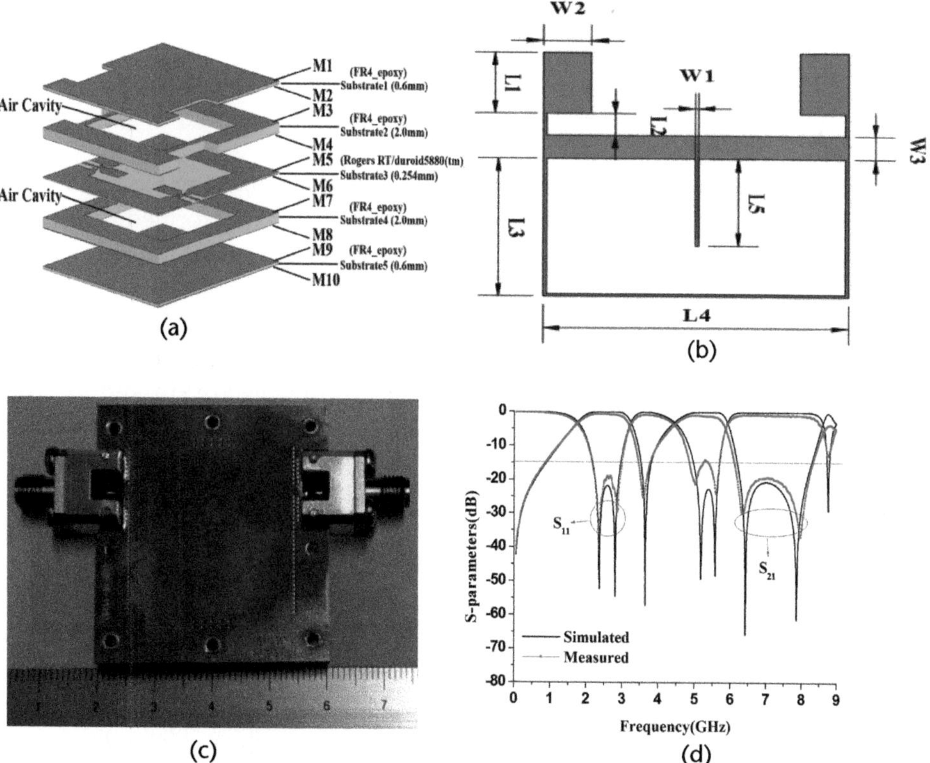

Figure 3.12 (a) The 3D diagram of SISL (M_1 to M_{10} are metal layers), (b) the layout of the ring resonator (the dimensions are $W_1 = 0.2$ mm, $W_2 = 3$ mm, $W_3 = 1.9$ mm, $L_1 = 5$ mm, $L_2 = 1.8$ mm, $L_3 = 11.3$ mm, $L_4 = 19$ mm, $L_5 = 7.1$ mm), (c) photograph of the self-packaged BPF, and (d) simulated and measured results of the dual-band bandpass filter [20].

a compact size with 0.23 $\lambda_g \times$ 0.25 λ_g (19.0 mm × 20.7 mm). The λ_g represents the guided wavelength of center frequency 2.48 GHz. Figure 3.12(d) shows the measured results, containing the center frequencies of 2.48/5.17 GHz, the 3-dB FBW of 59.4/27.3%, the insertion losses of 0.75/1.30 dB and the return losses of 19.0/14.2 dB, respectively. The actual insertion losses of the filter are only 0.45/0.83 dB due to the 0.30-dB insertion losses of SMA connectors and the 0.47-dB insertion losses of the transition structure of the two passbands.

Based on the SISL platform, a compact dual-band BPF is designed shown in Figure 3.13 [22]. Low-cost FR4 material and 3D metal connections technology with substrate cutout are used to reduce the loss. At 3.45 GHz/4.9 GHz, the improvement of insertion loss is 0.97 dB/1.6 dB. Two paths were formed between the source and load by paralleling quarter-wavelength step impedance resonators with different operating frequencies. The bandwidth and TZ location can be adjusted independently. Two added stubs can provide 20 dB out-of-band suppression up to 4.4 f_0. The two stubs are not only the stopband resonator but also the feed structure. In the first band of 3.45 GHz, the substrate cutoff reduced circuit loss by 0.97 dB. In the second band of 4.9 GHz, the circuit loss increases by 1.6 dB. The bandwidths of the circuit without cutout and with cutout are 17.7%/9.8% and 16.23%/7.7%, respectively. Before cutout, the insertion losses of the filter are 5.56 dB and 7.36 dB in the first and second passbands, respectively. After cutout, the insertion losses of the filter are 4.36 dB and 5.77 dB in the first and second passbands, respectively. However, the size of the circuit is almost the same. Figure 3.13(h) shows the measured suppression, which is up to 15.2 dB (4.4 f_0).

3.2.3 Wide Stopband Lowpass Filter

In this section, a wide stopband lowpass filter (LPF) using SISL is introduced [23]. The high-performance SISL LPF is composed of transformed radial stubs (TRS), SIR, and open stubs. The 3D structure of the proposed SISL LPF is shown in Figure 3.14(a). For low cost, we use 5 layers of FR4 Substrates 1–5 with a dielectric constant of 4.4 and a loss tangent of 0.02. The thicknesses of Substrate 1 to Substrate 5 are 0.6, 0.6, 0.2, 0.6, and 0.6 mm. Substrate 2 and Substrate 4 are hollowed out into a rectangle, and when the five-layer substrates are stacked in order, two air cavities of the suspended line circuit are formed. M_1 to M_{10} are metal layers. An electromagnetic shielding structure is formed by the metal grounding layers of M_2 and M_9 together with the via holes around the cavity. The SISL LPF also has the characteristic of self-packing property. The core circuits of SISL LPF are designed on M_5. The schematic diagram of the proposed SISL LPF is shown in Figure 3.14. The SISL LPF can be viewed as a cascade of three parts of a TRS (consisting of a coupled-line transformer paralleled by a radial stub), a pair of open stubs, and a pair of SIRs. Figure 3.14(c) shows the frequency responses of the filter. There exist six transmission zeros in the stopband, which achieves a great stopband rejection.

Figure 3.15(a) shows the composition structure of TRS, which is formed by the coupled-line transformer connected with a radial stub in shunt. Figure 3.15(b) shows the odd-mode circuit and even-mode equivalent circuit of TRS, where Z_t is the characteristic impedance of the suspended line section, Z_{co} and Z_{ce} represent the odd-mode and even-mode characteristic impedances of the parallel coupled-line, with

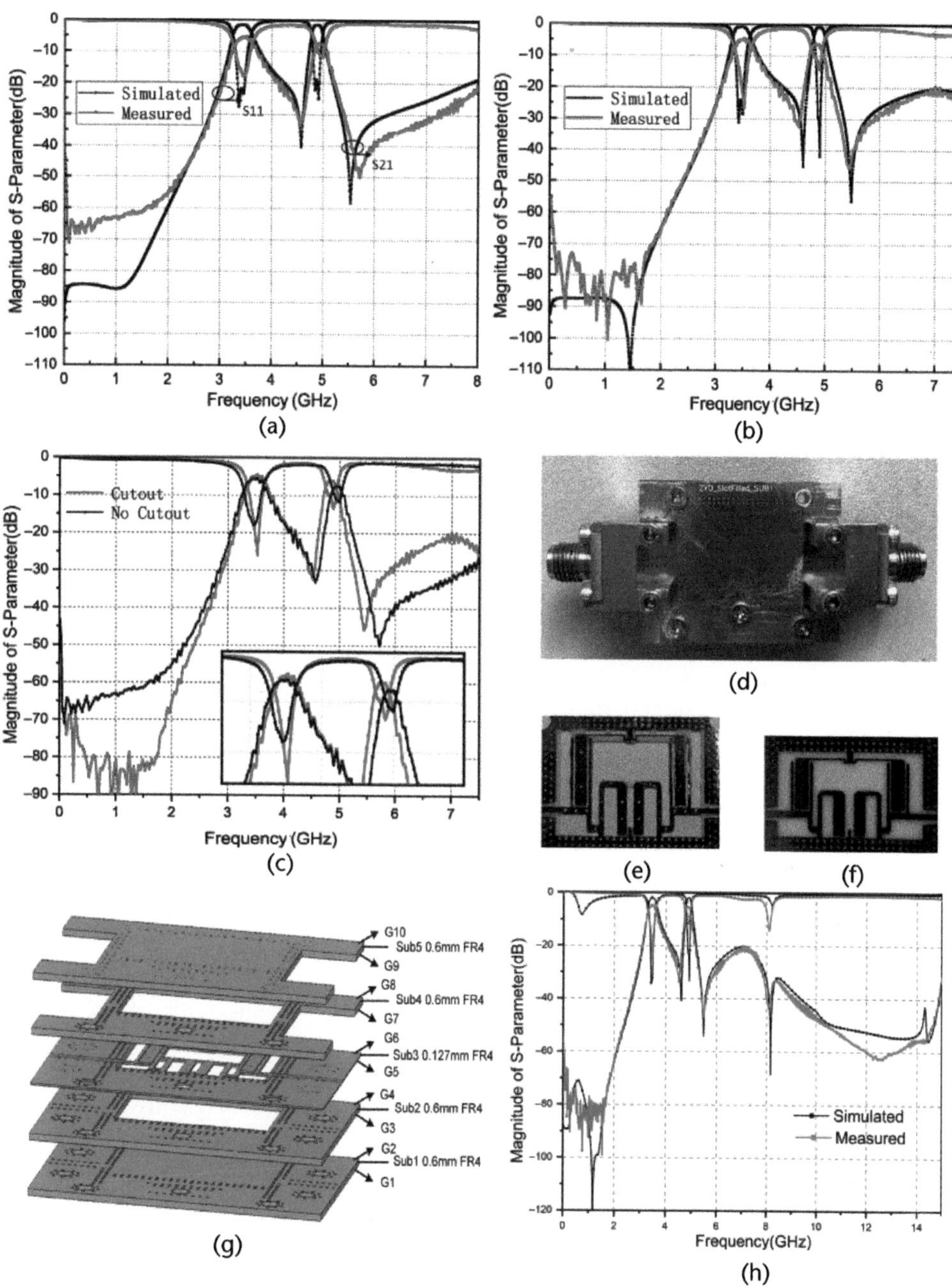

Figure 3.13 (a) Without substrate cutout; (b) with substrate cutout; (c) measured result compared between substrate cutout and no cutout; (d) complete filter with adapter; (e) photograph of substrate cutout circuit; (f) photograph of the circuit without substrate cutout; (g) 3D view of the proposed SISL structure, SISL, substrate integrates suspended line; and (h) simulated and tested results with substrate cutout [22].

θ_t and θ_c denoting the electrical lengths corresponding, respectively. The radial stub can be approximately modeled as a lumped capacitance [24], under the constraint of the effective electric length smaller than a quarter-wavelength. The capacitance of the radial stub can be extracted through one port input impedance using (3.28):

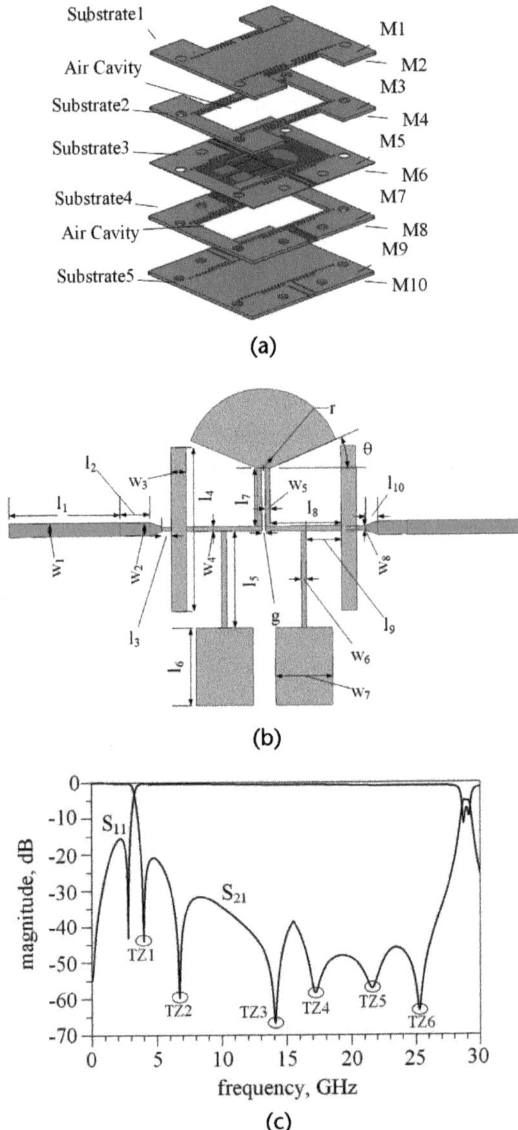

Figure 3.14 (a) The 3D structure of the proposed SISL LPF, (b) schematic diagram of the proposed SISL LPF, and (c) simulation results of the proposed SISL LPF [23].

$$C_g = \frac{-1}{2\pi f\left(\text{imag}\left(Z_{\text{in}}\right)\right)} \tag{3.28}$$

Figure 3.15(b)'s odd-mode and even-mode input impedance Z_{ino} and Z_{ine} can be expressed as follows:

$$Z_{\text{ino}} = jZ_t \frac{Z_{co}\tan\theta_c + Z_t\tan\theta_t}{Z_t - Z_{co}\tan\theta_t\tan\theta_c} \tag{3.29}$$

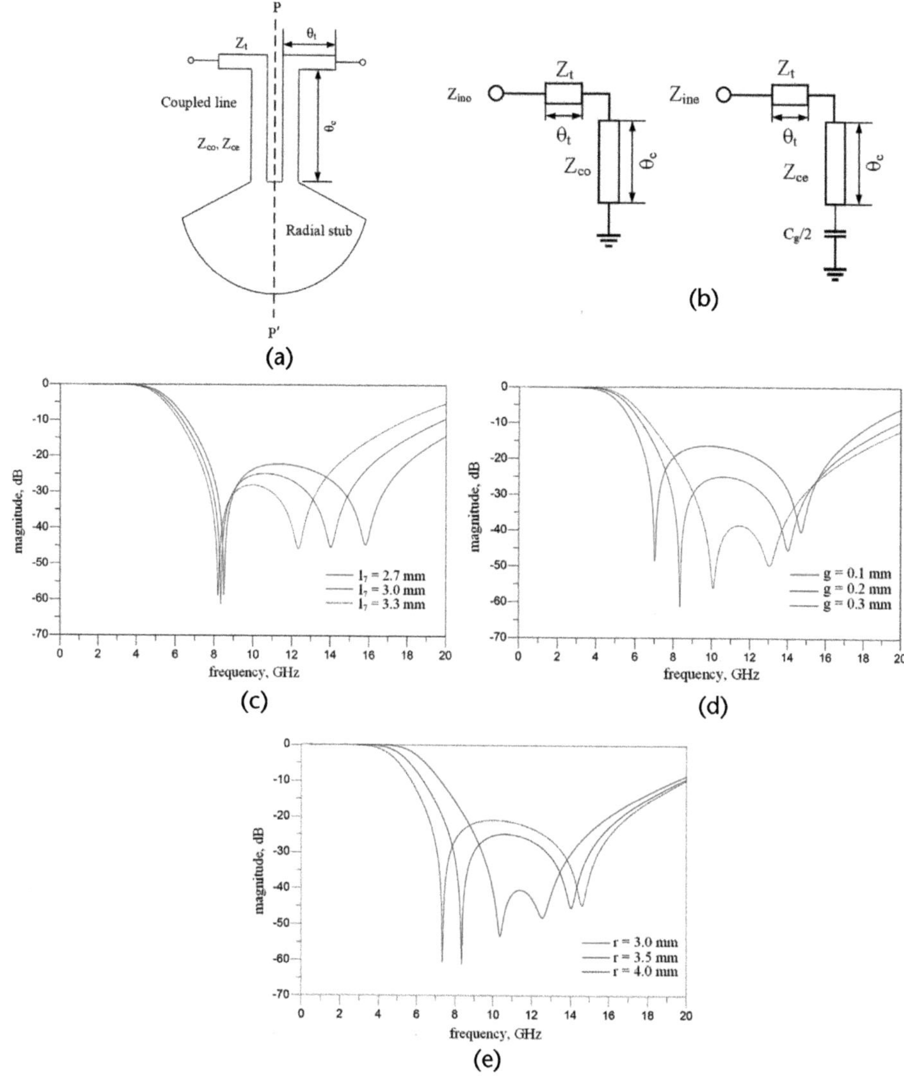

Figure 3.15 (a) The layout of transformed radial stub, (b) odd-mode and even-mode equivalent circuit of transformed radial stub, (c) transmission responses of transformed radial stub with different l_7, (d) transmission responses of transformed radial stub with different g, and (e) transmission responses of transformed radial stub with different r [23].

$$Z_{\text{ine}} = jZ_t \frac{-jZ_e + Z_t \tan\theta_t}{Z_t + jZ_e \tan\theta_t} \tag{3.30}$$

where

$$Z_e = jZ_{ce} \frac{\omega C_g Z_{ce} \tan\theta_c/2 - 1}{\omega C_g Z_{ce}/2 + \tan\theta_c} \tag{3.31}$$

The odd-mode and even-mode impedance can be used to calculate the S-parameter of TRS as shown in [11]:

$$S_{21} = \frac{z_{\text{ine}} - z_{\text{ino}}}{\left(z_{\text{ine}} + 1\right) \cdot \left(z_{\text{ino}} + 1\right)} \tag{3.32}$$

$$S_{11} = \frac{z_{\text{ine}} \cdot z_{\text{ino}} - 1}{\left(z_{\text{ine}} + 1\right) \cdot \left(z_{\text{ino}} + 1\right)} \tag{3.33}$$

where z_{ine}, z_{ino} represents the normalized value of Z_{ine}, Z_{ino}. When the normalized odd-mode impedance is equal to the normalized even-mode impedance, the transmission zeros can be obtained.

$$z_{\text{ine}} = z_{\text{ino}} \tag{3.34}$$

When the product of the normalized even-mode impedance and the normalized odd-mode impedance is equal to 1, the poles are generated.

$$z_{\text{ine}} \cdot z_{\text{ino}} = 1 \tag{3.35}$$

Figures 3.15(c–e) show the parameter analysis of TRS based on EM simulation. When the element values are changed, the different transmission responses of TRS are shown in Figure 3.15(c–e). From Figure 3.15, we can acknowledge that the values of l_7, g, and r (marked in Figure 3.14) dominantly determine the stopband rejection. As can be seen in Figure 3.15(c), the position of the second transmission zero is affected by the value of l_7, and the first transmission zero is hardly affected by l_7. As shown in Figure 3.15(d, e), the farther two transmission zeros can be separated when decreasing the value of g or increasing the value of r. As the analysis shows, a wide stopband and two TZs (TZ2 and TZ3 are marked in Figure 3.14(c)) can be produced by the TRS structure.

There are two purposes to introduce a pair of SIRs. The first purpose is to determine the cutoff frequency and improve the suppression of the stopband. The second purpose is to form two additional transmission zeros in the upper stopband which are marked as TZ5 and TZ6 in Figure 3.14(c).

As can be seen in Figure 3.14(c), the SIRs create a deep transmission zero near the passband, which is marked as TZ1, resulting in a very narrow transition from passband to stopband range of 0.5 GHz. When at low frequencies and the effective electric length smaller than a quarter-wavelength, the low impedance and high impedance lines of the SIR can be approximately modeled as capacitance and inductance, respectively. The position of TZ1 can be determined by the following formula [11]:

$$L = \frac{Z_h}{2\pi f_c} \sin\frac{2\pi l_5}{\lambda_{gL}} \tag{3.36}$$

$$C = \frac{1}{2\pi f_c Z_l}\sin\frac{2\pi l_6}{\lambda_{gC}} \tag{3.37}$$

$$f_{TZ1} = \frac{1}{2\pi\sqrt{LC}} \tag{3.38}$$

where l_5, l_6 are the lengths of high and low impedance line of SIR and Z_h, Z_l are the characteristic impedances of them, respectively. λ_{gL} and λ_{gC} are the waveguide lengths at the cutoff frequency when the impedance are Z_h and Z_l, respectively.

The high order mode of SIR and the weak coupling between the two symmetrical SIRs creates the two high frequency transmission zeros TZ5 and TZ6. Neglecting the weak coupling, when the input impedance of SIR is zero, the transmission zeros can be determined as:

$$Z_l \cot\theta_l - Z_h \tan\theta_h = 0 \tag{3.39}$$

where θ_h and θ_l are the high and low impedance lines of the SIR's respective electrical lengths.

The suggested LPF architecture uses a pair of open stubs to suppress the parasitic passband of TRS and widen the stopband. The open stubs are modeled as shunt connected series L-C branches that provide one transmission zero in the stopband (designated as TZ4 in Figure 3.14(c)) by shorting out transmission at their resonance frequency. One way to express the transmission zero position is [11]:

$$f = \frac{150}{l_4(\text{mm})\sqrt{\epsilon_{re}}}\text{GHz} \tag{3.40}$$

where ε_{re} is the SISL's effective dielectric constant and l_4 is the length of the open stub. The stopband bandwidth can be significantly increased by keeping the transmission zero close to the parasitic passband of the TRS. The open stubs function as parallel capacitors when they are not near the resonance frequency, suppressing high-frequency signals.

Figure 3.16(a) compares the simulated transmission response results of TRS. To better illustrate the working principle of the proposed filter, the comparison of the simulated transmission response results of TRS, LPF with TRS and SIRs, open stubs, and the final LPF is conducted and shown in Figure 3.16(a). It is confirmed that TZ2 and TZ3 are generated by TRS; TZ1, TZ5, and TZ6 are produced by SIRs; and TZ4 is realized by open stubs.

An SISL LPF with a cutoff frequency f_c at 3.27 GHz is designed, simulated, fabricated, and measured for verification. The photograph of the SISL LPF is shown in Figure 3.16(b) and the simulated and measured results are shown in Figure 3.16(c). The simulated results are in good agreement with the measured results. It can be observed from the measurement results that the upper stopband of the LPF can be extended to 24 GHz and the rejection of the stopband is better than 20 dB. The

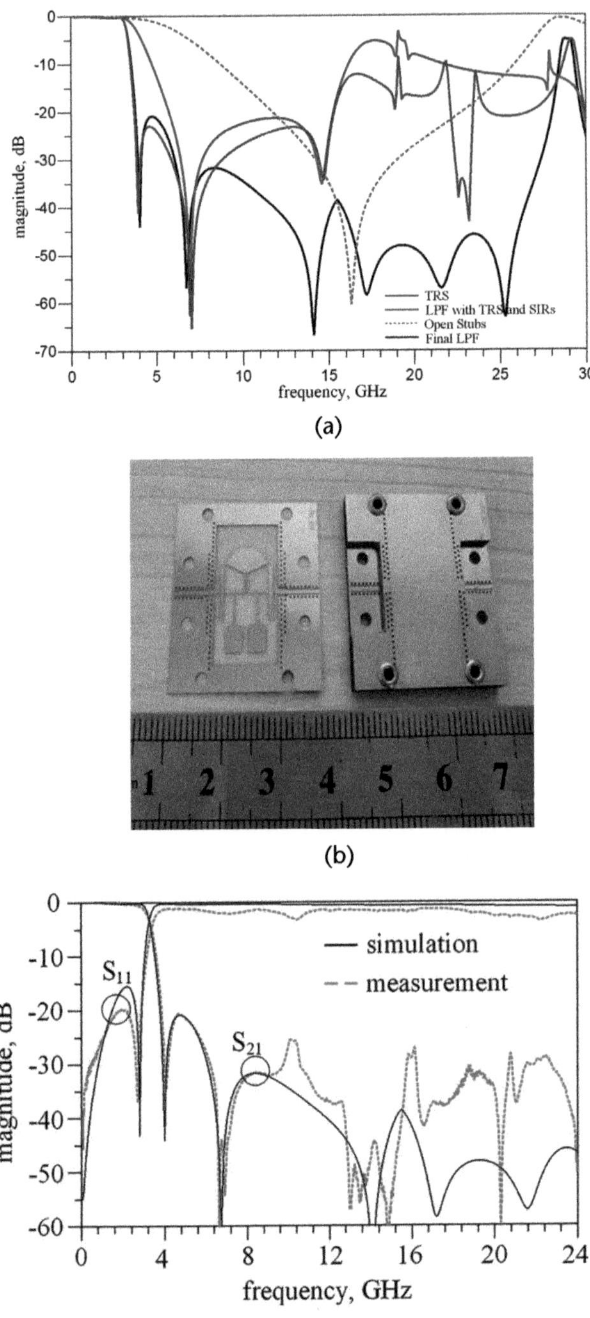

Figure 3.16 (a) Simulated transmission response results of TRS, LPF with TRS and stepped SIRs, open stubs, and final LPF; (b) photograph of fabricated substrate integrated suspended-line LPF; and (c) simulated and measured results of the proposed filter [23].

measured cutoff frequency is 3.27 GHz. The maximum insertion loss is 0.6 dB and the in-band return loss is 20 dB. The core size of LPF is 0.20 $\lambda_g \times 0.14\ \lambda_g$.

3.2.4 Triplexer

In recent years, wireless communication systems have developed rapidly. Multiplexer [25–27] plays a significant role for multistandards and multiservices to achieve combining or separating different filter channels. This section introduces an SISL triplexer [28]. The difference from the common construction methods of multiplexer is using multiple matching networks to connect the BPFs, and each matching network is only required to match the adjacent BPFs to a duplexer. This structure could be extended to the multiplexer and reduces the design difficulty of the matching network. Two adjacent BPFs in this triplexer compose a duplexer based on the design consideration. Each of the BPFs is composed of an LPF and a highpass filter (HPF), which have the advantages of high controllability and low insertion loss.

3.2.4.1 Investigation of the Triplexer

Figure 3.17(a) shows the whole schematic of the proposed triplexer. The triplexer consists of an input port and three output ports. When the BPFs operating on different frequencies are connected in parallel, a filter will introduce admittance in the other filter, resulting in poor transmission characteristic. In this case, we need to use a matching network circuit to connect the two BPFs to eliminate the loading effect between the different channel filters. The triplexer can be extended into a multiplexer with n BPFs based on this topology. Especially, any two adjacent filter channels could compose a duplexer (i.e., BPF 1, BPF 2, and BPF 3). Figure 3.17(a) shows that the BPFs are all composed of LPF and HPF. As shown in Figure 3.17(b), there exist three channels (i.e., channels 1, 2, and 3). The lowpass part in channel 1 consists of three LPFs that could extend the bandwidth of the stopband; the same pattern is seen for channel 2. In other words, LPFs are the common path of the channels. This design method simplifies the matching circuit design and makes the proposed design to be used for a multiplexer without space layout constraints.

In this triplexer design, the matching networks are only required to combine the two adjacent filters just as in a duplexer. As shown in Figure 3.17(b), the matching network consists of three ports corresponding to the two LPFs and an HPF as shown in Figure 3.17(c). The matching network needs to make the admittance of the passbands of the three filters a constant close to 0. The admittance relationship of it is calculated as follows:

$$y_{\text{in1}} = y_{\text{in2}} + y_{\text{in3}} \tag{3.41}$$

While y_{in1}, y_{in2}, and y_{in3} are the input impedance of the three ports, θ_1, θ_2, and θ_3 are the electric lengths of the transition lines, respectively. According to transmission line theory, in the first half-wavelength of the reactive load, there is always a value making the tail of the transmission line equivalent to the open end to the

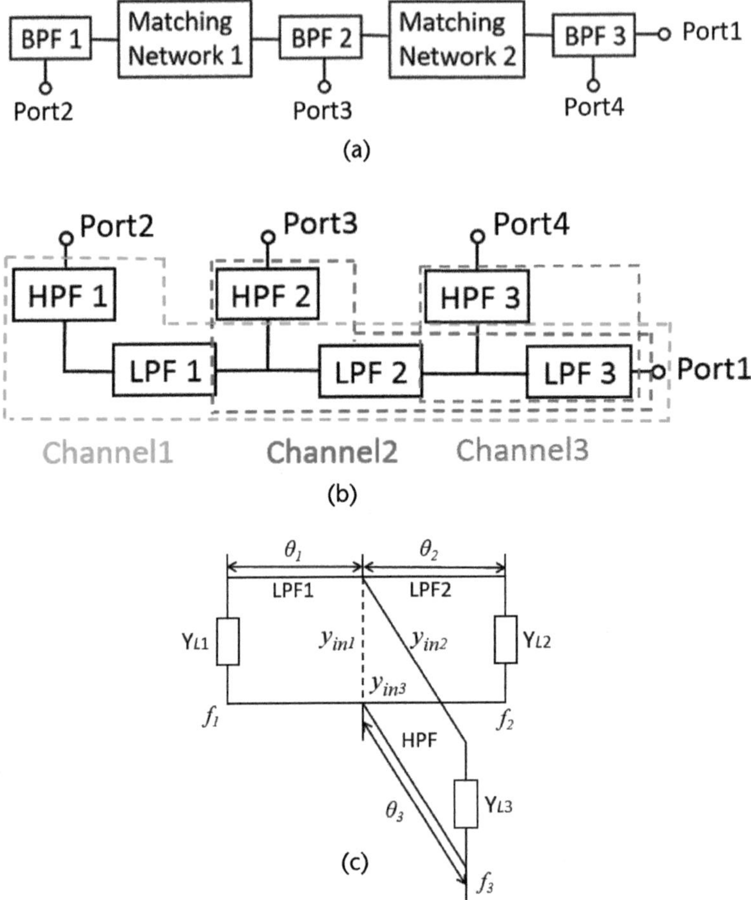

Figure 3.17 (a) Schematic 1 of the proposed triplexer, (b) schematic 2 of the proposed triplexer, and (c) matching network [28].

other filter, which offsets the damage to the channel characteristics. The electric length is a value of less than 90° corresponding to its own frequency.

3.2.4.2 LPF

The classical unit LPF cell is shown in Figure 3.18(a), which is composed of a straight stub and transmission lines. A section of distributed transmission line can be generally modeled as an inductance with the shunt capacitances at both ends. A straight stub for which the effective electric length is smaller than a quarter-wavelength is equivalent to a grounded capacitance. The equivalent-circuit model of Figure 3.18(a) is shown in Figure 3.18(b). The frequency response of the unit LPF cell under the different inductance are compared as shown in Figure 3.18(c). The used unit LPF cell and its equivalent circuit [24, 29] are shown in Figures 3.18(d, e). The unit LPF cell consists of capacitive element, coupled lines, and two transmission lines. The even-mode and odd-mode methods can be used to analyze the symmetrical model.

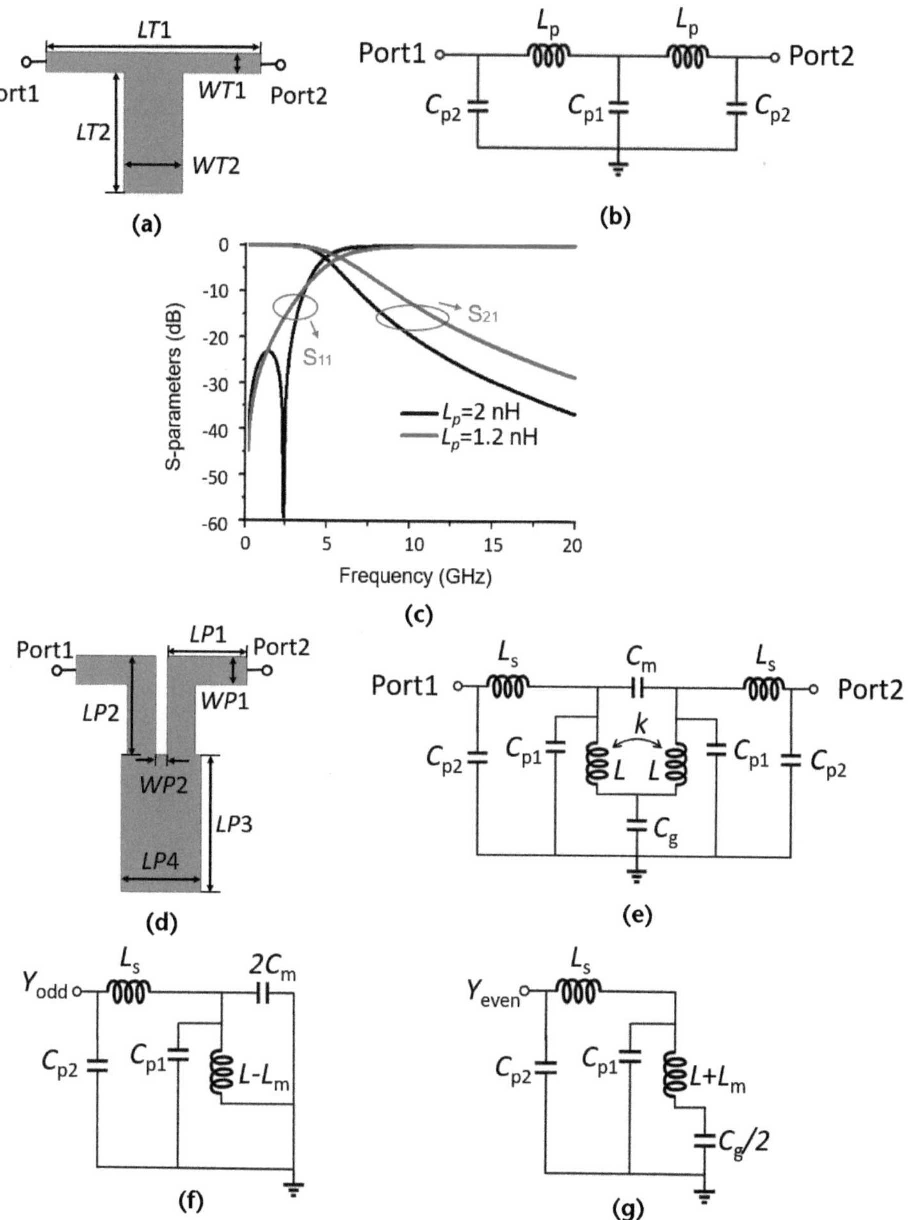

Figure 3.18 (a) Traditional unit LPF cell, (b) equivalent-circuit model of (a), (c) frequency response of the unit LPF cell with a different L_p, (d) proposed unit LPF cell, (e) equivalent-circuit model of (d), (f) odd-mode circuit of (e), and (g) even-mode circuit of (e) [28].

The even-mode and odd-mode equivalent circuits are demonstrated in Figures 3.18(f, g), respectively. The input admittances of the two modes can be calculated as:

$$Y_{odd} = j\omega C_{p2} + \cfrac{1}{j\omega L_s + \cfrac{1}{j\omega\left(C_{p1} + 2C_m\right) + \cfrac{1}{j\omega\left(L - L_m\right)}}} \tag{3.42}$$

$$Y_{\text{even}} = j\omega C_{p2} + \cfrac{1}{j\omega L_s + \cfrac{1}{j\omega C_{p1} + \cfrac{1}{j\omega\left(L + L_m\right) + \cfrac{2}{j\omega C_g}}}} \tag{3.43}$$

where $\omega = 2\pi f$ and $L_m = k \times L$. The S-parameter of the proposed equivalent circuits can be calculated as:

$$S_{21} = \frac{Y_0 Y_{\text{odd}} - Y_0 Y_{\text{even}}}{\left(Y_0 + Y_{\text{even}}\right)\left(Y_0 + Y_{\text{odd}}\right)} \tag{3.44}$$

$$S_{11} = \frac{Y_0^2 - Y_{\text{odd}} Y_{\text{even}}}{\left(Y_0 + Y_{\text{even}}\right)\left(Y_0 + Y_{\text{odd}}\right)} \tag{3.45}$$

When both even-mode and odd-mode circuits are in a state of resonance, the corresponding input admittance is 0:

$$Y_{\text{even}} = Y_{\text{odd}} = 0 \tag{3.46}$$

The resonant frequencies shown in Figure 3.18(f, g) are f_o and f_e, which could be derived from (3.42) and (3.43), and the functions are as shown in (3.47) and (3.48).

$$f_o = \frac{1}{2\pi}\sqrt{\frac{\sqrt{\left[\left(C_{p1} + 2C_m\right)\left(L - L_m\right) + C_{p2}\left(L - L_m\right) + L_S C_{p2}\right]^2 - 4\left(C_{p1} + 2C_m\right)\left(L - L_m\right)L_s C_{p2}} + \left[\left(C_{p1} + 2C_m\right)\left(L - L_m\right) + C_{p2}\left(L - L_m\right) + L_S C_{p2}\right]}{2\left(C_{p1} + 2C_m\right)\left(L - L_m\right)L_s C_{p2}}} \tag{3.47}$$

$$f_e = \frac{1}{2\pi}\sqrt{\frac{\sqrt{4C_g\left(L + L_m\right)C_{p1}L_s C_{p2}\left(C_g + 2C_{p1} + 2C_{p2}\right) - \left[C_g C_{p1}\left(L + L_m\right) + C_g C_{p2}\left(L + L_m\right) + C_g L_s C_{p2} + C_{p1} L_S C_{p2}\right]^2} - jC_g C_{p1}\left(L + L_m\right)C_{p1}L_S C_{p2}\left(C_g + 2C_{p1} + 2C_{p2}\right)}{2jC_g\left(L + L_m\right)C_{p1}L_s C_{p2}}} \tag{3.48}$$

The center frequency is calculated as

$$f = \frac{f_o + f_e}{2} \tag{3.49}$$

from which the physical size of the LPF unit could be determined.

The transmission zeros are generated when the even-mode admittance is equal to the odd-mode admittance:

$$Y_{\text{odd}} = Y_{\text{even}} \tag{3.50}$$

The functions of the zeros in (3.51) and (3.52) are obtained by solving (3.10).

$$f_{zp1} = \frac{1}{4\pi}\sqrt{-\frac{2\left(-2C_mL - C_gL_m + 2C_mL_m\right)}{C_gC_m\left(L^2 - L_m^2\right)} - 2\times\sqrt{\frac{\left(-2C_mL - C_gL_m + 2C_mL_m\right)^2}{C_g^2C_m^2\left(L^2 - L_m^2\right)^2} - \frac{4}{C_gC_m\left(L^2 - L_m^2\right)}}} \tag{3.51}$$

$$f_{zp2} = \frac{1}{4\pi}\sqrt{-\frac{2\left(-2C_mL - C_gL_m + 2C_mL_m\right)}{C_gC_m\left(L^2 - L_m^2\right)} + 2\times\sqrt{\frac{\left(-2C_mL - C_gL_m + 2C_mL_m\right)^2}{C_g^2C_m^2\left(L^2 - L_m^2\right)^2} - \frac{4}{C_gC_m\left(L^2 - L_m^2\right)}}} \tag{3.52}$$

The LPF unit can be cascaded up to *n* cells to form a high order filter. The proposed four-cell LPF is shown in Figure 3.19(a). Cell 1 and Cell 4 have the same physical parameter size, as do Cell 2 and Cell 3. The two different units improve the

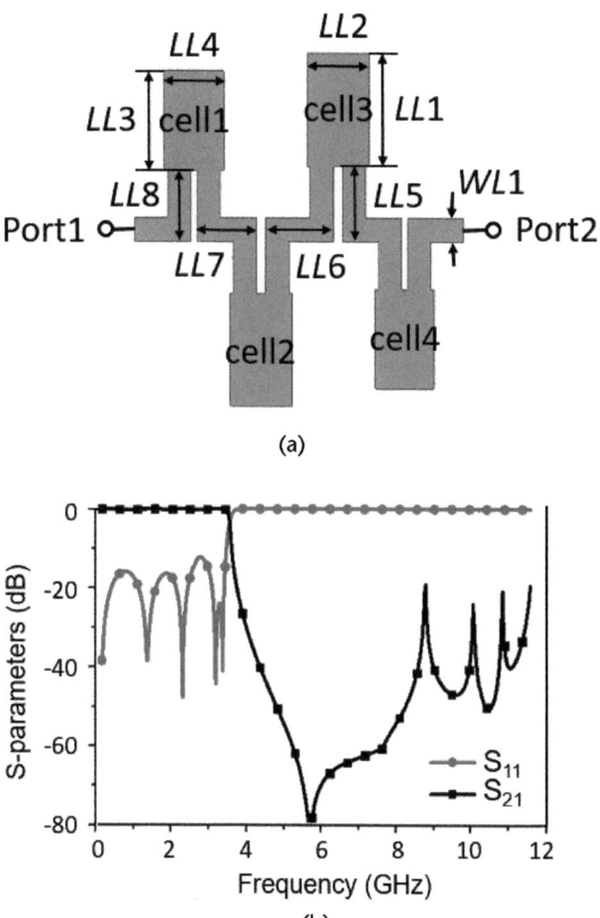

Figure 3.19 (a) Four-cell LPF structure, and (b) frequency response [28].

performance of the passband and stopband. In order to decrease the influence caused by coupling between the adjacent cells, the layout of the LPF adopts the method of interleaving the four units. Figure 3.19(b) shows the four-cell LPF's simulation result. LPF has good skirt selectivity and rejection depth.

3.2.4.3 HPF

As is shown in Figure 3.20(a), the seven-order HPF consists of four series capacitors and three shunt inductors. With the feature of suspended line, the broadside coupling

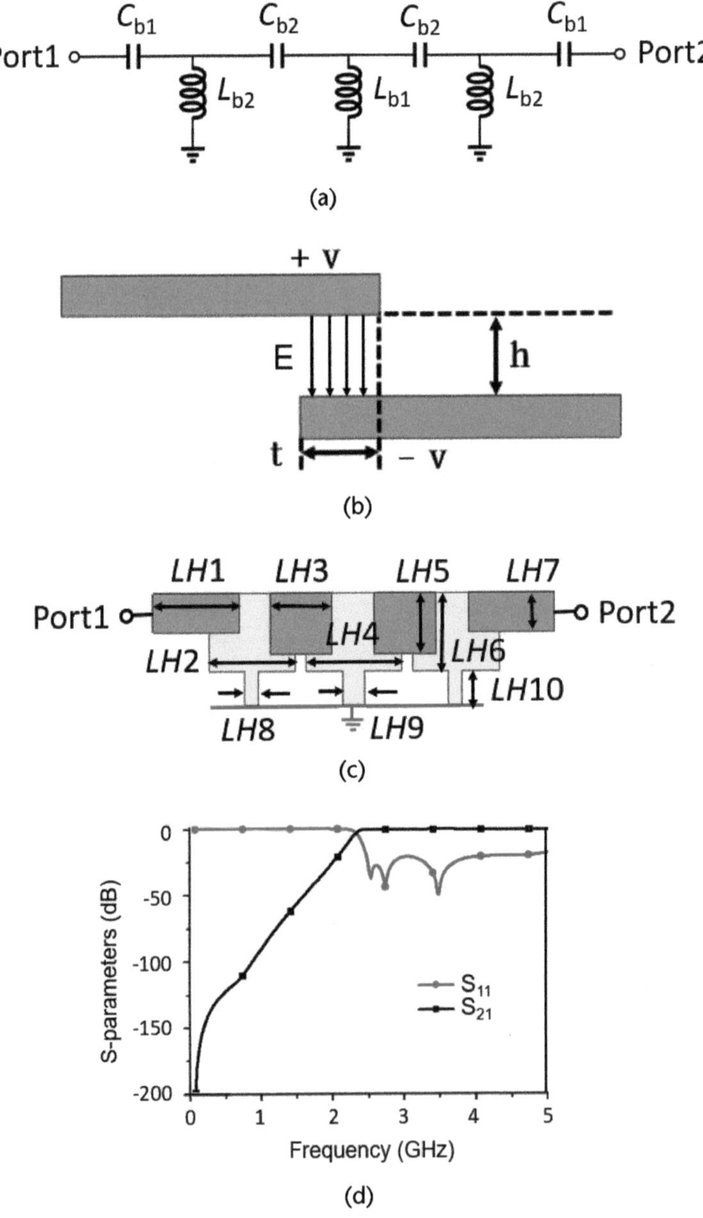

Figure 3.20 (a) Phototypes of the seven-order Chebyshev structure, (b) broadside coupling capacitance, (c) HPF structure, and (d) frequency response [28].

capacitance circuit is on both sides of the substrate, and there exists an overlap between the two metals as shown in Figure 3.20(b). The two metals are equivalent to two plates with positive and negative voltages, respectively. The capacitance of broadside coupling is larger than slot coupling.

Figure 3.20(c) shows the corresponding transmission line structure of HPF. The shorted shunt stub structure as the parallel inductance is used in HPF. According to the transmission-line theory, the input impedance of the terminal short transmission line is equivalent to an inductance under the constraint of the effective electric length of the short stub smaller than a quarter-wavelength. The frequency response of HPF is shown in Figure 3.20(d).

3.2.4.4 Bandpass Filter

A BPF can be composed of an LPF and an HPF whose frequency is lower than the LPF. All the three BPFs choose the same topology, which has the advantage of design controllability. Three BPF structures and their frequency responses are shown in Figure 3.21. All the BPFs centered at 0.8, 2.5, and 5.8 GHz are composed of LPF and HPF. As can be seen in Figure 3.21(a), the structure of BPF 1 is composed of an LPF, an element in the blue circle, and an HPF. The frequency response of the BPF 1 of 0.8 GHz is shown in Figure 3.21(b), the stopband rejection reaches 20 dB at 1.25 GHz, and up to 9 GHz. BPF 2 which is at 2.5 GHz is shown in Figures 3.21(c, d). The stopband more than 20-dB attenuation is from 4 to 12.2 GHz. The third BPF of 5.8 GHz shown in Figure 3.21(e) is also composed of an LPF and an HPF. The frequency response is shown in Figure 3.21(f). The method of combining LPF and HPF to get BPF obtains good rejection depth and high controllability.

3.2.4.5 Duplexer 1 of Channels 1 and 2

Figure 3.22(a) shows the planar view of duplexer 1 of channels 1 and 2 with substrate excavation [30]. As can be seen, Tx and Rx are connected through a matching network to the antenna. For the duplexer using just one antenna, higher isolation for signals independently transmitted by the Tx/Rx region is necessary. A high isolation of the duplexer could mostly suppress the leakage signal from the transmitter to the receiver. The design of the matching network is shown in Figure 3.17(c). Moreover, to reduce the influence of coupling between the two LPFs, there exists via holes to form the metallic wall between two BPFs.

Substrate 3 is partly hollowed to make the main electromagnetic field distribute in the air cavities to reduce dielectric loss, and then results in a higher Q-factor. The transition structure consists of three parts including microstrip, stripline, and SISL. The input and output of the signal are on the metal layer G_5. The characteristic impedance of each part is 50Ω. The parameters shown in Figure 3.22(a) are: $W_{A1} = W_{A2} = 2.4$ mm, $W_{A3} = 0.6$ mm, $W_{A4} = 1$ mm, $W_{A5} = 1.3$ mm, $W_{A6} = 2$ mm, $L_{A1} = 27$ mm, $L_{A2} = 24$ mm, $L_{A3} = 8.4$ mm, $L_{A4} = 14$ mm, $L_{A5} = 11.4$ mm, $L_{A6} = 10.2$ mm, $L_{A7} = 5.5$ mm, $L_{A8} = 6$ mm, $L_{A9} = 10$ mm, and $L_{A10} = 3$ mm.

The simulated and measured results of this duplexer are shown in Figures 3.22(b, c). The center frequencies of the duplexer are 0.8 and 2.5 GHz, respectively. The measured FBWs are 55% and 43% for the two bands, respectively. Removing the influence of the connector, the measured insertion losses are 1.26/1.21 dB, while

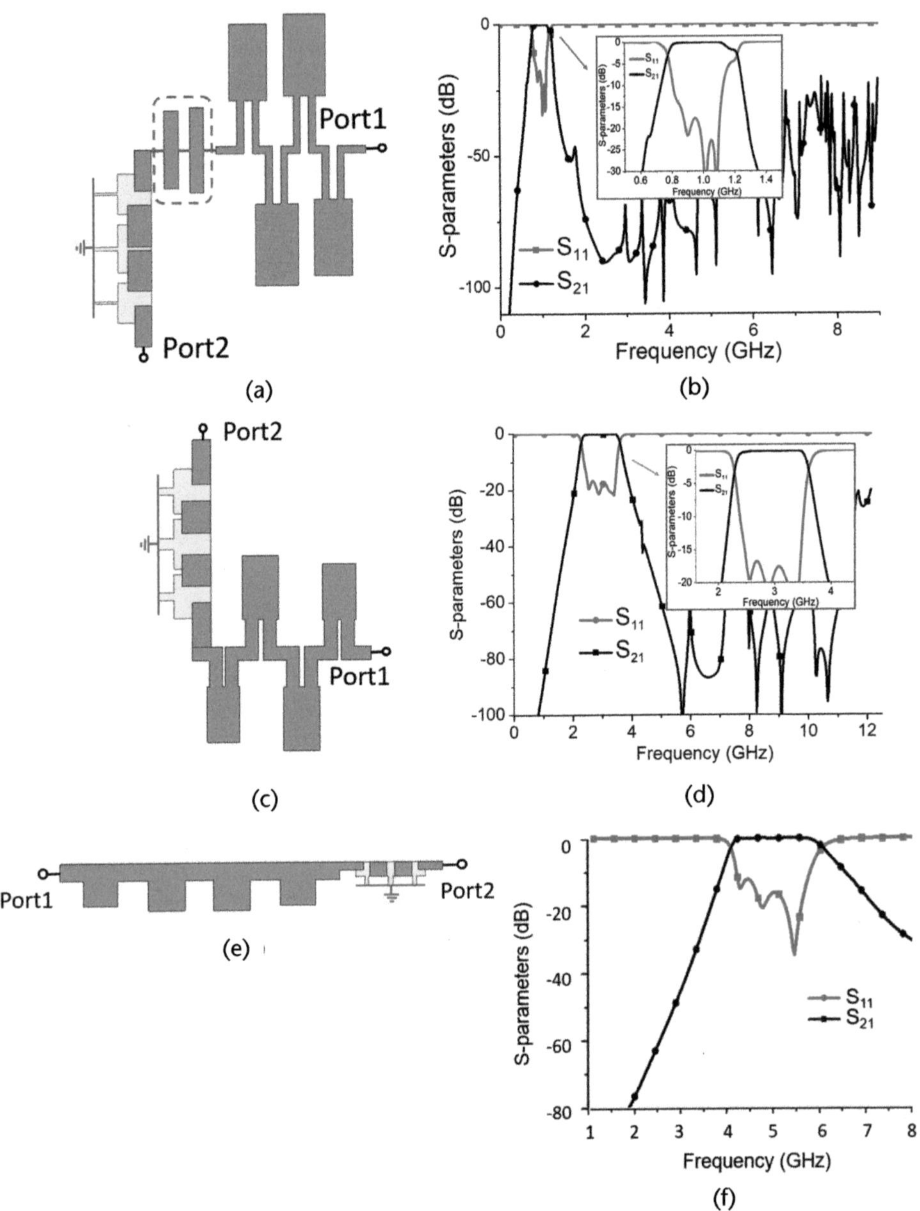

Figure 3.21 (a) Structure of BPF 1, (b) frequency response of BPF 1, (c) structure of BPF 2, (d) frequency response of BPF 2, (e) structure of BPF 3, and (f) frequency response of BPF 3 [28].

the modified design with substrate excavation reduces the insertion loss at the two bands by 1.03 dB/0.96 dB. The stopband characteristic of the two passbands and the isolation between Rx and Tx signals reach more than 60 dB.

3.2.4.6 Duplexer 2 of Channels 2 and 3

The structure of the duplexer 2 of channels 2 and 3 is shown [28]. The two BPFs are connected through the matching network of the same method above. The

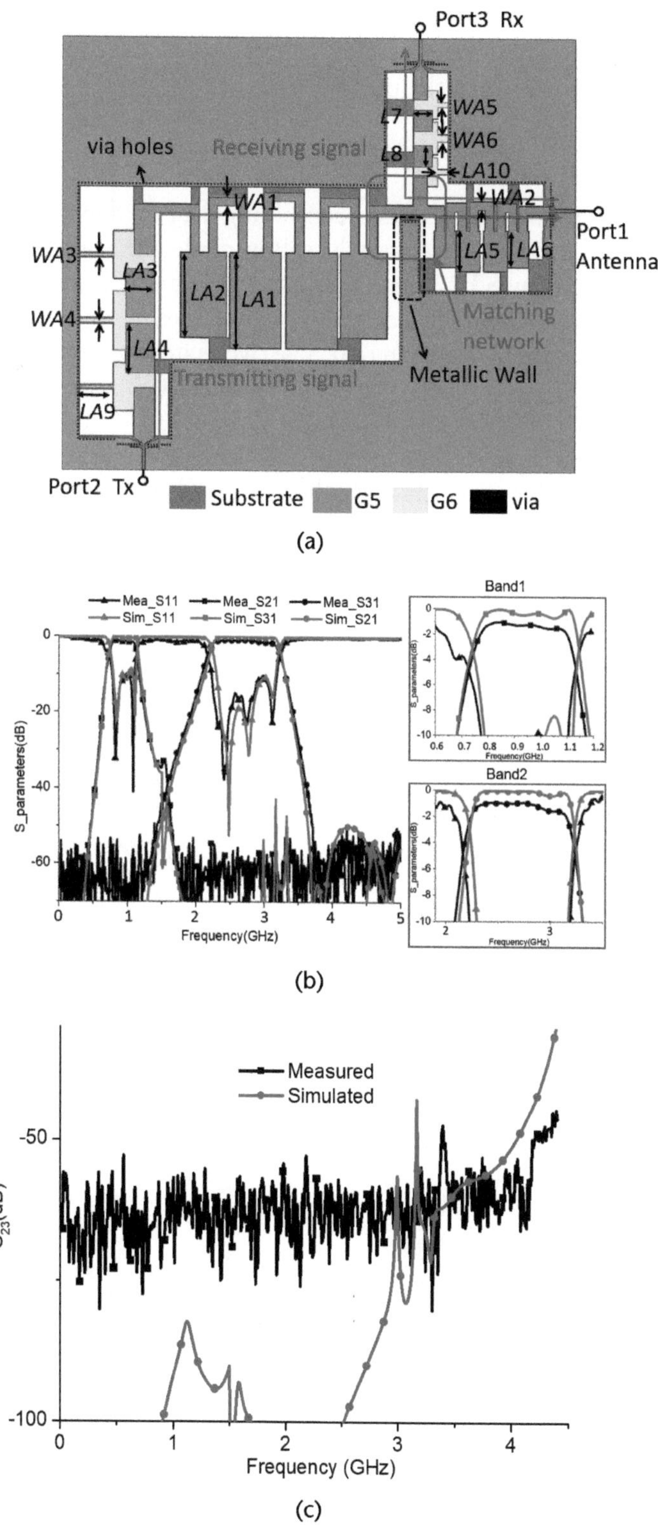

Figure 3.22 (a) Top-side view of the duplexer of channel 1 (0.8 GHz) and channel 2 (2.5 GHz) with substrate excavation, (b) simulated and measured results of the duplexer in (a), and (c) simulated and measured results of Tx-Rx isolation of the duplexer in (a) [28].

center frequency of operation is located at 2.9 and 6 GHz. The measured FBWs are 44.82%/40.02% and the measured insertion losses are 1.4 dB/0.84 dB, respectively. The isolation between the ports achieves more than 40 dB.

3.2.4.7 Verification of the Triplexer

With the design method above, combining the two duplexers can obtain a triplexer. In Figure 3.23(a), the circuit in the yellow circle is the duplexer 1 of channels 1 and 2 described in Figure 3.22(a), in which the LPFs are improved to expand the stopband just as BPFs shown in Figures 3.21(a, c), and the circuit in the green circle duplexer 2 of channels 1 and 2 shown in [28]. The two matching networks only need to consider combining the two adjacent filters. Channels 1, 2, and 3 are the paths from port 1 to ports 4, 3, and 2, respectively. Figure 3.23(b) shows the 3D view of the SISL triplexer. To prove the concept, the implemented triplexer has been fabricated. As can be seen in Figures 3.23(c, d), the simulated and measured results of the triplexer with substrate excavation are in good agreement. The center frequencies of these circuits are 1, 3, and 5.8 GHz. The measured FBWs are 50.10%/43.33%/43.10% for the three bands, respectively. With the influence of the modified design with

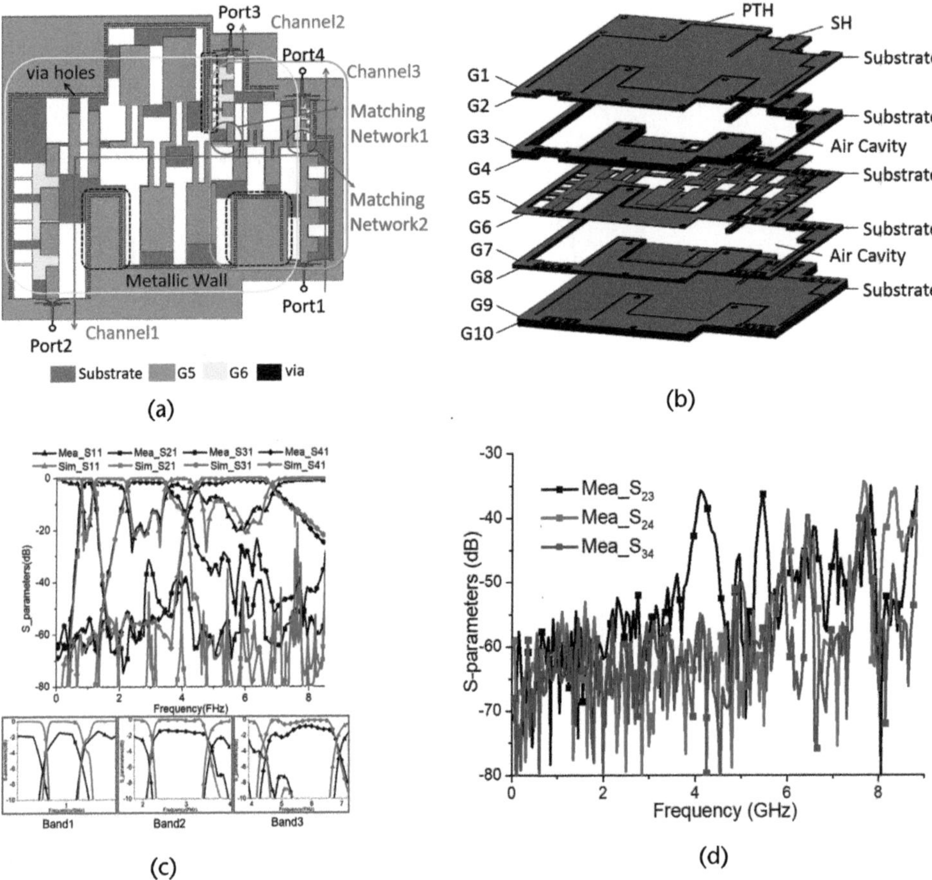

Figure 3.23 (a) Top-side view of the triplexer; (b) 3D diagram of the SISL triplexer, G_1–G_{10} are the metal layers; (c) simulated and measured results of the triplexer; and (d) measured isolation results of the triplexer [28].

substrate excavation, the insertion losses at the three bands are 1.4 dB/1.3 dB/0.8 dB. The isolation among the ports is all more than 35 dB.

3.3 Lumped-Element Filters and Tunable Filters

3.3.1 Lumped-Element Filter

3.3.1.1 Introduction of Lumped-Element Filter

A lumped element filter is an RF circuit that incorporates lumped elements, such as inductors and capacitors, and is often used to selectively attenuate or pass the signal within specific bands of frequencies. The lumped filters can realize kinds of responses, including Butterworth, Chebyshev, and Gaussian responses. Due to the miniaturization advantage of lumped element circuits, lumped element filters can achieve a smaller size compared to filters with the distributed element.

Most filters are often designed according to the synthesis of a lowpass filter equivalent called a lowpass prototype. The ladder networks are often employed to design lowpass prototype filters as shown in Figure 3.24, where n indicates the count of reactive elements and denotes either the series inductance of an inductor or the shunt capacitance of a capacitor. g_i can be obtained by looking up the table of element values for filters of specific response types in [7]. The practical circuits of the LPFs and HPFs can be transformed from the lowpass prototype.

The specific capacitance and inductance values of LPFs can be obtained with g_i by

$$L = \frac{Z_0 g}{2\pi f_c} \quad \text{for } g \text{ representing the inductance} \tag{3.53}$$

$$C = \frac{g}{2\pi f_c Z_0} \quad \text{for } g \text{ representing the capacitance} \tag{3.54}$$

The specific capacitance and inductance values of HPFs can be obtained by

$$C = \frac{1}{2\pi f_c Z_0 g} \quad \text{for } g \text{ representing the inductance} \tag{3.55}$$

$$L = \frac{Z_0}{2\pi f_c g} \quad \text{for } g \text{ representing the capacitance} \tag{3.56}$$

where Z_0 is the source impedance and f_c is the cutoff frequency of the filter.

Based on the SISL platform, we can improve the performance of lumped element capacitors and inductors, thereby improving the performance of the lumped element filters. In addition, the SISL platform can provide a flexible layout for lumped element filters to achieve a smaller size. SISL also has the advantage of low loss, thus the lumped filter can achieve better performance. This section will mainly introduce the lumped LPF [31–33] and HPF [31] on the SISL platform.

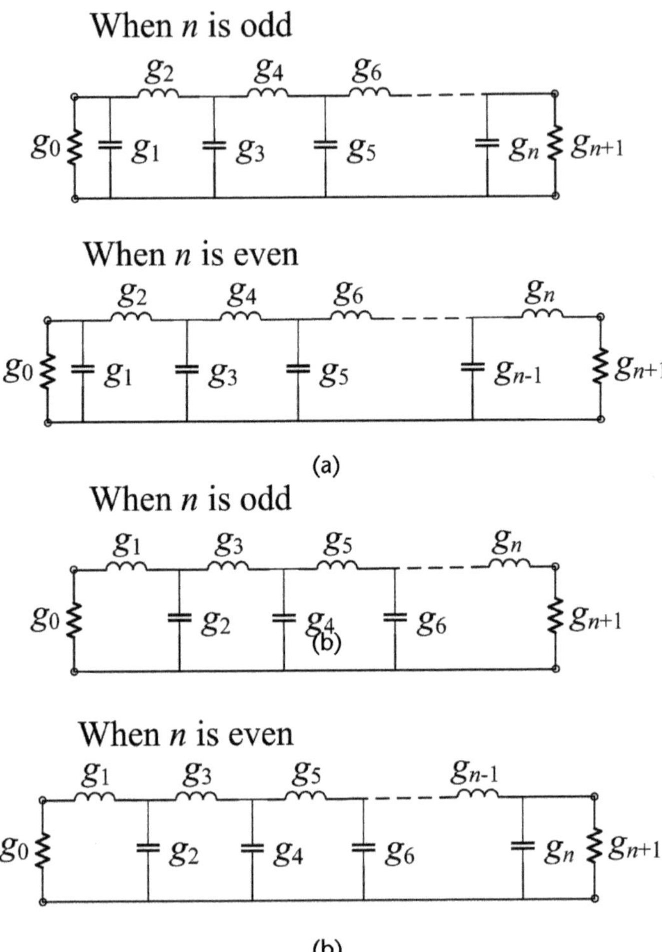

Figure 3.24 The circuits of lowpass prototype filters with the ladder network: (a) a ladder network structure and (b) its dual.

3.3.1.2 Lumped LPF

The lumped LPF, as reported in [31], achieves an ultrawide stopband and a compact size by a double interdigital capacitor (DIDC) and double-sided SISL inductor. The DIDC and DSISL inductor have already been analyzed in Chapter 2. The circuit shown in Figure 3.24(a) was chosen for this design according to the circuit size, stopband suppression, and loss. The values of the inductors and capacitors were calculated using (3.53) and (3.54): $L_1 = L_7 = 6.34$ nH, $C_2 = C_6 = 4.43$ pF, $L_3 = L_5 = 13.91$ nH, and $C_4 = 5.20$ pF.

The LPF is implemented on a classical five-substrates SISL structure. The topology of the lumped SISL LPF and the planar view of M_5 and M_6 are presented in Figures 3.25(a–c). The circuit size of the filter is 53.9 mm × 24.5 mm, corresponding to 0.180 $\lambda_g \times 0.082\ \lambda_g$, where λ_g represents the guide wavelength at 1 GHz. The normalized circuit size (NCS) is 0.015 λ_g^2. The lumped SISL LPF is simulated and measured from dc to 8 GHz, and the cutoff frequency of the LPF is 1 GHz. The passband and stopband performance exhibit good characteristics. The insertion loss within the passband is less than 1 dB, accompanied by a return loss better than 25 dB. The stopband attenuation is better than 20 dB over 24 GHz (24 f_c) starting at 1.36 GHz. Group delay is a crucial parameter in the evaluation of the communication system linearity and data rate. From dc to 1 GHz, the measured group delay of this filter ranges is 0.9 to 1.5 ns.

Based on the previous design, the stopband suppression of the circuit is affected by the presence of many harmonics within the stopband range. To improve the structure of the SISL LPF and extend the stopband while reducing the loss, several

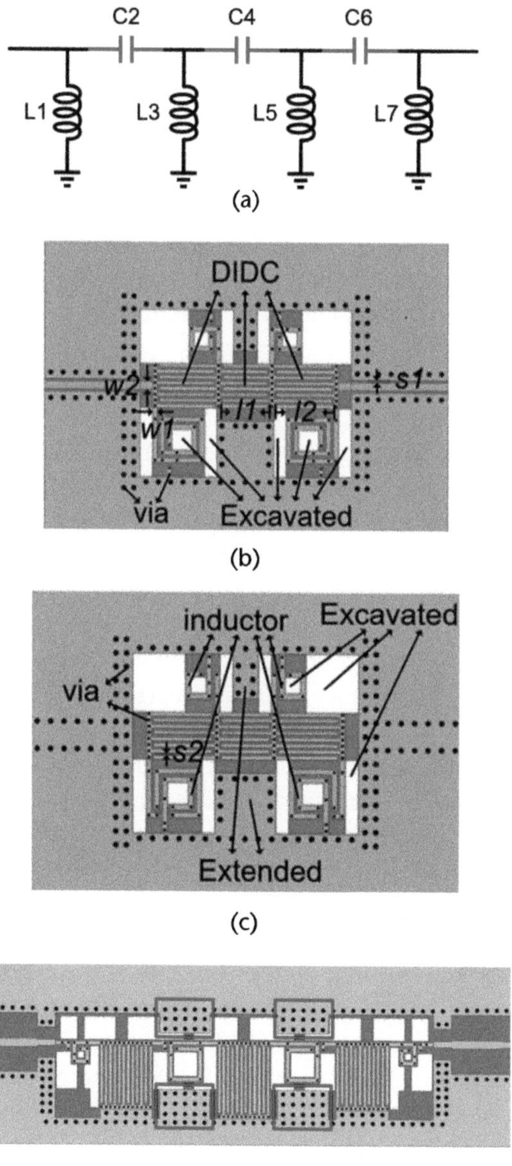

Figure 3.25 The SISL lumped element lowpass filter: (a) the topology of the SISL LPF, (b) the planar view of M_5, (c) the planar view of M_6, and (d) the planar view of M_5 of the SISL LPF with substrate excavation [31].

factors are considered. First, the self-packaged structure of the SISL may lead to the cavity mode effect on the expansion of the stopband. Therefore, the metallic walls adjacent to the inductor are extended to mitigate this issue in the design of LPF. Additionally, the substrate is excavated, except for necessary parts for mechanical support, to further decrease dielectric loss. The planar configuration of M_5 within the SISL LPF, featuring metallic walls and substrate excavation, is depicted in Figure 3.25(d). The dimensions of the filter measure 53.8 mm × 23.6 mm, corresponding to 0.180 λ_g × 0.079 λ_g, with an NCS of 0.014 λ_g^2. This research presents a comparison between simulated and measured results of the lumped SISL LPF utilizing substrate excavation across the dc to 10-GHz range. The cutoff frequency of the designed LPF is 1 GHz. Within the passband, insertion loss remains below 1 dB, and the return loss is better than 25 dB. At 1.38 GHz, the designed LPF achieves a stopband rejection of 20 dB. The measured group delay from dc to 1 GHz ranges from 0.9 to 1.5 ns. The measured stopband suppression better than 20 dB is improved from 24 GHz (24 f_c) to 32 GHz (32 f_c). The results demonstrate that substrate excavation and packaging optimization significantly extend the stopband.

Employing a novel dielectric-filled capacitor with high-capacitance density mentioned in Chapter 2, the lumped-element LPF in [32] achieves a smaller size as shown in Figure 3.26(a). The inductors and capacitor in Figure 3.26(b) can also be calculated by (3.53) and (3.54) with g_1 = 0.8516 and g_2 = 1.1032. The cutoff frequency f_c is selected as 120 MHz. Consequently, L = 56.47 nH and C = 29.26 pF are calculated. Based on SISL platform, a compact circuit size of 0.005 λ_g × 0.013 λ_g × 0.0025 λ_g is also obtained. The insertion loss is measured between 0.02 and 0.32 dB from dc to 120 MHz, and the in-band return loss is better than 19.5 dB. A wide 20-dB rejection stopband is achieved from 0.35 to 2.27 GHz. The measured group delay in the passband is 1.9 to 3.4 ns.

A lumped LPF in [33] achieves a compact size and wide stop with multiplate miniaturized capacitor mentioned in Chapter 2 and honeycomb cavity.

In traditional LPFs, all the capacitors and inductors are placed in one common SISL air cavity, which may result in electromagnetic field interaction and performance degradation. To address this issue, this research adopts the SISL honeycomb concept with multiple cavities in our design, as shown in Figures 3.27(a, b). In this design, each inductor and capacitor are placed in an individual cavity and isolated from others by metal holes and patterned substrates. A circle of compact metal holes is used to surround each capacitor and inductor, penetrating all the substrate and metal layers to provide electromagnetic isolation. In the individual cavity of the inductor, the substrate is patterned to reduce the area of the inductor, thereby improving the design flexibility and extending the stopband of the lumped LPF.

The circuit and the planar view of the LPF are shown in Figure 3.27(c). The capacitors and inductors are calculated using (3.53) and (3.54), where $g_1 = g_7$ = 0.7970, $g_2 = g_6$ = 1.3924, $g_3 = g_5$ = 1.7481, and g_4 = 1.6331. The cutoff frequency (f_c) is selected as 0.6 GHz in the design. The final topology parameters are as follows: $L_1 = L_7$ = 10.57 nH, $C_2 = C_6$ = 7.39 pF, $L_3 = L_5$ = 23.18 nH, and C_4 = 8.66 pF.

The designed LPF is measured and simulated. The measured insertion loss is less than 0.75 dB from dc to 0.3 GHz and the measured return loss within the passband is better than 25.4 dB. A wide stopband with a suppression level over 20 dB is achieved from 0.82 to 67 GHz. The size of the LPF is 0.085 λ_g × 0.015 λ_g ×

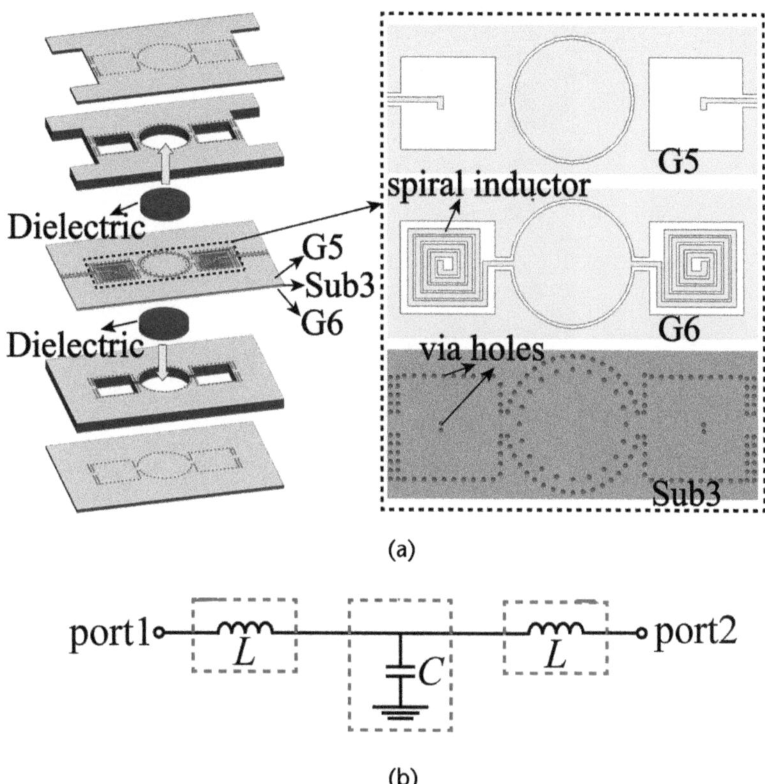

Figure 3.26 The compact SISL LPF using filled dielectric capacitor: (a) 3D view, and (b) its equivalent circuit [32].

$0.012\ \lambda_g$. The proposed SISL LPF demonstrates advantages in terms of compact size, self-packaging, and wide stopband.

3.3.1.3 Lumped HPF

In [31], a lumped HPF on SISL platform is also proposed. The HPF also employs techniques similar to the LPF proposed in this article, namely, using DIDC and DSISL spiral inductors and substrate excavation techniques. The circuit is shown in Figure 3.28(a) and the capacitance and inductance are calculated with (3.53) and (3.54): $L_1 = L_7 = 9.99$ nH, $C_2 = C_6 = 2.3$ pF, $L_3 = L_5 = 4.55$ nH, and $C_4 = 1.95$ pF.

The planar view of M_5 and M_6 of the proposed SISL HPF are shown in Figure 3.28(b, c). To facilitate a comparison between different designs, two types of SISL HPFs were fabricated: one with substrate excavation and the other without. The measured results demonstrate that both designs exhibit a sharp skirt characteristic and excellent 20-dB return loss from 1 to 5 GHz. Furthermore, both SISL HPFs can achieve a 20-dB rejection up to 720 MHz. Notably, the SISL HPF with substrate excavation shows a remarkable minimum insertion loss of 0.1 dB, which is 0.2 dB lower than that of the SISL HPF without substrate excavation. The designed HPF achieves a compact size of 32.4 mm × 25.6 mm, corresponding to $0.108\lambda_g \times 0.085\ \lambda_g$, with an NCS of $0.009\ \lambda_g^2$.

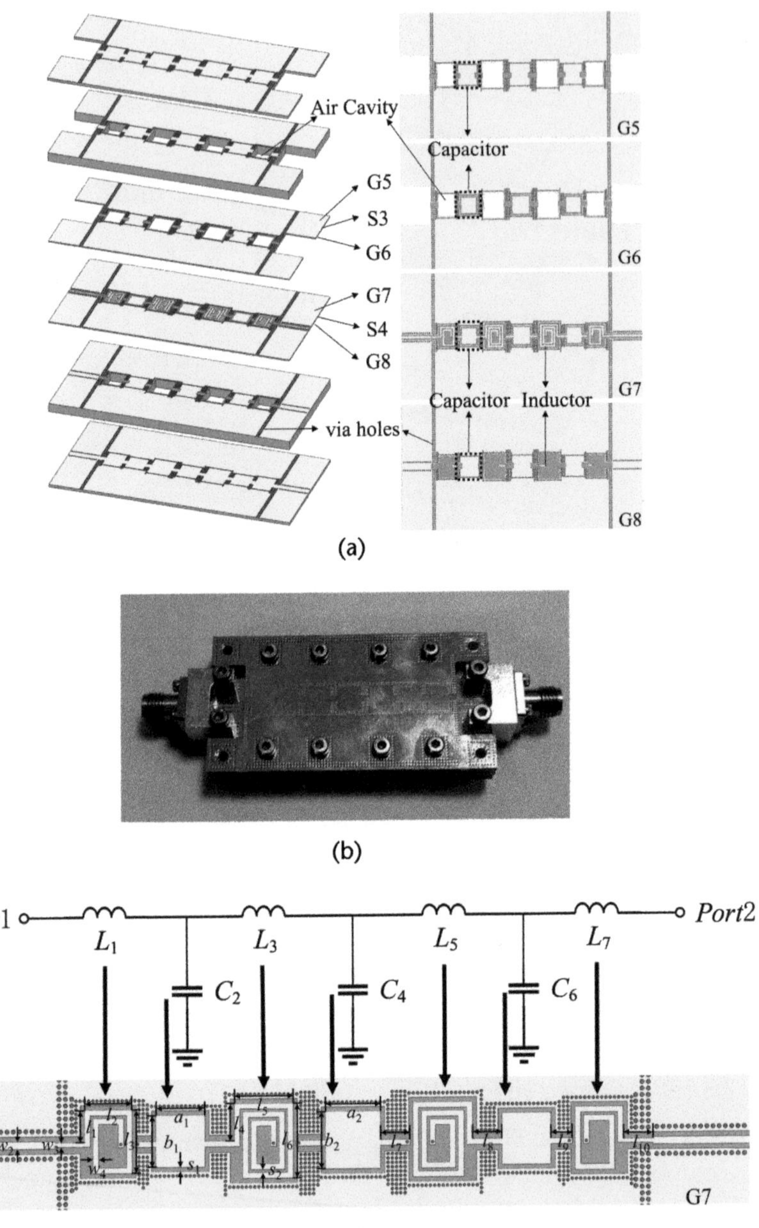

(a)

(b)

(c)

Figure 3.27 The compact substrate integrated suspended-line LPF: (a) 3D view, (b) fabricated circuit, and (c) its equivalent circuit and dimensions of the structure [33].

3.3.2 Tunable and Reconfigurable Filters

3.3.2.1 Introduction of Tunable and Reconfigurable Filters

Tunable and reconfigurable filters can provide control over parameters such as frequency and bandwidth while reducing the need for several switches sandwiched between electrical components.

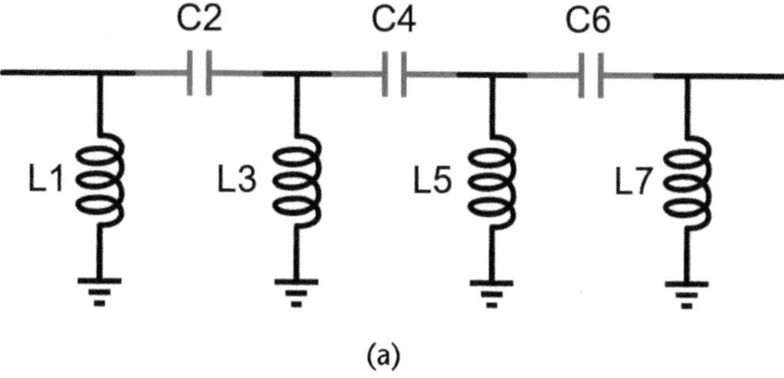

(a)

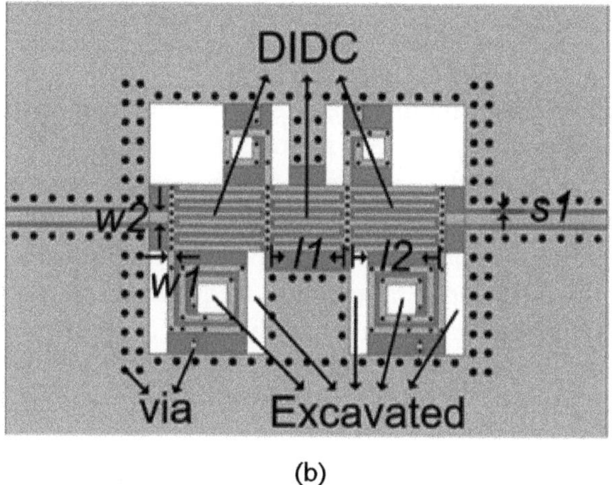

(b)

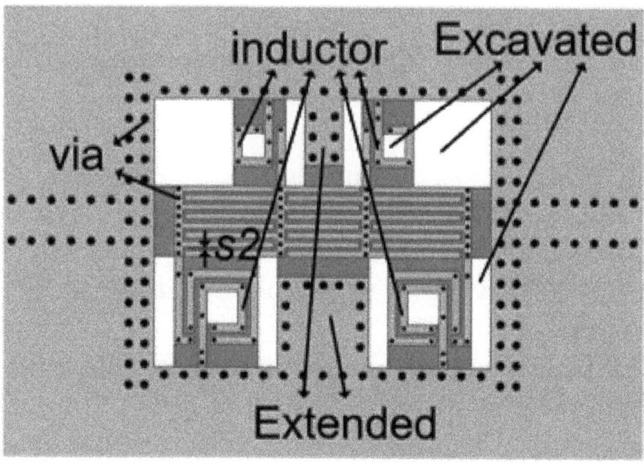

(c)

Figure 3.28 The lumped SISL HPF: (a) the seven-order topology of the HPF, (b) planar view of M_5, and (c) planar view of M_6 [31].

Tunable BPFs come in three main types: frequency-tunable, bandwidth-tunable, and combined frequency and bandwidth-tunable. Before we discuss the specific circuit, there are several performance parameters about tuned bandpass filters that should be explained.

The center frequency f_0 of the bandpass filter can be expressed as follows:

$$f_0 = \frac{f_L + f_H}{2} \tag{3.57}$$

where f_L and f_H are the frequencies on both sides of the passband of the bandpass filter that attenuates by 1 dB or 3 dB relative to f_0.

The center frequency of an adjustable filter is an interval range, commonly measured as a percentage. The frequency tuning range (FTR) can be expressed as follows:

$$\text{FTR} = \frac{2\left(f_{\text{max}} - f_{\text{min}}\right)}{f_{\text{max}} + f_{\text{min}}} \times 100\% \tag{3.58}$$

where f_{max} is the maximum center frequency that can be tuned and f_{min} is the minimum center frequency that can be tuned.

To fully describe the bandwidth characteristics of filters, bandwidth indicators include absolute bandwidth (ABW) and relative bandwidth (FBW), with the former focusing on the absolute value of bandwidth and the latter focusing on the percentage of bandwidth relative to its center frequency.

The expression for absolute bandwidth is as follows:

$$\text{ABW} = f_H - f_L \tag{3.59}$$

The expression for relative bandwidth is as follows:

$$\text{FBW} = \frac{f_H - f_L}{f_0} \times 100\% \tag{3.60}$$

Based on SISL platform, a tunable quad-operation-mode BPF with wide tuning range, low loss, compact size, and self-packaging is presented in [34]. Besides, the tunable SISL BPF adopts several techniques to improve the performance, including combining SEMCPs with varactor loaded and introducing source-load coupling and stub-loaded.

Different from the tunable filters on the other platform, tunable filters on SISL platform have their characteristics, such as wide tuning range, low loss, compact size, and self-packaging, which will be shown in detail in this section.

3.3.2.2 Quad-Operation-Mode Tunable SISL BPF

The operating mechanism of each part of the SISL BPF is described below. To generate TZs and adjust the external quality factor, the filter employs heterogeneous feed lines. Meanwhile, the varactor-loaded resonators are responsible for setting the operating frequency of the filter. The coupling coefficient between the resonators is fine-tuned by utilizing SEMCPs. Finally, the bias voltage of the varactors is the sole parameter necessary to achieve the tunable SISL BPF with quadruple operating modes.

The layout of the self-packaging tunable SISL BPF is depicted in Figure 3.29(a). The structure of feed lines and resonators with SEMCP are considered carefully. The feed lines of the filter consist of two parts. The first part, shown as L_1 in Figure 3.29(b), employs double metal layers via holes to reduce the required width and utilize design transition structure carefully to achieve impedance matching. This section carries stubs $L_5 + L_6$ and L_7 of the M_6 layer, which introduce two TZs in the stopband. Notably, the ends of these two stubs are broadside-coupled to the resonators, and their length is shorter than a normal stub-loaded line. The second part, shown as L_9 in Figure 3.29(c), also utilizes broadside coupling to provide consistent coupling energy to the resonators across a broad frequency range. The source-load coupling is formed by employing these structures and two TZs are introduced. Between the two feed-line sections, the varactors D_1, along with their bias voltage V_1, dominate the external quality factor of the filter. When operating in different modes, impedance mismatch may occur, but V_1 can be tuned for compensation to maintain an excellent match.

Varactors D_2 is loaded on the resonators to modify the equivalent electrical length of the tunable SISL BPF. As the bias voltage V_2 increases, the capacitance value of D_2 decreases, causing the BPF to be tuned to a higher frequency and vice versa. To effectively tune the coupling coefficient of the BPF, SEMCPs are employed between the resonators. The SEMCP technique involves using a high-impedance line to provide the magnetic coupling path and varactors (D_3) to provide the electrical coupling path. The coupling coefficient is influenced by the combined magnetic and electrical couplings. After the filter is fabricated, the magnetic coupling cannot be changed, while the electrical coupling can be tuned by changing the bias voltage (V_3) of the varactors. The path counteraction of SEMCP creates a controllable TZ, offering the flexibility to position it either slightly above or below the passband to enhance the selectivity of the filter. By avoiding the need for additional resonators, such as in cross-coupling techniques, SEMCP allows for the miniaturization of the tunable BPF and a reduction in insertion loss. SEMCP enables the adjustment of the coupling coefficient and tuning of the TZ in proximity to the passband, serving as a pivotal element for achieving a tunable BPF with quad-operation modes.

The measured and simulated results of the tunable SISL BPF are given in [34]. The frequency of the designed BPF can be tuned across the range of 0.6 to 1.32 GHz with a constant absolute bandwidth (CABW) of 70 MHz. The tuning range and insertion loss are 75% and 2.6 to 2.8 dB, respectively. The frequency tuning of the BPF with constant fractional bandwidth (CFBW) is 11.7%. In this case, the frequency is tunable from 0.6 to 1.3 GHz, encompassing a tuning range of 73.7%, along with an insertion loss ranging from 1.5 to 2.8 dB. At a central frequency of 1 GHz, the 1-dB bandwidth exhibits a tunable range spanning 50 to 190 MHz, accompanied by a tuning ratio of 3.8 dB and an insertion loss that changes from 1.6 to 3.8 dB. The BPF is in the all-off state, with S_{21} between than −24 dB, demonstrating signal blocking. The measured results show a wideband response of the BPF at 0.6 GHz, with a stopband better than 18 dB, which is more than 6 f_c. The BPF features five TZs: TZ_1 and TZ_3 are created by source-load coupling, TZ_2 is created by SEMCP, and TZ_4 and TZ_5 are created by stub ($L_5 + L_6$) and stub L_7. The influence of the interfaces and transition structure is considered, and the insertion loss of them is

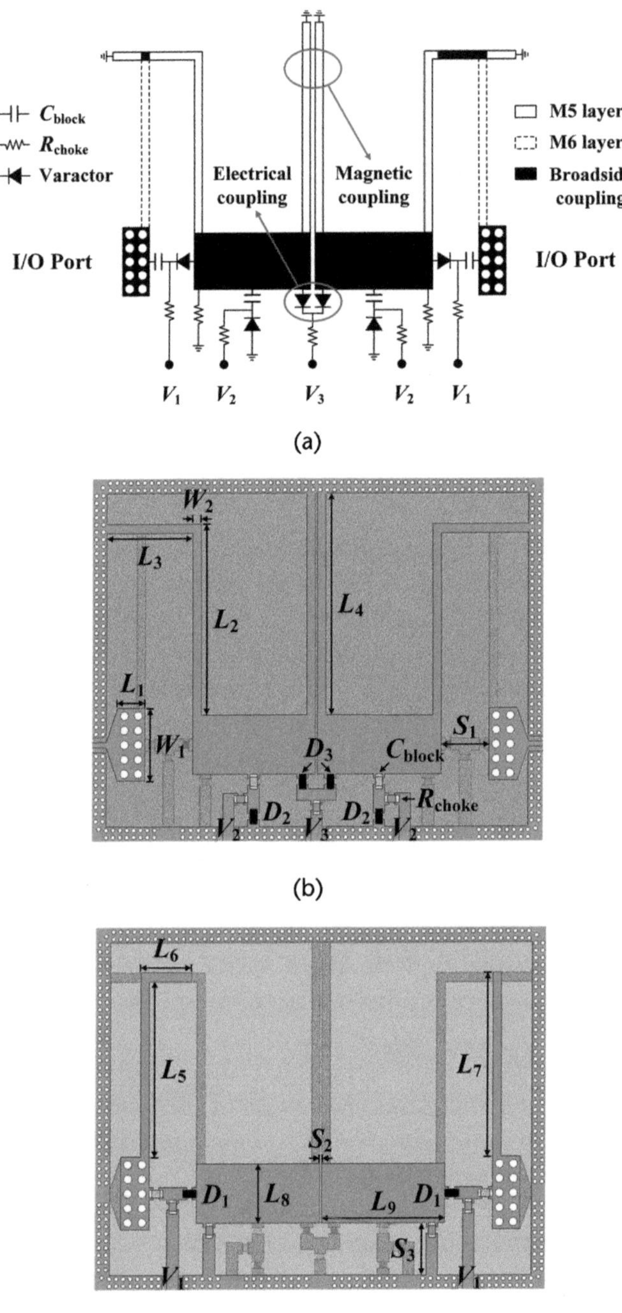

Figure 3.29 The tunable SISL BPF: (a) layout, (b) M_5 layer, and (c) M_6 layer [34].

calibrated in advance, and the return loss is better than 15 dB. The slight deviations between the measured and simulated results are primarily attributed to varactor manufacturing tolerances, soldering, and fabrication errors.

3.4 Metasubstrate Filter on the SISL Platform

Due to its periodic patterns, the metasubstrate has been shown to have a high permittivity, aiding in the downsizing of microwave circuits and systems [35]. Based on a metasubstrate SISL platform, compact self-packaged second/fourth-order quarter-wavelength hairpin filters have been designed in [36].

Two parallel linked wires are constructed into the dumbbell-shaped metasubstrate unit and are joined via metalized via-holes. In Figure 3.30(a), $a = 2$ mm, $l = 2$ mm, $d = 0.3$ mm, $w = 0.3$ mm, and $t = 0.035$ mm. The NRW algorithm is used to calculate the effective permittivity ε and permeability μ [37, 38].

$$\varepsilon = \frac{\left[-\gamma^2\left(\frac{\lambda_0}{2\pi}\right)^2 + \frac{\lambda_0}{\lambda_c}\right]}{\mu} \tag{3.61}$$

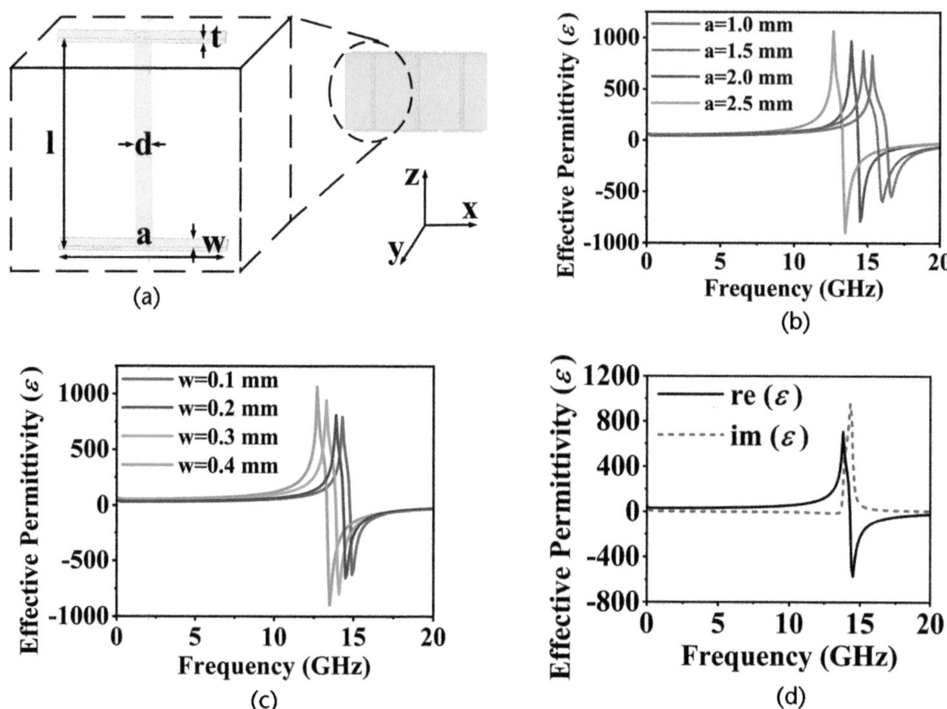

Figure 3.30 (a) Schematic diagram of dumbbell-shaped metasubstrate unit, (b, c) curves of effective permittivity *e* with frequency for different *a* and *w*, and (d) curves of effective permittivity *e* with frequency for dumbbell-shaped metasubstrate unit [36].

$$\mu = -\frac{j\lambda_0\gamma}{2\pi\sqrt{1-\left(\frac{\lambda_0}{\lambda_c}\right)^2}}\left(\frac{1+\Gamma_c}{1-\Gamma_c}\right) \tag{3.62}$$

ε and μ can be calculated by

$$\frac{S_{11}^2 - S_{21}^2 + 1}{2S_{11}} = \frac{1+\Gamma_c^2}{2\Gamma_c} \tag{3.63}$$

$$K = \frac{S_{11}^2 - S_{21}^2 + 1}{2S_{11}} \tag{3.64}$$

$$\Gamma_c = K \pm \sqrt{K^2 - 1} \tag{3.65}$$

$$T_l = \frac{S_{11} + S_{21} - \Gamma_c}{1-\left(S_{11}+S_{21}\right)\Gamma_c} \tag{3.66}$$

$$\gamma = -\frac{1}{l}\ln\left(T_l\right) \tag{3.67}$$

l is the thickness of a single unit structure, λ_c is the cutoff wavelength, and λ_0 is the free-space wavelength.

As the wave, polarized with an electric field vector in the y direction, interacts with the metasubstrate structure, a substantial accumulation of surface charges accumulate on the arm of the metallic patches. Charge accumulation contributes to a large effective permittivity.

The relationship between the real part of effective permittivity and frequency is shown in Figures 3.30(b, c). The increase of a and w causes the increase of the effective permittivity. a and w are 2 mm and 0.3 mm according to the design requirements. In Figure 3.30(d), the designed dumbbell-shaped unit shows a metasubstrate characteristic with high permittivity of 50 in the range of 2 to 5 GHz due to a strong electric resonance.

In Figure 3.31, the 3D view of transmission line structure based on the six-layer SISL metasubstrate platform is displayed. The metasubstrate structure is manufactured on the upper and lower layers of the substrate M. Six substrates are manufactured with the low-cost FR4 material.

One can calculate the slow-wave factor (SWF) of the metasubstrate as shown in [39]

$$\text{SWF} = \frac{\beta}{\beta_0} \tag{3.68}$$

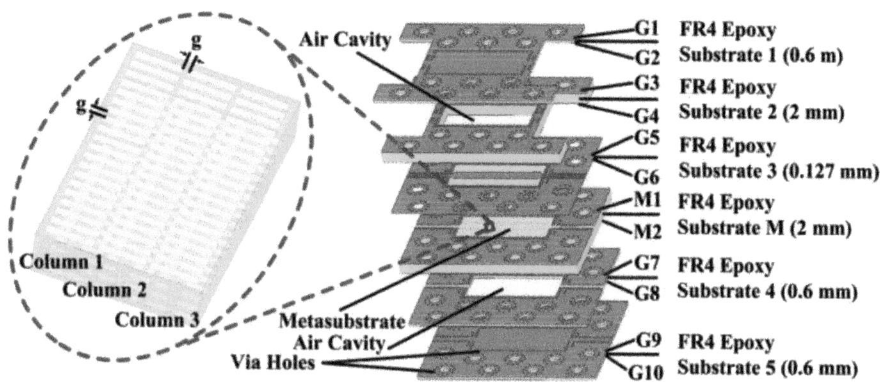

Figure 3.31 The 3D view of SISL metasubstrate platform [36].

where β and β_0 are the phase constant and phase constant in free space, respectively. According to (3.68), the β increases with the increase of the SWF.

In Figure 3.32, the measured S-parameters and S_{21} phase curves of the SISL transmission line can be obtained. The insertion loss and return loss exceed 0.6 dB and 20 dB, respectively, across the frequency range from dc to 10 GHz. The electromagnetic wave shows a better match and lower loss and transmits smoothly in this frequency band. From 2 to 5 GHz, the phase changes of 70° and 131° are obtained for the SISL transmission lines without and with metasubstrate, which

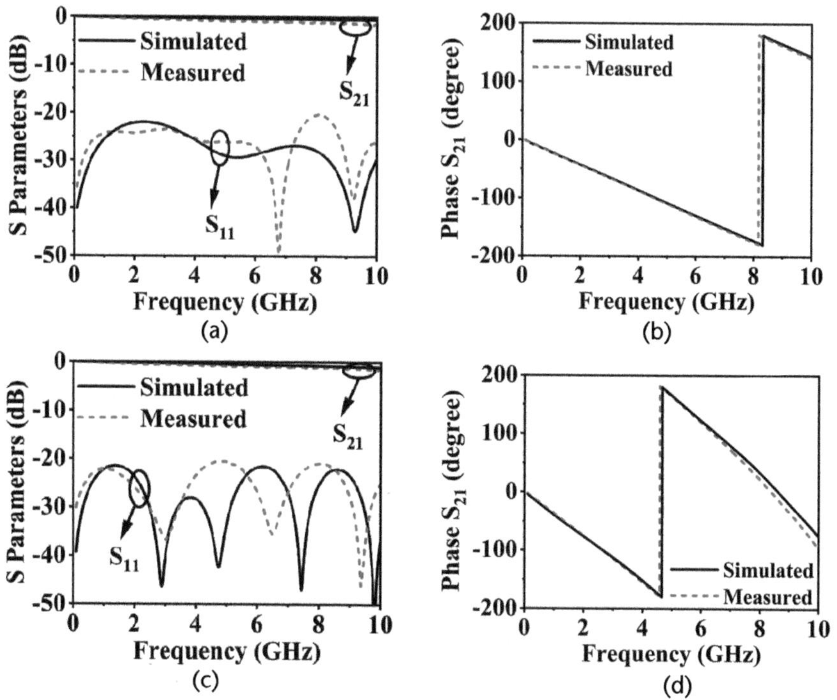

Figure 3.32 Simulated and tested S-parameters and S_{21} phase curves of SISL transmission line: (a, b) without metasubstrate, and (c, d) with metasubstrate [36].

means that the β is increased by 1.87 times. Therefore, the calculated permittivity is increased by 3.5 times.

$$V_p = \frac{\omega}{\beta} \tag{3.69}$$

$$V_p = \frac{c}{\sqrt{\varepsilon}} \tag{3.70}$$

FR4 material shows a relative permittivity of 4.4. The proposed periodic metasubstrate exhibits a high relative permittivity value of 15.4. Based on the full-wavelength resonator method [40], the effective permittivity of the SISL metasubstrate platform is calculated by (3.71).

$$\varepsilon_{\text{eff}} = \left(\frac{ncf_0}{2\pi r f_{\text{eff}}}\right)^2 \tag{3.71}$$

where f_0 is the fundamental frequency of unit, c is the light speed in vacuum, n is the mode number or number of wavelengths on the dumbbell-shaped unit, f_{eff} is the measured frequency, and r is the mean radius of the unit.

According to the proposed SISL metasubstrate platform, miniaturized self-packed second-order quarter-wavelength hairpin filters are designed and manufactured. In Figure 3.33(a), the feed lines and two quarter-wavelength resonators of R1 and R2 generate the passband in 3.3–3.6 GHz, f_0 of 3.45 GHz, 3-dB FBW of 17.4%, and return loss better than 20 dB. According to the coupling scheme in Figure 3.33(b), W_4 is the equivalent gap capacitance in the electric (E_l) coupling. Short stub L_2 is the equivalent inductance. They form the E_l and magnetic (M_a) coupling path, respectively. The expression for the coupling coefficient K is given in quarter-wavelength filters with SEMCPs [6].

$$K = \frac{\omega_0^2 - \omega_e^2}{\omega_0^2 + \omega_e^2} = B\left(Y_c L_m - C_m Z_c\right) = M_a - E_l \tag{3.72}$$

where

$$B = \frac{4\left(D + Y_c L_m + Z_c C_m\right)}{\left(2Y_c L_m + D\right)^2 + \left(2Z_c C_m + D\right)^2} \tag{3.73}$$

$$M_a = BY_c L_m \tag{3.74}$$

and

$$E_l = BZ_c C_m \tag{3.75}$$

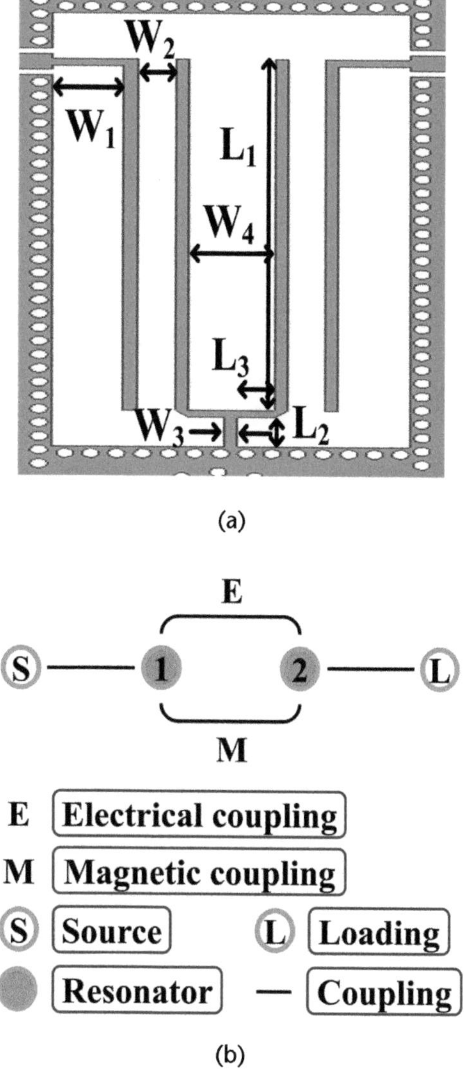

Figure 3.33 (a) The core second-order filter circuit on G_5 layer of SISL, and (b) proposed second-order filter topology [36].

$$\omega_o = \frac{\pi}{2(2C_m Z_c + D)} \tag{3.76}$$

and

$$\omega_e = \frac{\pi}{2(2L_m Y_c + D)} \tag{3.77}$$

The odd-mode resonant angular frequencies are denoted as ω_o, while the even-mode resonant angular frequencies are denoted as ω_e. Z_c represents the characteristic impedance of the resonator.

$$Z_c = \frac{1}{Y_c} \tag{3.78}$$

and

$$D = \frac{\sqrt{\varepsilon_{re}}\,l}{c} \tag{3.79}$$

c and l are the light speed in vacuum and the length of resonator, respectively.

The difference value of M_a and E_l determines K. Magnetic coupling plays domination while K is greater than zero. In Figure 3.34, a TZ marked TZ2 in the upper stopband is produced due to the cancelling of E_l and M_a coupling. M_a coupling will be gradually enhanced with the increase of the length of the short stub L_2. Furthermore, the TZ2 progressively shifts towards higher frequency. The TZ1 in the lower stopband hardly changes, which is created by the parallel coupling feed structure [41].

The above circuit design is validated through the manufacture and fabrication of the BPF. W_1, W_2, W_3, W_4, L_1, L_2, and L_3 are 1.65 mm, 0.9 mm, 0.3 mm, 2 mm, 15.5 mm, 1.29 mm, and 0.85 mm, respectively. In Figure 3.35, the photographs of the core second-order filter circuit, multilayer PCB structures, and fabricated filter are displayed. The core size of the second-order filter with metasubstrate is $0.26\ \lambda_g \times 0.19\ \lambda_g$. The core size of the second-order filter without metasubstrate is $0.46\ \lambda_g \times 0.20\ \lambda_g$. A size reduction (SR) of 46% is achieved by the above proposed dumbbell-shaped metasubstrate technique. According to Figure 3.36, the measured center frequency and FBW of the SISL metasubstrate second-order filter are 3.49 GHz and 30.5%, respectively. The insertion loss is lower than 1.13 dB, while the return loss is better than 20 dB.

Miniaturized self-packaged second/fourth-order quarter-wavelength hairpin filters are developed and produced based on the novel proposed SISL dumbbell-shaped metasubstrate platform. The designed filters with metasubstrate show more compact

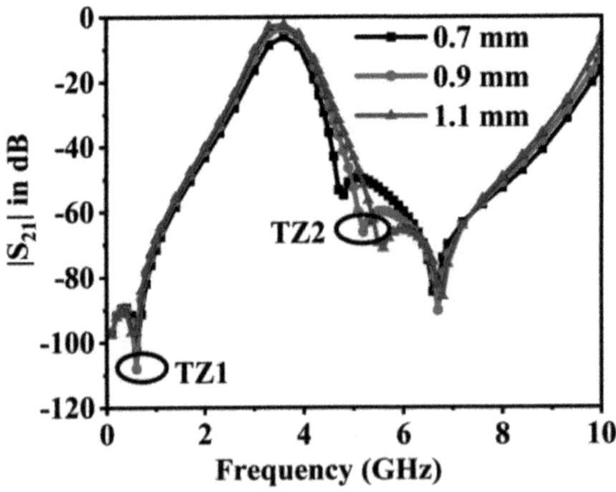

Figure 3.34 S-parameters of the SISL second-order filter with frequency for a different L_2 [36].

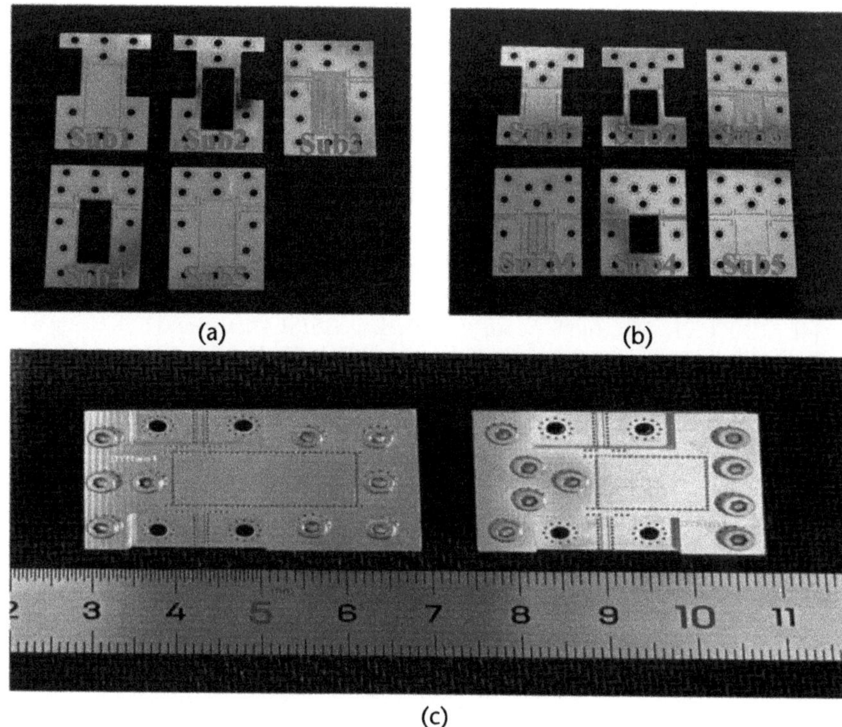

Figure 3.35 Photographs of multilayer PCB structures of substrates S_1–S_5, M, and fabricated SISL second-order filter: (a) without metasubstrate, (b) with metasubstrate, left of (c) without metasubstrate, and right of (c) with metasubstrate [36].

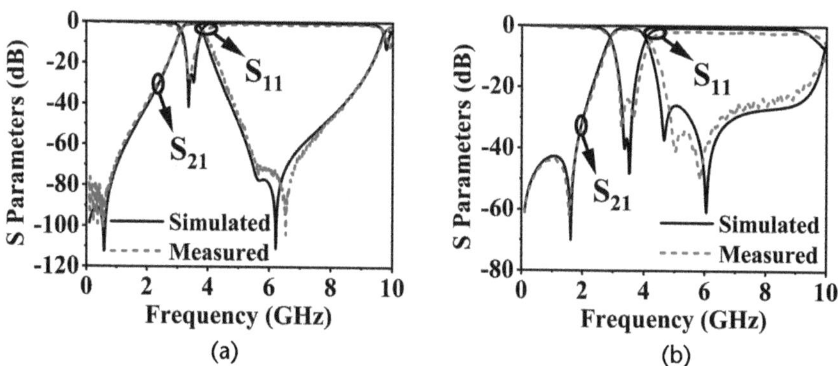

Figure 3.36 Simulated and tested S-parameters of SISL second-order filter: (a) without metasubstrate, and (b) with metasubstrate [36].

size, which is facilitated by the multilayer SISL and periodic metasubstrate structures, contributing to the development of advanced miniaturized self-packaged filters.

Acknowledgments

We would like to thank Qingling Zhang, Zehua Yue, Shuangxu Li, Haiyang Zhang, Haozhen Huang, and Yufeng Ding for their valuable support to this chapter.

References

[1] Xu, W., K. Ma, and Y. Guo, "Design of Miniaturized 5G SISL BPFs with Wide Stopband Using Differential Drive Inductor Resonators," *IEEE Transactions on Microwave Theory and Techniques*, Vol. 70, No. 6, 2022, pp. 3115–3124.

[2] Du, C., et al., "A Self-Packaged Bandpass Filter with Controllable Transmission Zeros Using Substrate Integrated Suspended Lines," *2016 IEEE International Conference on Microwave and Millimeter Wave Technology (ICMMT)*, No. 1, 2016, pp. 317–319.

[3] Yue, Z., K. Ma, and Y. Wang, "FR-4 Based Ka-Band SISL Bandpass Filters Using Folded Half-Wavelength Resonators," *Microwave and Optical Technology Letters*, 2023.

[4] Du, C., K. Ma, and S. Mou, "Self-Packaged SISL Dual-Band BPF Using a New Quadruple-Mode Resonator," *Electronics Letters*, Vol. 52, No. 13, 2016, pp. 1143–1145.

[5] Xu, W., K. Ma, and C. Du, "Design and Loss Reduction of Multiple-Zeros Dual-Band Bandpass Filter Using SISL," *IEEE Transactions on Circuits Syst. II*, Vol. 68, No. 4, April 2021, pp. 1168–1172.

[6] Zhang, W., et al., "Design of a Compact SISL BPF with SEMCP for 5G Sub-6 GHz Bands," *IEEE Microw. Wireless Compon. Lett.*, Vol. 30, No. 12, December 2020, pp. 1121–1124.

[7] Chu, Y., et al., "A Self-Packaged Low-Loss and Compact SISL DBBPF with Multiple TZs," *IEEE Microw. Wireless Compon. Lett.*, Vol. 29, No. 3, March 2019, pp. 192–194.

[8] Gao, W., et al., "Design of Novel Compact SISL Filters with Multiple TZs Based on Patch Resonators," *Microwave and Optical Technology Letters*, December 2021.

[9] Du, C., K. Ma, and S. Mou, "A Miniature SISL Dual-Band Bandpass Filter Using a Controllable Multimode Resonator," *IEEE Microw. Wireless Compon. Lett.*, Vol. 27, No. 6, June 2017, pp. 557–559.

[10] Ma, K., et al., "A Compact Size Coupling Controllable Filter with Separate Electric and Magnetic Coupling Paths," *IEEE Transactions on Microwave Theory and Techniques*, Vol. 54, No. 3, March 2006, pp. 1113–1119.

[11] Hong, J. -S., *Microstrip Filters for RF/Microwave Applications*, 2nd ed., New York: Wiley, 2011.

[12] Chu, Q. -X., and H. Wang, "A Compact Open-Loop Filter with Mixed Electric and Magnetic Coupling," *IEEE Transactions on Microwave Theory and Techniques*, Vol. 56, No. 2, February 2008, pp. 431–439.

[13] Rosenberg, U., and S. Amari, "Novel Coupling Schemes for Microwave Resonator Filters," *IEEE Transactions on Microwave Theory and Techniques*, Vol. 50, No. 12, December 2002, pp. 2896–2902.

[14] Djaiz, A., and T. A. Denidni, "A New Compact Microstrip Two-Layer Bandpass Filter Using Aperture-Coupled SIR-Hairpin Resonators with Transmission Zeros," *IEEE Transactions on Microwave Theory and Techniques*, Vol. 54, No. 5, May 2006, pp. 1929–1936.

[15] Kuo, J. T., and E. Shih, "Microstrip Stepped Impedance Resonator Bandpass Filter with an Extended Optimal Rejection Bandwidth," *IEEE Transactions on Microwave Theory and Techniques*, Vol. 51, No. 5, May 2003, pp. 1554–1559.

[16] Chen, C. H., et al., "Highly Miniaturized Multiband Bandpass Filter Design Based on a Stacked Spiral Resonator Structure," *IEEE Transactions on Microwave Theory and Techniques*, Vol. 60, No. 5, May 2012, pp. 1278–1286.

[17] del Castillo Velazquez-Ahumada, M., et al., "Design of a Dual Band-Pass Filter Using Modified Folded Stepped-Impedance Resonators," *IEEE MTT-S Int. Microw. Symp. Dig.*, June 2009, pp. 857–860.

[18] del Castillo Velazquez-Ahumada, M., et al., "Application of Stub Loaded Folded Stepped Impedance Resonators to Dual Band Filters," *Progr. Electromagn. Res.*, Vol. 102, No. 2, 2010, pp. 107–124.

[19] Eisenstadt, W. R., and Y. Eo, "S-Parameter-Based IC Interconnect Transmission Line Characterization," *IEEE Transactions on Components, Hybrids, and Manufacturing Technology*, Vol. 15, No. 4, August 1992, pp. 483–490.

[20] Du, C., K. Ma, and S. Mou, "A Controllable and Low Loss Dual-Band Bandpass Filter by Using a Simple Ring Resonator," *2016 IEEE MTT-S International Microwave Workshop Series on Advanced Materials and Processes for RF and THz Applications (IMWS-AMP2016)*, 2016.

[21] Xu, J., W. Wu, and C. Miao, "Compact and Sharp Skirts Microstrip Dual-Mode Dual-Band Bandpass Filter Using a Single Quadruple-Mode Resonator (QMR)," *IEEE Transactions on Microwave Theory and Techniques*, Vol. 61, No. 3, March 2013, pp. 1104–1113.

[22] Zhang, Y., K. Ma, and X. Chen, "A Compact Low-Cost Dual-Band Bandpass Filter with Enhanced Suppression Using Substrate Integrates Suspended Line Technology," *Microw. Opt. Technol. Lett.*, Vol. 64, No. 2, February 2022, pp. 276–282.

[23] Zhang, H., et al., "Design of Wide Stopband Lowpass Filter Using Transformed Radial Stub Based on Substrate Integrated Suspended Line Technology," *Microwave and Optical Technology Letters*, 2020.

[24] Ma, K., and K. Yeo, "New Ultra-Wide Stopband Low-Pass Filter Using Transformed Radial Stubs," *IEEE Transactions on Microwave Theory and Techniques*, Vol. 59, No. 3, 2011, pp. 604–611.

[25] Verdú, J., et al., "Synthesis Methodology for the Design of Acoustic Wave Stand-Alone Ladder Filters, Duplexers and Multiplexers," *Proc. IEEE Int. Ultrason. Symp. (IUS)*, Washington, D.C., September 2017, pp. 1–4.

[26] González-Andrade, D., et al., "Ultra-Broadband Mode (De)Multiplexer Based on a Sub-Wavelength Engineered MMI Coupler," *Proc. IEEE Photon. Conf. (IPC)*, Orlando, FL, October 2017, pp. 423–424.

[27] González-Andrade, D., et al., "Ultra-Broadband Mode Converter and Multiplexer Based on Sub-Wavelength Structures," *IEEE Photon. J.*, Vol. 10, No. 2, April 2018, Art. No. 2201010.

[28] Chu, Y., K. Ma, and Y. Wang, "A Novel Triplexer Based on SISL Platform," *IEEE Transactions on Microwave Theory and Techniques*, Vol. 67, No. 3, 2019, pp. 997–1004.

[29] Rhodes, J. D., "Suspended Substrate Filters and Multiplexers," *Proc. 16th Eur. Microw. Conf.*, Dublin, Ireland, September 1986, pp. 8–18.

[30] Li, L., K. Ma, and S. Mou, "A Novel High Q Inductor Based on Double Sided Substrate Integrated Suspended Line Technology with Patterned Substrate," *Proc. IEEE MTT-S Int. Microw. Symp. (IMS)*, Honolulu, HI, June 2017, pp. 480–482.

[31] Ma, Z., et al., "Quasi-Lumped-Element Filter Based on Substrate-Integrated Suspended Line Technology," *IEEE Transactions on Microwave Theory and Techniques*, Vol. 65, No. 12, December 2017, pp. 5154–5161.

[32] Wang, Y., M. Yu, and K. Ma, "A Compact Low-Pass Filter Using Dielectric-Filled Capacitor on SISL Platform," *IEEE Microw. Wirel. Compon. Lett.*, Vol. 31, No. 1, January 2021, pp. 21–24.

[33] Zhang, H., Y. Wang, and K. Ma, "A Compact Low-Pass Filter Using a Miniaturized Capacitor and Honeycomb Cavity on Substrate Integrated Suspended Line Platform," *Microw. Opt. Technol. Lett.*, 2023, pp. 1–7.

[34] Chen, Z., et al., "A Quad-Operation-Mode Tunable SISL Bandpass Filter with Compact Size and Self-Packaging," *IEEE Microwave and Wireless Technology Letters*, Vol. 33, No. 4, April 2023, pp. 379–382.

[35] Kim, D., "Planar Magneto-Dielectric Metasubstrate for Miniaturization of a Microstrip Patch Antenna," *Microw. Opt. Techn. Lett.*, Vol. 54, No. 12, December 2012, pp. 2871–2874.

[36] Ding, Y., et al., "Novel Miniaturized Self-Packaged Filters Based on Metasubstrate in SISL

Platform," *IEEE Transactions on Circuits Syst. II, Exp. Briefs*, Vol. 70, No. 8, August 2023, pp. 2879–2883.

[37] Nicolson, A. M., and G. F. Ross, "Measurement of the Intrinsic Properties of Materials by Time-Domain Techniques," *IEEE Transactions on Instrumentation and Measurement*, Vol. IM-19, No. 4, November 1970, pp. 377–382.

[38] Weir, W. B., "Automatic Measurement of Complex Dielectric Constant and Permeability at Microwave Frequencies," *Proc. IEEE*, Vol. 62, No. 1, January 1974, pp. 33–36.

[39] Chang, W. S., and C. Y. Chang, "A High Slow-Wave Factor Microstrip Structure with Simple Design Formulas and Its Application to Microwave Circuit Design," *IEEE Transactions on Microwave Theory and Techniques*, Vol. 60, No. 11, November 2012, pp. 3376–3383.

[40] Liu, Z. Y., K. Ma, and N. N. Yan, "Study of Effective Permittivity and Loss Reduction of SISL Using Ring Resonator," *Microw. Opt. Technol. Lett.*, Vol. 64, No. 3, March 2022, pp. 482–488.

[41] Kuo, J. T., S. P. Chen, and M. Jiang, "Parallel-Coupled Microstrip Filters with Over-Coupled End Stages for Suppression of Spurious Responses," *IEEE Microw. Wireless Compon. Lett.*, Vol. 13, No. 10, October 2003, pp. 440–442.

CHAPTER 4

SISL Power Dividers and Couplers

This chapter describes the passive components including the power dividers, and couplers, which are widely used in RF circuits and microwave circuits. With the proposed SISL platform with the advantages of low loss, weak dispersion, and self-packaging, these passive devices have been studied for new structures and excellent performance. Hence, this chapter will describe the development of these passive components on the SISL platform.

4.1 Power Dividers on the SISL

In recent years, with the deterioration of the electromagnetic environment, the design of power dividers not only needs to realize power dividing but also give some additional functions such as filter response, phase shifting, broadband, multiband, miniaturization, harmonic suppression, and so on.

4.1.1 Architecture of the SISL Power Dividers

The 3D structure of the SISL is shown in Figure 4.1, including 5 substrate layers and 10 metal layers. The power divider circuit is designed on M_5 and M_6 of sub3. M_2 and M_9 are the signal ground of the suspension circuit. Sub2 and Sub4 are partially removed and become the cavities of the SISL. Usually, many via holes around the air cavity are arranged on each layer of substrates to achieve connection of all metal

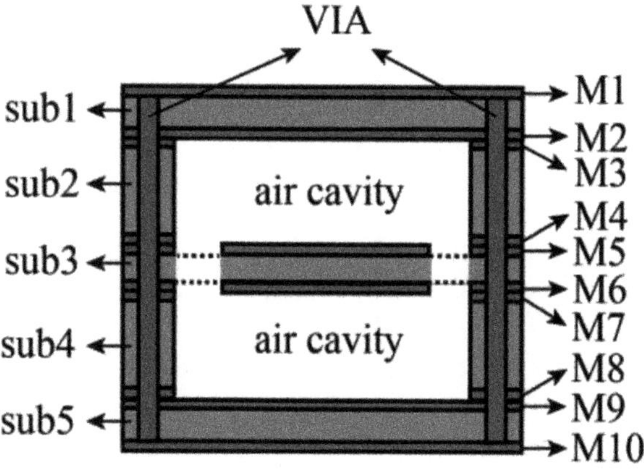

Figure 4.1 The power divider circuit on SISL.

layers of the five-layer substrates. Via holes and copper layers give the air cavity good electromagnetic shielding, which can reduce the radiation loss of the circuit.

4.1.2 Design Examples of the SISL Power Dividers

This chapter presents five design cases of power dividers for miniaturization, low loss, and multifrequency bands. The five design cases include lumped parameter elements, distributed parameter elements and waveguide elements, demonstrating the design flexibility of the SISL power divider.

4.1.2.1 A Miniaturized Bandpass Filtering Power Divider Using Quasi-Lumped Elements

The proposed power divider consists of two series resonators L_1C_1, two parallel resonators L_2C_2, two series inductors L_3, and an isolation resistor R. The parallel resonator L_2C_2 produces a transmission zero (TZ) at 0 GHz, and the series resonator L_1C_1 produces a TZ at the third harmonic frequency, thus forming a bandpass filtering response.

The circuit schematic is shown in Figure 4.2(a). The symmetrical structure of the circuit is analyzed by the even-mode/odd-mode method. The equivalent circuits of the even mode and odd mode are given in Figures 4.2(b, c), respectively. The theoretical S-parameter of the proposed power divider can be obtained by the equivalent circuit of the even/odd mode in Figure 4.2.

$$|S_{11}| = |S_{11}^e| \tag{4.1}$$

$$|S_{12}| = |S_{13}| = \sqrt{\frac{\left(1 - |S_{11}|^2\right)}{2}} \tag{4.2}$$

$$|S_{22}| = |S_{33}| = \frac{|S_{22}^e + S_{22}^o|}{2} \tag{4.3}$$

$$|S_{23}| = |S_{32}| = \frac{|S_{22}^e - S_{22}^o|}{2} \tag{4.4}$$

The relationship between TZ and resonator LC can be calculated according to S_{12}, and the isolation resistance value can be calculated according to S_{23}. The 3D view of the proposed SISL power divider is shown in Figure 4.3 and the circuit layout is shown in Figure 4.4. As shown in Figure 4.4, the main circuit is designed on layers G_5 and G_6 of Sub3, which is embedded inside the electromagnetic shielding box.

Based on lumped parameters, the power divider uses an LC parallel resonator to generate a TZ at the zero frequency and an LC series resonator to generate another TZ at the third harmonic frequency, thus forming a bandpass filtering response. Two additional LC series resonators are used to generate another two

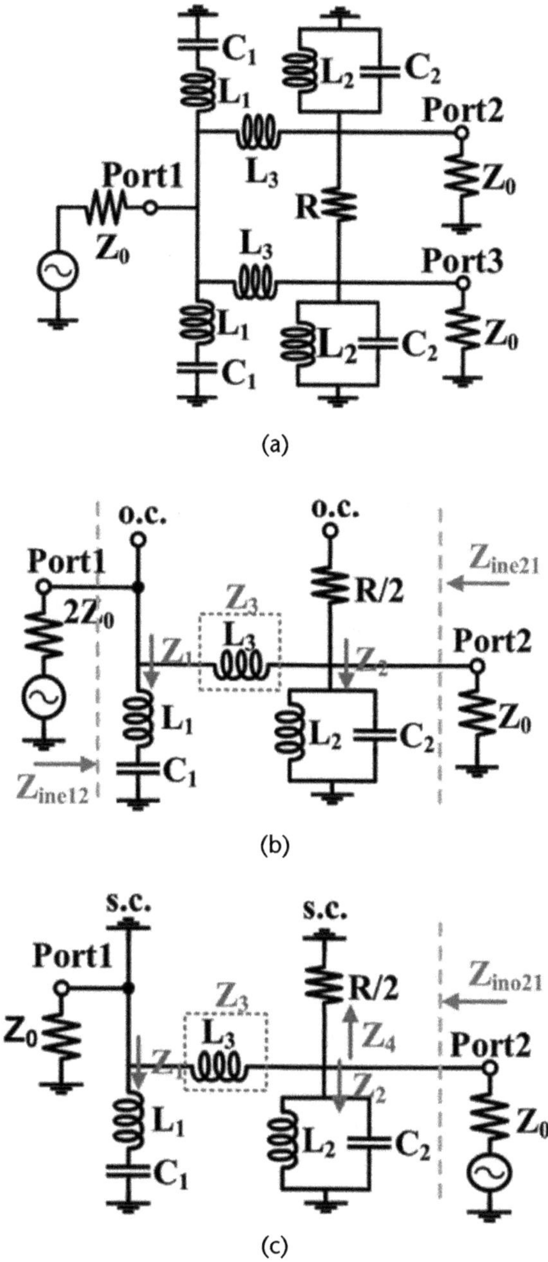

Figure 4.2 (a) The equivalent circuit, (b) even-mode, and (c) odd-mode [1].

TZs, to enhance upper stopband rejection. The lumped elements are realized by SISL-based dual-layer interdigital capacitors and dual-layer spiral inductors, for miniaturization and low insertion loss.

The measured results agree well with the simulated results; the measured input and output port return losses are better than 20 dB from 0.8 GHz to 1.1 GHz, with a fractional bandwidth (FBW) of 31.6%. The minimum in-band insertion loss is 0.31 dB (the theoretical 3-dB loss is subtracted). The in-band isolation is better than 18 dB.

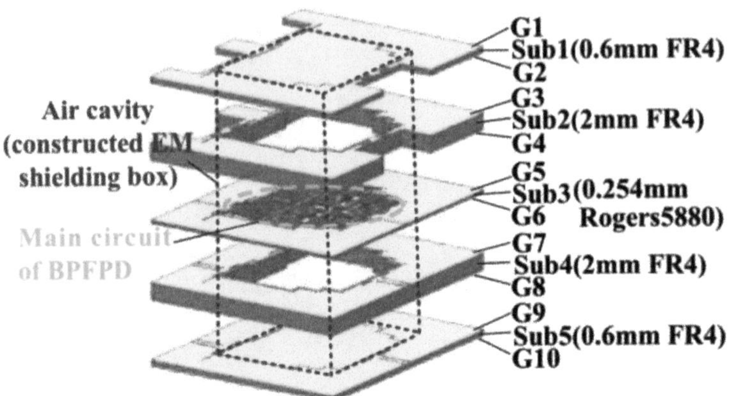

Figure 4.3 The 3D view of the proposed power divider [1].

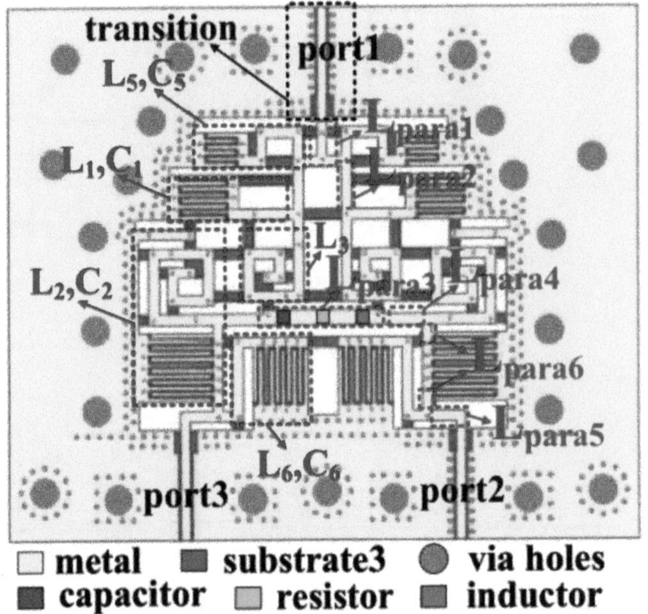

Figure 4.4 The circuit layout of the proposed power divider [1].

4.1.2.2 A Self-Packaged Power Divider with Compact Size and Low Loss

In this design, a coupled linear power divider is proposed by combining the advantages of SISL multilayer PCB integration. The proposed power divider consists of two transmission lines, a pair of coupled lines, and a resistor.

The topology of the proposed power divider is shown in Figure 4.5(a); the resistance and the coupling coefficient can be obtained by solving the S-parameter. To make the circuit layout compact, we bend the two output paths forming a spiral shape and place the isolation resistor at the center of the spiral. The circuit layout of the proposed spiral-shaped power divider is shown in Figure 4.5(b). The two

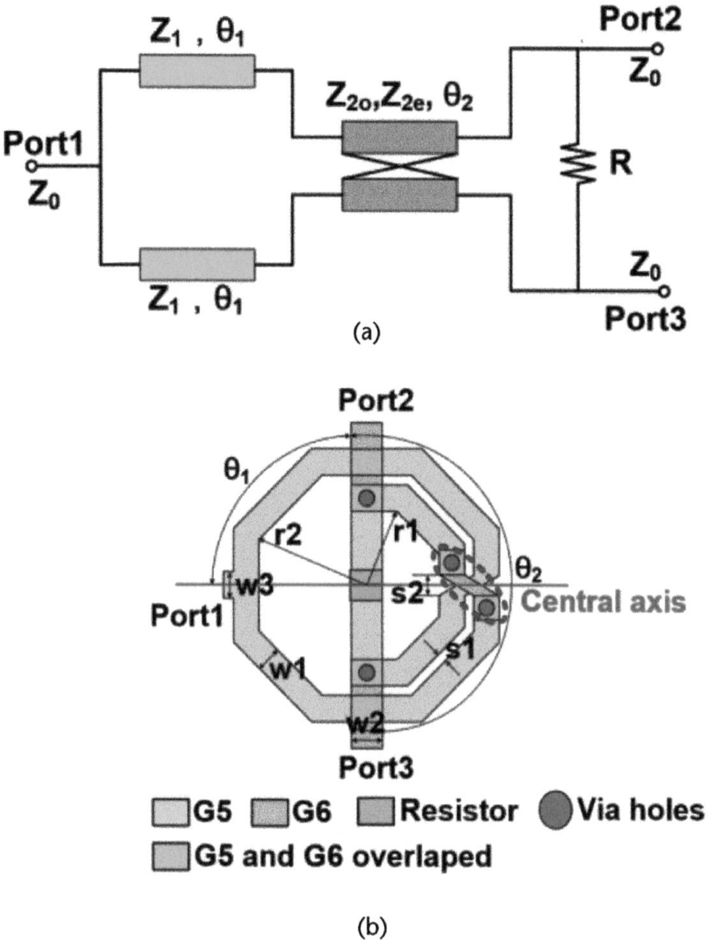

Figure 4.5 The proposed power divider [2]: (a) the equivalent circuit, and (b) topology.

output paths are designed mutually coupled, by properly increasing the coupling, the total electrical length can be reduced as compared with the conventional Wilkinson power divider.

The SISL structure used in this design is based on the PCB process and consists of five substrate layers (from Sub1 to Sub5) and ten metal layers (from G1 to G10). Each substrate layer is covered with metal. Sub1 and Sub5 act as the cover board, and Sub2 and Sub4 are hollowed to form air cavities. The designed plate through via holes around the air cavity, together with G_2 and G_9 metal layers, construct an electromagnetic shielding box. The 15-dB bandwidth of the patterned SISL SSPD is 55.1% (from 1.75 GHz to 3.08 GHz), which is determined by the worst performance. The insertion loss of the SISL power divider without the patterned substrate is less than 0.3 dB from 1.25 GHz to 2.8 GHz, and that of the patterned SISL power divider is less than 0.2 dB from 1.23 GHz to 3.35 GHz. It can be seen that the SISL power divider with the patterned substrate achieves a loss reduction of up to 0.1 dB as compared with that without the patterned substrate.

4.1.2.3 A Miniaturized and Self-Packaged Gysel Power Divider with Embedded Metamaterials on the SISL Platform

Through the combination of metamaterials and SISL, a Gysel PD with embedded high-dielectric constant metamaterials (HDCM) in the SISL platform is designed.

The multilayer SISL structure is shown in Figure 4.6. The SISL is realized by five substrates marked S_1 to S_5 with ten faces marked M_1 to M_{10}. We make S_2 and S_4 hollow into rectangles to form two cavities, which not only reduces dielectric loss and realizes suspension, but also can easily place the required 3D components such as resistors in the cavity. The core circuit is designed on S_3, and the designed metal patterns can be etched on M_5 and M_6. Each substrate can be designed and manufactured separately and then be pressed together with rivets. The upper and lower metamaterial layers of the PCB are, respectively, plated with metal patches, and the middle is connected by metal through-holes. The characteristics of a high dielectric constant are realized through the periodic arrangement.

To illustrate that after embedding metamaterials, the relative permittivity can be increased, so that the RF circuit can be miniaturized. Taking Gysel PD as an example, SISL Gysel PD without and with embedded metamaterials are designed. The core circuits as shown in Figure 4.7. It can be seen from Figure 4.7 that the SISL power divider based on metamaterial is smaller than the SISL power divider. The center frequency is fixed to 3.45 GHz, for the 5G application, and the metamaterial-based Gysel power divider achieves miniaturization while achieving almost the same performance.

4.1.2.4 A Dual-Band Coupled-Line Power Divider Using SISL Technology

A coupled-line-based dual-band power divider using SISL for wireless local area network (WLAN) application has been proposed. Two-section cascaded coupled

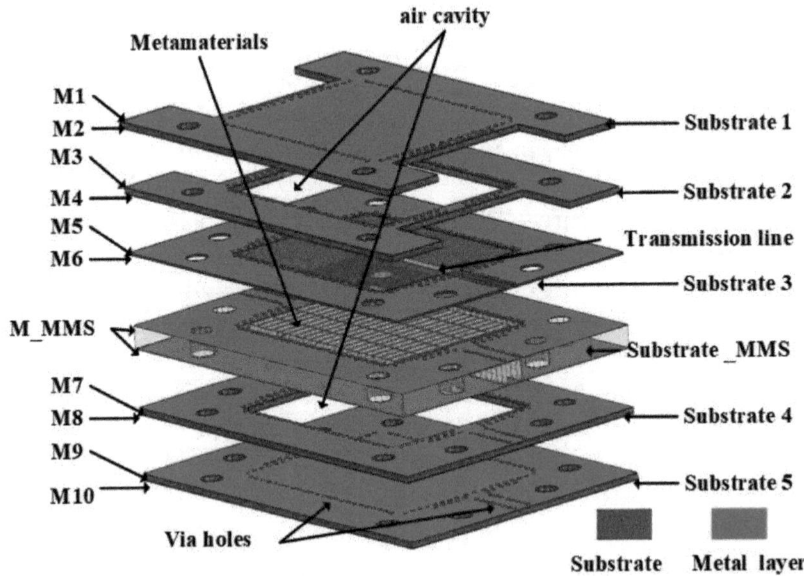

Figure 4.6 Design of the proposed HDCM power divider [3].

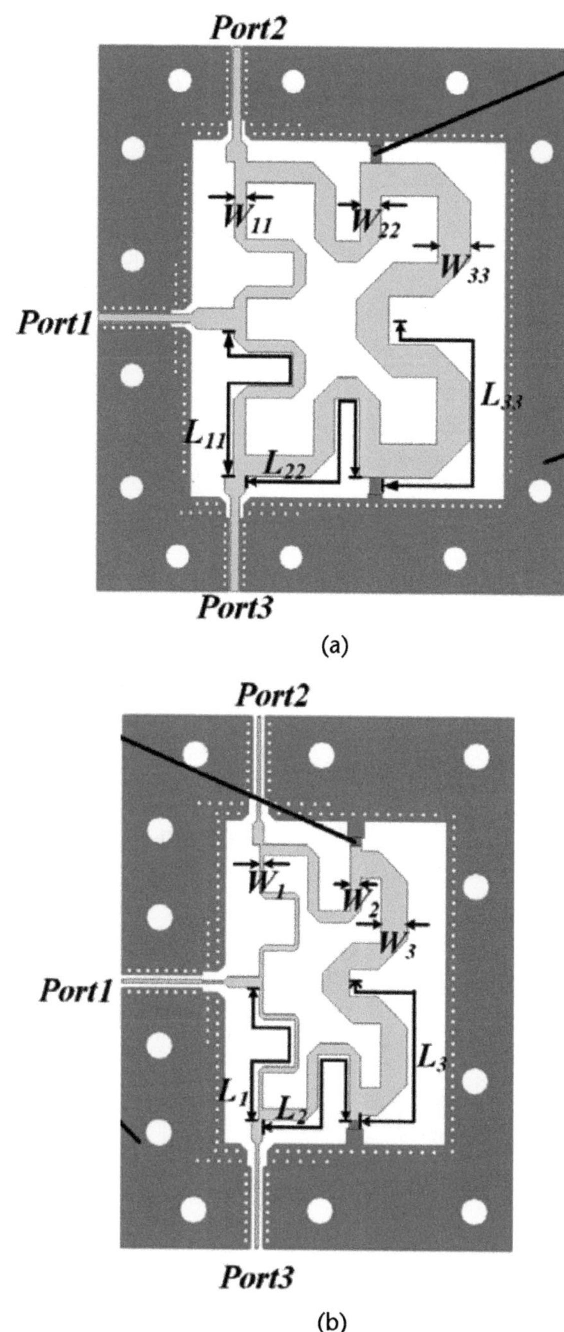

Figure 4.7 The core circuit on the M_5 layer of the proposed SISL Gysel PD [3]: (a) without metamaterials, and (b) when embedding metamaterials.

line in each output path is used for dual-band response. One coupled-line loaded at the input port is to improve the frequency selectivity. Two lumped resistors placed between two pairs of the two-section coupled line are used for isolation enhancement. To easily achieve tight coupling, a multilayer SISL structure is introduced, and the coupling strength can be adjusted by changing the overlapped line width of the broadside-coupled offset stripline. By properly adjusting the coupling strength, the desired frequency ratio can be tuned. To further reduce the dielectric loss of the proposed double-band power divider based on the SISL platform, all the suspended substrate except for the metal covered and for supporting is cut out.

The schematic of the proposed power divider is shown in Figure 4.8(a). Each output path is a two-section cascaded coupled-line structure, two resistors R_1 and R_2 are placed between the two two-section coupled lines. The S-parameter of the power divider can be calculated according to the equivalent circuit in Figure 4.8(a). The S-parameter is related to the coupling coefficient of the coupling line, and the coupling coefficient can be adjusted properly to make the power divider have good dual passband response.

Figure 4.8(b) shows the 3D view of the proposed dual-band power divider, the used SISL structure consists of five substrate layers (Sub1 to Sub5) and ten metal layers (G_1 to G_{10}), which are stacked in order, and each substrate layer is covered with metal on both sides. Sub1 and Sub5 act as the cover boards, Sub2 and Sub4

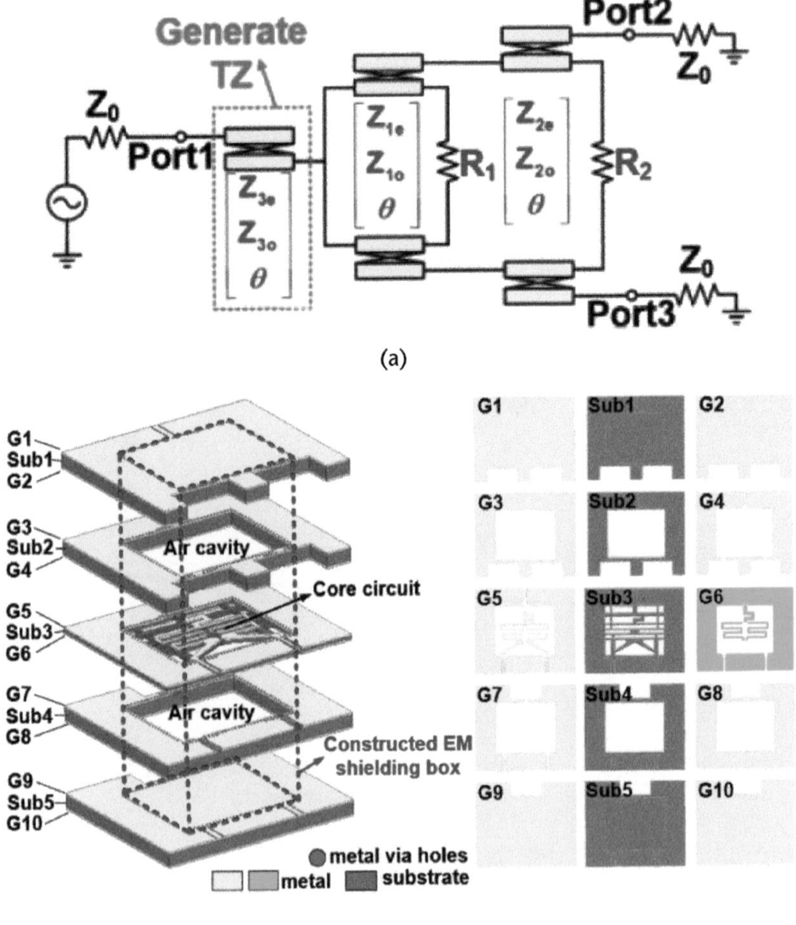

Figure 4.8 The proposed power divider [4]: (a) the equivalent circuit, and (b) the multilayer circuit structure.

are hollowed to form air cavities, and the core circuit of the DBPD is designed on the G_5 and G_6 layers of Sub3. The designed via holes around the air cavity, together with G_2 and G_9, construct an electromagnetic shielding box, the core circuit is embedded inside the electromagnetic shielding box, forming a good electromagnetic shielding environment.

The measured results agree well with the simulated ones; the center frequencies of the two passbands are 2.45 GHz and 5.13 GHz, respectively. The 15-dB FBW for the worst return loss is 64% (from 1.7 GHz to 3.3 GHz) for the lower band and 17.8% (from 4.6 GHz to 5.5 GHz) for the upper band. Removing the transition loss and connector loss, the measured minimum insertion losses for the two passbands are 0.47 dB and 0.8 dB, respectively. The 15-dB FBW for isolation is 78.7% (from 1.35 GHz to 3.1 GHz) and 55.6% (from 3.9 GHz to 6.9 GHz). Due to the precision of PCB fabrication and the substitution isolation resistors, the measured S_{23} shifts a little to lower frequency. A TZ at 3.9 GHz is generated by the coupled line loaded at the input.

4.1.2.5 An FR4-Based Self-Packaged Full Ka-Band Low-Loss 1:4 Power Divider Using SISL to Air-Filled SIW T-Junction

A Ka-band (26–40 GHz) power divider with broadband, low loss, and low cost using SISL technology has been proposed and its power-handling capability has been studied. To achieve broadband performance, an iris structure is adopted in the SISL to air-filled SIW T-junction. The iris structure will flatten the input impedance changes versus frequency, which can introduce multiple resonance points and achieve broadband matching. FR4 substrates are used to design the entire circuit, which can significantly reduce circuit costs. Because of using double metal layers and substrate-patterned SISL to reduce dielectric loss, this power divider achieves a relatively low insertion loss of 1.2 dB, despite using high-loss substrates. In addition, it can handle 7-W input power at 40 GHz, which can meet the needs of medium-power applications. Moreover, a power divider with 15-dB output isolation is proposed by adding three ports connected to the load resistor.

The configuration of the proposed power divider is shown in Figure 4.9. It is laminated by 15 layers of substrate, each sandwiched between two metal layers of copper. The substrates are partly hollowed to form air-filled SIW and the cavity of SISL.

The double-layer metal signal line and substrate patterned SISL (DSISL) is a kind of SISL with minor insertion loss. In DSISL, two metal layers of signal lines are interconnected through metallic vias, and most of the substrates on both sides of signal lines are cut out except for the necessary support. Therefore, the electromagnetic field is mainly distributed in the air, and the substrate material has little effect on the circuit loss, which also means that DSISL on FR4 substrates can be used to design low-loss circuits.

The DSISL to air-filled SIW T-junction with iris is shown in Figure 4.10(a). A difficulty is matching the frequency-dependent impedance of the air-filled SIW (Z_w) to the constant impedance of DSISL (Z_s). The flatter the input normalized impedance (Z_{in}) changes versus frequency, the more conducive it is to achieve broadband matching. When there is no iris, the amount of normalized reactance change is 0.17

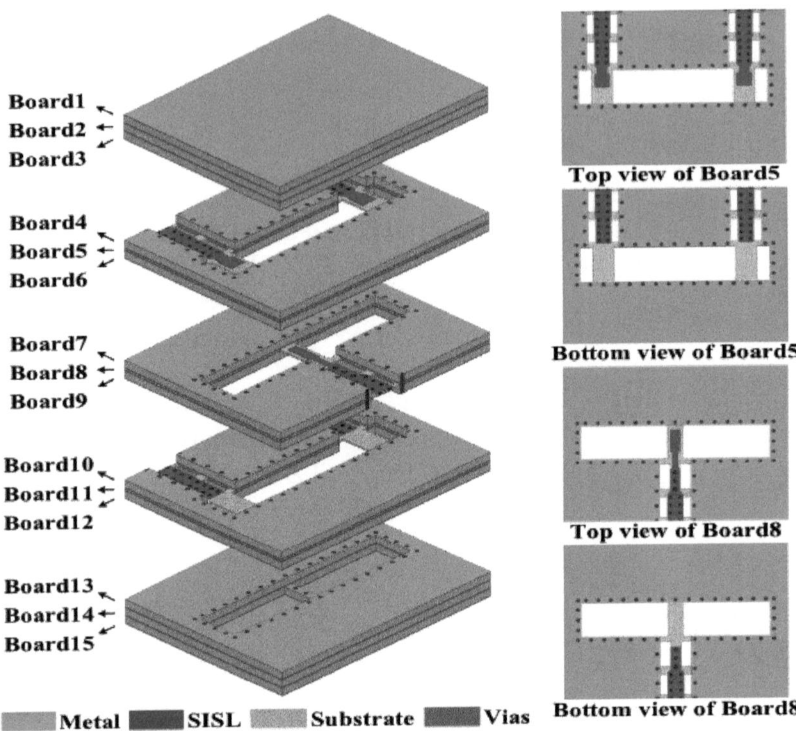

Figure 4.9 Configuration of the proposed 1:4 power divider [5].

in the bandwidth of 24–43 GHz, which is not good enough, $h_i = 0$ mm. While the iris is introduced, the curve of input normalized impedance versus frequency will be flatter. To ensure the symmetry of the electric field distribution, the iris is also designed symmetrically. The normalized input impedance changes versus frequency with different sizes of iris. For example, when adopting an iris with a size of 0.9 mm × 0.6 mm, the amount of normalized reactance change is reduced from 0.17 to 0.04 in the bandwidth of 25–43 GHz. Then Z_{in} can be matched to 50Ω through a high-resistance line and a quarter-wavelength line. The fractional bandwidth of the T-junction with iris improves by 20% compared with that without iris and achieves 61.7% (i.e., from 24.4 to 46.2 GHz), with a return loss of better than 20 dB and an insertion loss of less than 0.15 dB. In addition, the iris can push the resonance point caused by a high-order mode to a higher frequency.

Based on the above analysis, we first design the DSISL to air-filled SIW T-junction and air-filled SIW to DSISL probe-pair transition, then cascade two stages and adjust the distance between the center probe and the periphery probe which is set to $3\lambda/4$ to achieve the best power transmission in this design, so that the 1:4 power divider design can be completed.

As a design case, two power dividers are connected back-to-back to achieve a 1:4 passive power divider. The power divider with standard SISL was fabricated as a control group to quantify the reduction of dielectric loss. The effect of cavity

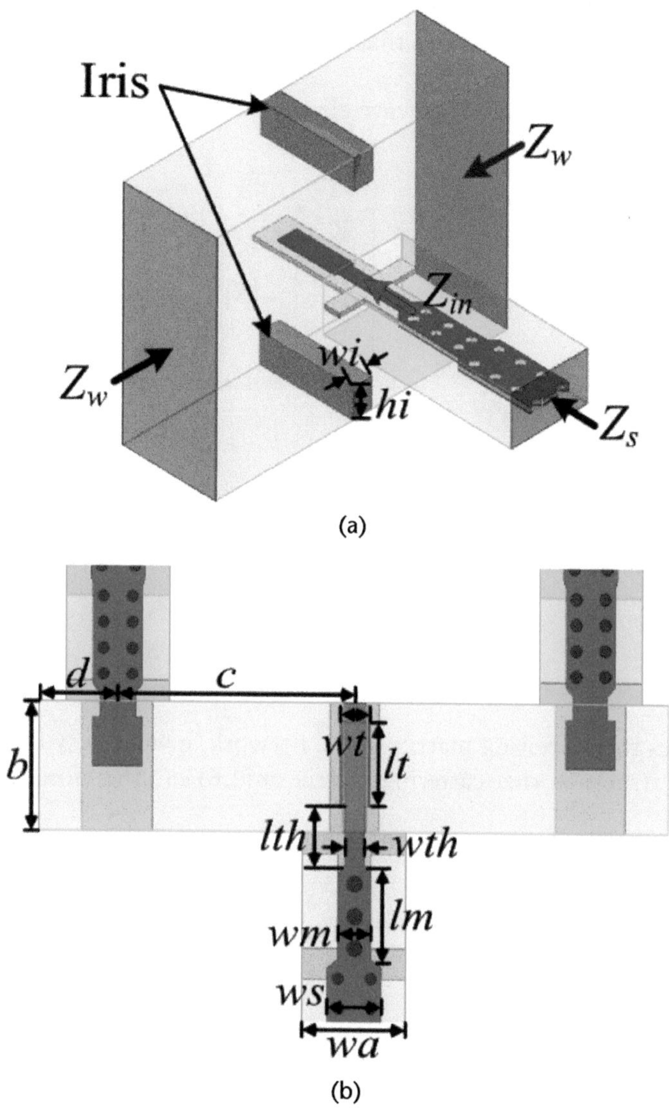

Figure 4.10 (a) T-junction [5], and (b) geometric parameters of the power divider [5].

height deviation and probe layers offset on circuit performance is studied, and the circuit can work well under certain assembly errors.

4.2 Couplers on the SISL

4.2.1 Coupled-Line Coupler on the SISL Platform

As an essential passive component of the RF front end, a coupler plays a crucial role in achieving a specific power distribution ratio while maintaining a certain phase difference between the output ports. This output signal can be utilized as a source for signal power, frequency detection, power synthesis, and distribution. The coupler is a four-port device that includes an input port, a through port, a coupled port, and an isolated port.

According to the microwave circuit network theory, a four-port network can be expressed as follows [6]:

$$\begin{bmatrix} V_1^- \\ V_2^- \\ V_3^- \\ V_4^- \end{bmatrix} = [S] \begin{bmatrix} V_1^+ \\ V_2^+ \\ V_3^+ \\ V_4^+ \end{bmatrix} \tag{4.5}$$

where

$$[S] = \begin{bmatrix} S_{11} & S_{12} & S_{13} & S_{14} \\ S_{21} & S_{22} & S_{23} & S_{24} \\ S_{31} & S_{32} & S_{33} & S_{34} \\ S_{41} & S_{42} & S_{43} & S_{44} \end{bmatrix} \tag{4.6}$$

denotes the scattering matrix of the network, given the symmetry and reciprocity of the structure, the scattering matrix of (4.6) can therefore be expressed as

$$[S] = \begin{bmatrix} S_{11} & S_{21} & S_{31} & S_{41} \\ S_{21} & S_{22} & S_{41} & S_{42} \\ S_{31} & S_{41} & S_{11} & S_{21} \\ S_{41} & S_{42} & S_{21} & S_{22} \end{bmatrix} \tag{4.7}$$

The following quantities are commonly used to characterize a directional coupler:

$$\text{Coupling} = C = 10\log\frac{P_1}{P_3} \text{ (dB)} \tag{4.8}$$

$$\text{Directivity} = D = 10\log\frac{P_3}{P_4} \text{ (dB)} \tag{4.9}$$

$$\text{Isolation} = I = 10\log\frac{P_1}{P_2} \text{ (dB)} \tag{4.10}$$

$$\text{Insertion loss} = L = 10\log\frac{P_1}{P_4} \text{ (dB)} \tag{4.11}$$

4.2.1.1 The Theory of the Coupled Line

When two unshielded transmission lines are close together, due to the interaction of the electromagnetic field of the transmission lines, there is energy coupling between the transmission lines, which is called a coupled transmission line. Traditionally, coupled lines are symmetrical, the coupling modes include even and odd modes, and the coupling between symmetrical lines can be determined according to phase velocity and characteristic impedance. The analysis of symmetrical coupled lines usually uses odd-mode and even-mode analysis.

First, consider two special excitation types: even mode, that is, the current amplitude and directivity on the two transmission lines are the same; and odd mode, that is, the current amplitude on the two transmission lines is the same, but the direction is opposite.

For the even mode, the electric field is evenly symmetrical about the center line, and no current flows between the two transmission lines. For the even mode, the resulting capacitance from each wire to the ground is:

$$C_e = C_{11} = C_{22} \tag{4.12}$$

The characteristic impedance of the even mode is:

$$Z_{0e} = \sqrt{\frac{L_e}{C_e}} = \frac{1}{v_p C_e} \tag{4.13}$$

For the odd mode, the power line is oddly symmetrical about the center line, and there is zero voltage between the two transmission lines. The capacitance generated from each line to the ground is:

$$C_o = C_{11} + 2C_{12} = C_{22} + 2C_{12} \tag{4.14}$$

The characteristic impedance of the odd mode is:

$$Z_{0o} = \sqrt{\frac{L_o}{C_o}} = \frac{\sqrt{L_o C_o}}{C_o} = \frac{1}{v_p C_o} \tag{4.15}$$

4.2.1.2 Coupled-Line Coupler

The literature [7] has proposed a coupled-line coupler based on the SISL platform. As shown in Figure 4.11(a), the main circuit of the coupler is located on the upper and lower metal layers of Substrate 3, specifically on metal layers G_5 and G_6, while the signal ground is established on G_2 and G_9. The coupled-line coupler is implemented by offset parallel coupled striplines, shown in Figure 4.11(b), where w represents the line width of the coupled line and *offset* denotes the line spacing.

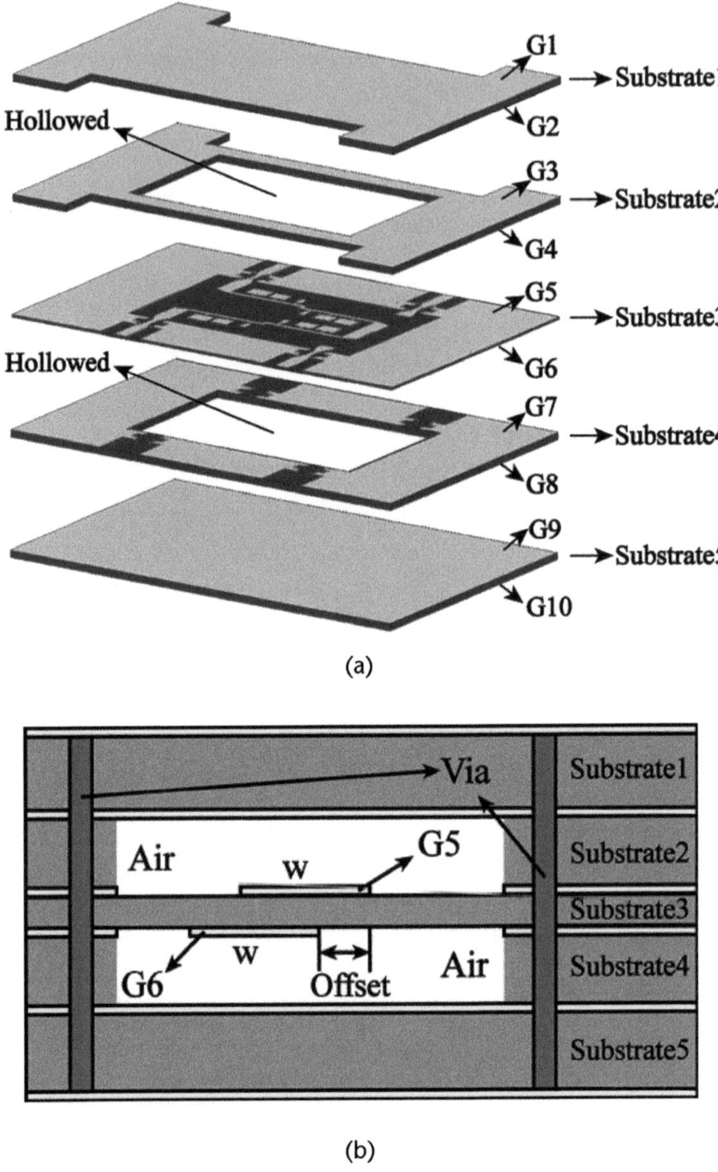

Figure 4.11 (a) A 3D view of the proposed SISL tandem coupler [7], and (b) a cross-sectional view of the proposed SISL tandem coupler [7].

For general asymmetrically coupled line couplers, the analytical methods of inductive and capacitive coupling coefficients can be applied. In contrast, odd-mode and even-mode analysis can be used for symmetrically coupled-line couplers. As the symmetrical coupler can be considered a specific instance of the asymmetrical coupler, this chapter employs a more comprehensive analytical approach based on the inductance-capacitance coupling coefficient. This method allows for a more generalized analysis, which enhances the accuracy and clarity of this chapter. For the ideal coupled-line coupler:

$$K_L = \frac{L_m}{\sqrt{L_1 L_2}} \tag{4.16}$$

$$K_C = \frac{C_m}{\sqrt{C_1 C_2}} \tag{4.17}$$

$$Z_0 = \sqrt{\frac{L_i}{C_i}} \tag{4.18}$$

$$K_C = K_L \tag{4.19}$$

The circuit consists of two coupled striplines with self-inductance and self-capacitance represented by L_1, L_2, C_1, and C_2, respectively. The mutual inductance and mutual capacitance between the two striplines are represented by L_m and C_m, respectively. K_L and K_C denote the inductive and capacitive coupling coefficients, respectively, which represent the degree of coupling between the two lines. Ideally, the port characteristic impedance (Z_0) is 50Ω, and $K_L = K_C$. In this case, the coupler exhibits optimal directivity, and the standing wave ratio and isolation curve are ideal. However, because the SISL employs a nonuniform dielectric in the form of upper and lower air cavity intermediate substrates, K_L and K_C are not equal, which can affect the directivity of the coupler.

The 3D electromagnetic simulation software, Sonnet, is utilized to simulate and analyze the RLC matrix of the coupling line in the SISL. This software enables the calculation and extraction of inductance and capacitance coupling coefficients for the SISL coupling line. In this design, the thickness of the dielectric Substrate 3 remains constant, and the coupling coefficient of the coupled line is solely dependent on the line width (w) and line spacing offset (*offset*). By altering the line width and line spacing offset of two offset parallel coupled lines, the impact of these dimensions on the inductance and capacitance coupling coefficients is analyzed.

Fix the line spacing *offset* parameter (0 mm), and scan the line width w (0.2–0.8 mm) of two coupled lines at the same time to obtain the capacitive coupling coefficient K_C and inductive coupling coefficient K_L, and the ratio of the two K_C/K_L. When the line width w increases, both the capacitive coupling coefficient K_C and the inductive coupling coefficient K_L increase, but the growth rate of K_C is slower than that of K_L, which shows that K_C/K_L decreases with the increase of w.

Fix the line width w parameter (0.5 mm), and scan the line spacing *offset* (0.2–0.8 mm) of the two coupled lines at the same time to obtain the capacitive coupling coefficient K_C and the inductive coupling coefficient K_L, and the ratio of the two K_C/K_L. When the line spacing offset increases, both the capacitive coupling coefficient K_C and the inductive coupling coefficient K_L decrease, but the rate of decrease of K_L is faster than that of K_C, which is shown as K_C/K_L with the increase of *offset*. The simulation results indicate that in the SISL coupled-line structure, the inductive coupling coefficient is consistently smaller than the capacitive coupling coefficient K_C which can negatively impact the directional property of the coupler designed using this structure.

In order to improve the directionality of the coupler, the open-circuit stub can be loaded along the coupling line to create a capacitive effect on the ground. Assuming that the loading capacitance is C_a, then the capacitive coupling coefficient after loading the capacitive stub is:

$$K_C' = \frac{C_m}{\sqrt{C_1' C_2'}} = \frac{C_m}{C_1'} = \frac{C_m}{C_1 + C_a} \tag{4.20}$$

As the loading capacitance increases, the inductive coupling coefficient remains relatively constant, while the capacitive coupling coefficient gradually decreases until it reaches the value of the inductive coupling coefficient K_L, which is the coupling coefficient of the coupled line.

To achieve a coupled linear coupler circuit with a 3-dB coupling degree based on this SISL platform, the total coupling coefficient is required to be 0.707. However, according to the inductive coupling coefficient curve, it can be deduced that a larger w value is required to obtain a coupling coefficient of 0.707, which will also lead to problems for port matching.

After considering both strong coupling and directivity, it was decided to cascade multiple weak couplers with a total coupling degree of 8.34 dB and load open-circuit stubs simultaneously. The circuit plane structure and equivalent circuit diagram of the coupler on the dielectric substrate 3 are shown in Figure 4.12.

The specific steps for designing a highly directional cascade coupler are:

1. To achieve 3-dB strong coupling, use two 8.34-dB couplers for cascading, where 8.34 dB is converted into a value of 0.383.
2. According to the relationship curve in (4.18), adjust the size of w and *offset* to make the inductive coupling coefficient K_L equal to 0.383, and at the same time control the impedance of the coupling line (using (4.18)), to slightly exceed the standard impedance value of 50Ω, as this will help reduce the impedance value after loading the compensation branch.
3. Load the open branch at a uniform position along the coupling line and adjust the size of the open branch so that K_C is ultimately reduced to be equal to K_L.
4. Optimize the electromagnetic simulation software to determine the final size parameters. Figure 4.13 shows the complete simulation model of the design.

To test SISL internal circuits, it is necessary to design a transition structure from SISL to other planar circuits. In this design, the transition structure from SISL to CPW as reported before has been adopted. The loaded open-circuit stubs adopt different sizes, with corresponding loaded capacitance values of C_{a1}, C_{a2}, and C_{a3}. which also provides a higher degree of design freedom for electromagnetic simulation and tuning of the circuit.

In the frequency range of 1.36 GHz to 3.04 GHz, the measured S_{21} and S_{31} parameters were -3.32 ± 0.6 dB and -3.38 ± 0.63 dB, respectively. Over a relative bandwidth of 111% from 1 GHz to 3.5 GHz, the measured return loss (negative

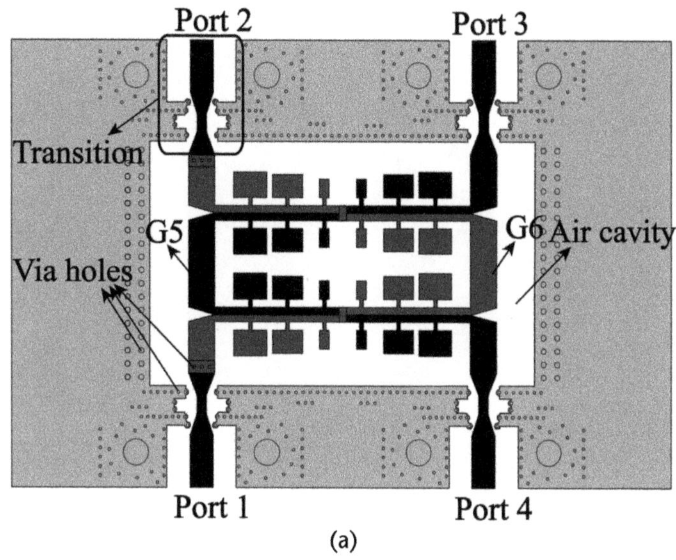

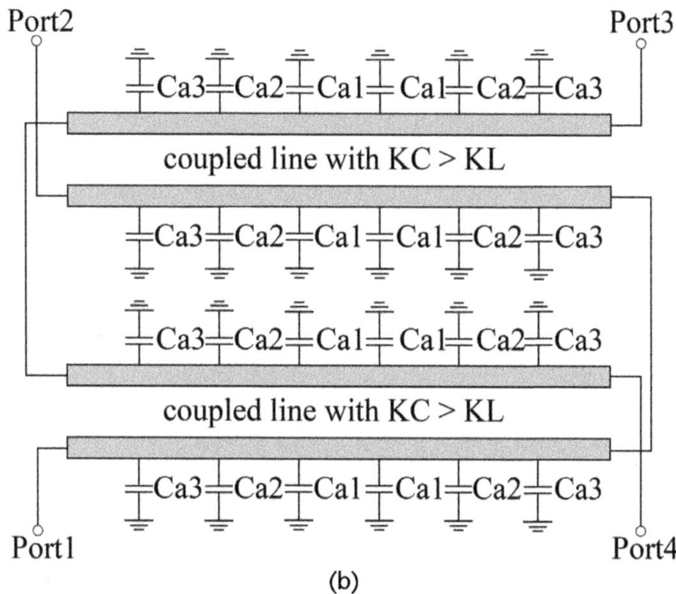

Figure 4.12 (a) Circuit pattern of the proposed SISL tandem coupler [7], and (b) the theoretical schematic of the circuit pattern [7].

value of S_{11}) and isolation (negative value of S_{41}) were both better than 20 dB, and the phase difference of the output port was 90.5° ± 2.5°. In order to accurately calculate the overall loss of the four-port network due to dielectric loss, radiation loss, and conductor loss, while also considering the matching and isolation of the ports, the loss energy can be expressed as a percentage using (4.21). For this design, at a center frequency of 2.25 GHz, the percent loss tested was only 5%.

$$Loss_{\text{four port}} = 1 - \left|S_{11}\right|^2 - \left|S_{21}\right|^2 - \left|S_{31}\right|^2 - \left|S_{41}\right|^2 \tag{4.21}$$

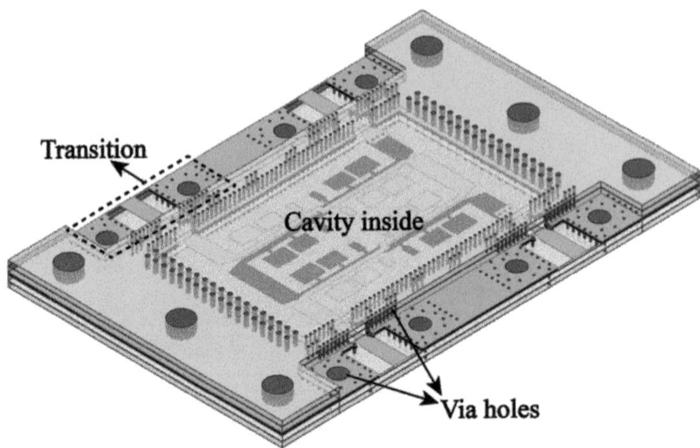

Figure 4.13 Electromagnetic simulation software model of the self-packaged SISL tandem coupler [7].

Thanks to the self-encapsulation structure of the multilayer circuit, it offers several advantages, including low cost, lightweight, and compact size.

4.2.1.3 Multiple Inner Boards Coupled-Line Coupler

A new form of SISL platform with multiple inner board layers is proposed [8], whose structure is made up of two inner boards, a hollow board for separation and support between the inner boards, and two hollow boards forming a cavity, and two upper and lower cover boards, as shown in Figure 4.14. DB3 and DB5 are two inner boards, DB4 is the inner hollow board, DB2 and DB6 are cavity hollow boards, and DB1 and DB7 are cover boards. This structure has the advantage of effectively reducing the dielectric loss of the broadside coupled-line structure circuit, especially for FR4 substrate with large loss angle tangent, which has a wide range of application prospects. As shown in Figure 4.14, the two inner boards DB3 and DB5 are capable of inter-board coupling as a set of broadside coupled lines, and their electric field is concentrated in the air, which avoids the large dielectric loss of single-layer substrate broadside coupled line structure.

To compare the loss of two SISL structures applied to broadside coupled line circuits, this section simulates a group of coupled lines using both structures. The plate thickness and dielectric materials used in the two structures are listed in Table 4.1. The simulated coupling lines are all in the form of broadside coupling without any horizontal offset. Each coupling line has a length of 15 mm and a width of 3 mm. Assuming a negligible thickness of the metal layer, the vertical distance between the coupled lines in both cases is approximately equal.

After analyzing the electric field of coupling lines of two different SISL structures, it is clear that the coupling electric field between the coupling lines in the SISL structure with a single inner layer is mostly distributed in the lossy dielectric substrate, resulting in higher dielectric loss. However, in the SISL double inner layer structure, the coupling electric field between the coupling lines is primarily distributed in the middle cavity, with minimal dielectric loss.

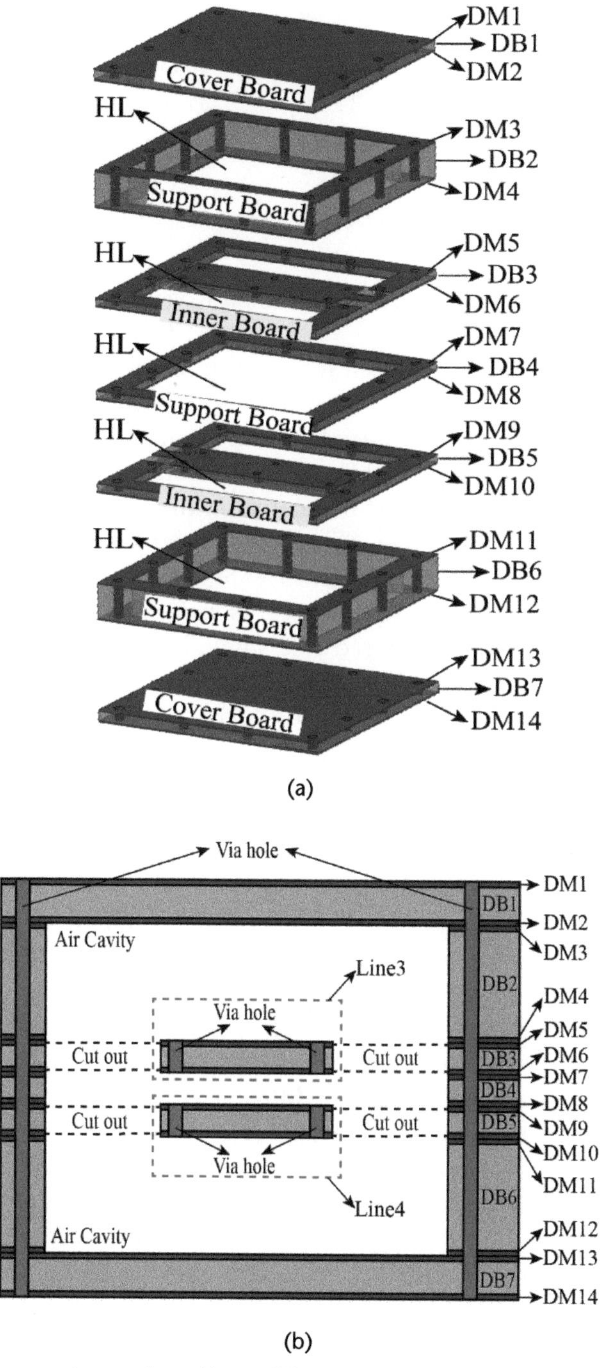

Figure 4.14 (a) Proposed seven-board layers [8], and (b) proposed seven-board layer SISL [8].

Dielectric loss in a substrate is typically characterized by the loss of tangent value. For instance, the commonly used FR4 dielectric material has a loss tangent value of 0.02 in electromagnetic simulation software by default, whereas high-frequency plates like Rogers 5880 have a much lower value of 0.0009. Consequently,

Table 4.1 Thickness of the Boards Used in the Two Types of the SISL Structure

Board	*Thickness*	*Board*	*Thickness*	*Board*	*Thickness*
SB1	0.6 mm	SB2	3 mm	SB3	0.34 mm
SB4	3 mm	SB5	0.6 mm	DB1	0.6 mm
DB2	3 mm	DB3	0.34 mm	DB4	0.34 mm
DB5	0.34 mm	DB6	3 mm	DB7	0.6 mm

FR4 exhibits greater dielectric loss than Rogers 5880, especially at higher operating frequencies.

The relationship between loss and different loss tangent values for two SISL coupling-line structures in Figure 4.14 is also analyzed. Assuming both structures use the same dielectric material with the same loss tangent value, electromagnetic simulation can be used to obtain the changing relationship of the loss percentages. To avoid the impact of port mismatch on the simulated loss value, the percentage of energy loss is calculated. When the loss tangent value is 0.02, the loss percentages of the simulated single-inner-layer SISL and double-inner-layer SISL are 0.37% and 0.19%, respectively. In other words, using a double inner layer SISL can reduce circuit loss by 95% compared to a single-inner-layer structure. Moreover, as the loss tangent of the substrate gradually increases, the low-loss characteristics of the double inner layer SISL in the broadside coupled line circuit become more prominent.

Therefore, when selecting a low-cost material with high dielectric loss, such as FR4, a dual inner-plate SISL structure can significantly reduce circuit loss for broadside coupled-line circuits. Similarly, this conclusion applies to three-inner-layer SISL or SISL circuits with more inner-layers as well.

In this section, a miniaturized coupled line coupler is proposed based on the SISL double-inner plate structure presented earlier, as Figure 4.15(a) shows. The coupler is fed directly from the transition structure to the internal coupling line. Additionally, the distance S between the side signal ground and the side of the internal coupling line in the horizontal direction is very small. This proximity generates a ground compensation capacitance and reduces the discontinuity of the port feed structure.

By analyzing the influence of different S values on the electric field distribution. It can be concluded that a large S value causes discontinuity in the connection between the transition feeder and the internal coupling line, which reduces the return loss and isolation of the coupler. When the S value is reduced, the electric field energy in the gap increases, resulting in a larger compensation capacitance to the side ground. Furthermore, the parasitic inductance caused by the high-impedance line at the port feeder is reduced, achieving better port impedance matching.

Figure 4.15(b) depicts the relationship between the simulated return loss and isolation and S. It shows that reducing S from 0.5 mm to 0.2 mm significantly improves the return loss and isolation in a broadband range. Additionally, the circuit size of the coupler is smaller due to the removal of the chamfer gradient line portion.

The center frequency of the coupler is set to 4 GHz. Through electromagnetic simulation and optimization, the parameter dimensions in Figure 4.15(a) are as follows: $l_3 = 19.38$ mm, shift = 0.47 mm, $w = 2.73$ mm, $d = 0.8$ mm, and $S = 0.16$ mm. The actual processing is shown in Figure 4.16, where all substrates are made of FR4 boards, making the cost of production very low.

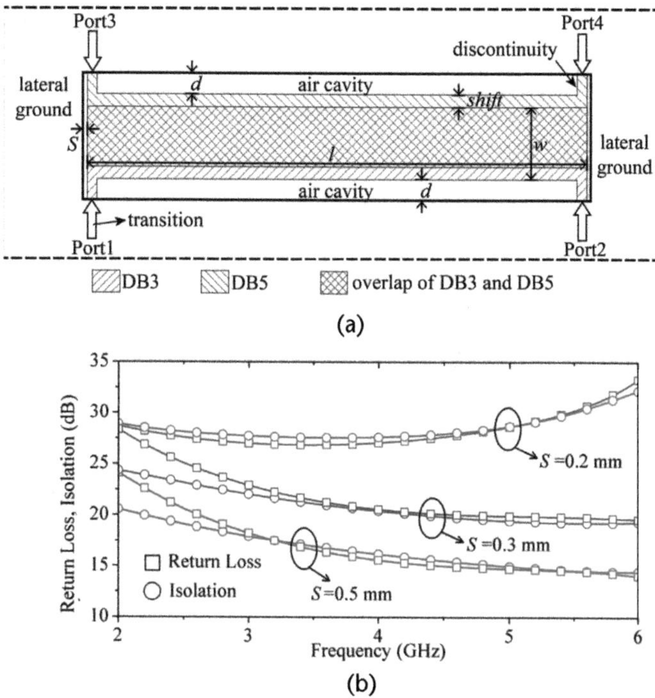

Figure 4.15 (a) Simplified planar view of the coupled-line coupler implemented on the double inner board SISL platform [8], and (b) simulated return loss and isolation change versus S [8].

The simulated return loss and isolation are better than 30 dB throughout the frequency range from 2 GHz to 6 GHz, while the measured values are about 5 dB to 10 dB lower. At the center frequency of 4 GHz, the measured S_{21} and S_{31} are −3.28 dB and −3.2 dB, respectively, resulting in an insertion loss of around 0.25 dB. Additionally, the output port phase difference of the simulation and test is close to the ideal 90° throughout the entire frequency band.

4.2.2 Lumped-Element Coupler on the SISL Platform

At lower operating frequencies, the lumped-element coupler is widely used for size miniaturization. Based on the SISL platform, the lumped-element coupler has also progressed in research [9].

The topology of the second-order set total parameter circuit is shown in Figure 4.17(a), including four equivalent inductances L, two capacitances C_2, and one capacitance C_1. Port 1 is the input port, Port 2 and Port 3 are the through and coupling port for outputting quadrature signals of equal amplitude, and Port 4 is the isolation port.

The parameters of each lumped element can be calculated using the odd-even mode analysis method:

$$L = \frac{Z_0}{\omega} \tag{4.22}$$

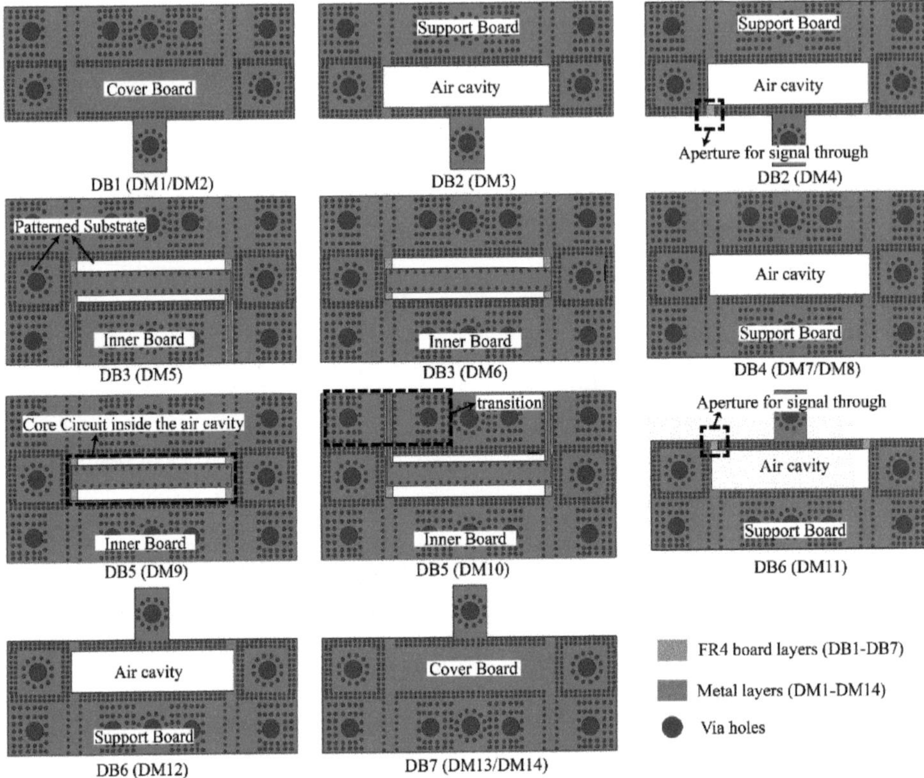

Figure 4.16 Planar view of the coupled-line coupler implemented on the double inner board SISL platform [8].

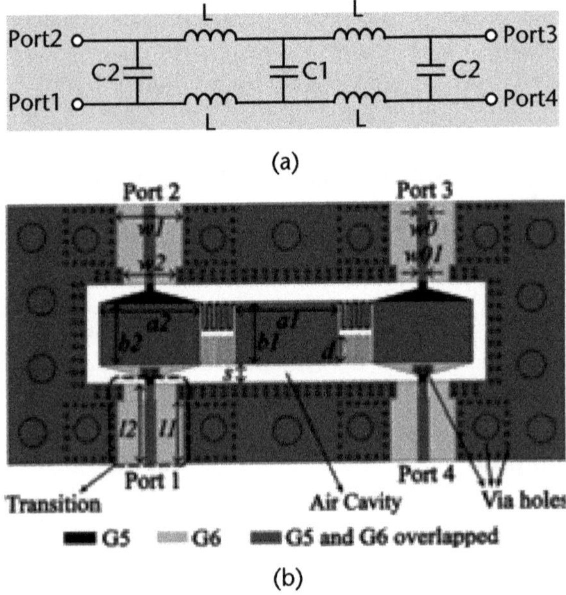

Figure 4.17 (a) Schematic diagram of the proposed SISL lumped coupler [9], and (b) planar view of the proposed lumped SISL coupler [9].

$$C_1 = \frac{1}{\omega Z_o} \tag{4.23}$$

$$C_2 = C_1 + \Delta C \tag{4.24}$$

where Z_0 is the characteristic impedance of the port and ω is the angular frequency.

When designing the sizes of C_1 and C_2, this chapter simulates the frequency response of the coupler whether ΔC is equal to 0 or not, respectively. When ΔC is equal to 0, there is only one intersection point between the S_{21} and S_{31} curves, that is, the coupler only achieves a relatively ideal response at the center frequency. When ΔC is 0.1 pF, there are two intersection points between the curves of S_{21} and S_{31}, which increases the bandwidth of amplitude and phase balance and can also reduce the influence of capacitance error and processing accuracy on circuit performance.

Figure 4.18 shows the coupler frequency response for different ΔC. Figure 4.17(b) is the planar view of the lumped parameter coupler based on the SISL platform. The circuit is mainly in the upper and lower metal layers of Substrate 3, where the capacitor is implemented in the form of a plate capacitor, and the inductor is implemented with a narrow line in a meander lie shape to reduce the circuit area. When $f_0 = 1.18$ GHz, it is obtained by calculation: $L = 6.74$ nH, $C_1 = 2.7$ pF, and $C_2 = 2.8$ pF.

In the case where the thickness and dielectric constant of dielectric substrate 3 are fixed, the capacitance value of the plate capacitance is mainly determined by its area, which is represented by the edge lengths a_1 and b_1, and a_2 and b_2 as shown in Figure 4.17(b).

Due to the fact that the wire width of the curve inductance can affect conductor loss and circuit area, a line spacing and curve inductance cash of 0.15 mm is chosen for comprehensive consideration. The inductance and capacitance values are extracted using the two-port equivalent circuit method and the open-short-circuit extraction method. Other parameters of the coupler are fine-tuned and simulated to optimize performance.

The substrate consists of five layers: FR4 (0.6 mm), FR4 (2 mm), Rogers 5880 (0.254 mm), FR4 (0.6 mm), and FR4 (2 mm). The 2-mm layer of FR4 allows for a larger cavity height, enhancing coupling and increasing plate capacitance density. The Figure 4.17(b) parameters are as follows: $w_1 = 5$ mm, $w_2 = 4$ mm, $w_0 = 0.78$ mm, $w_{01} = 0.6$ mm, $l_1 = 4.6$ mm, $l_2 = 5.8$ mm, $a_1 = 7.8$ mm, $a_2 = 7.5$ mm, $b_1 = 4.5$ mm, $b_2 = 4.5$ mm, $s = 1.4$ mm, and $d = 2$ mm.

The actual size of the circuit is $0.025\lambda_g \times 0.11\lambda_g$, where λ_g represents the waveguide wavelength. The simulation and test results are in good agreement, across the frequency range of 1 GHz to 1.4 GHz, the measured amplitude unbalance is within ±0.85 dB. Between 1.02 GHz and 1.34 GHz, the tested S_{21} and S_{31} values are −3.6 ± 0.4 dB and −3.8 ± 0.4 dB, respectively. At the center frequency of 1.18 GHz, the tested return loss and isolation are better than 24 dB, and the tested output port phase difference is 89.8°.

4.2.3 Differential Coupler on the SISL Platform

A differential coupler can be realized by connecting a balun at the port of a single-ended coupler, but this will increase the circuit area and loss. Another fully integrated

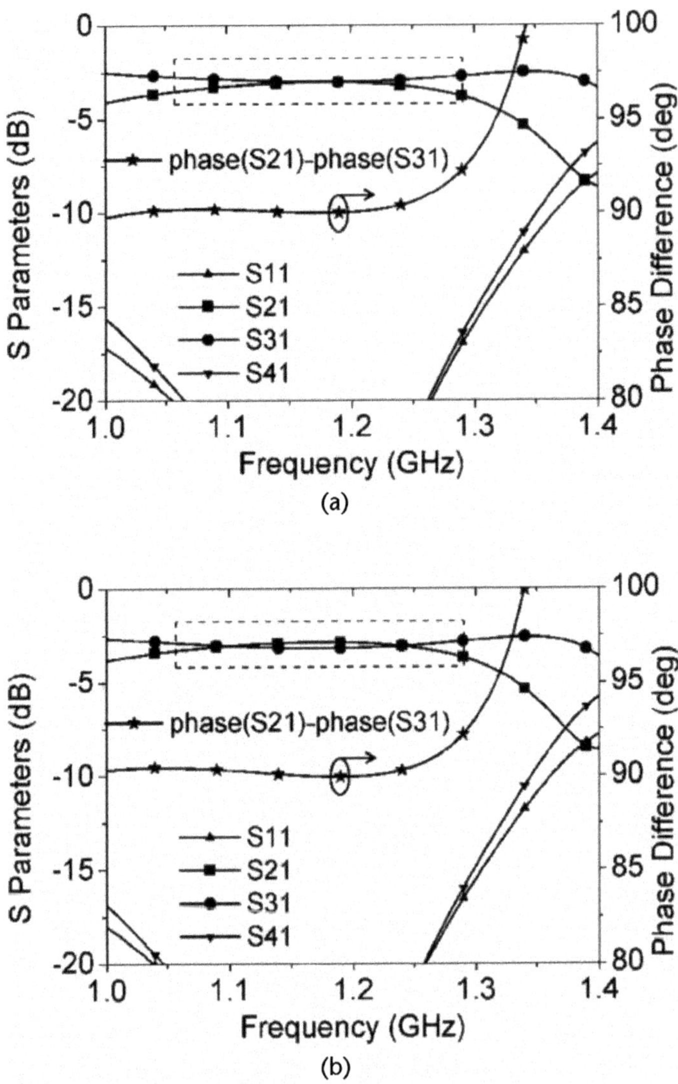

Figure 4.18 Simulated results of the lumped coupler given in Figure 4.17 [9]: (a) $\Delta C = 0$ pF, and (b) $\Delta C = 0.1$ pF ($L = 6.74$ nH, $C_1 = 2.7$ pF, center frequency is 1.18 GHz).

differential coupler includes four differential ports: A, B, C, and D. Each differential port consists of two single-ended ports, Ports 1 and 5, Ports 2 and 6, Ports 3 and 7, and Ports 4 and 8. For a 3-dB quadrature differential coupler, when differential Port A inputs a differential signal, differential Ports B and C will respectively output equal-amplitude and quadrature differential signals, while differential Port D is an isolated port, and the differential coupler can be used as a feed. The electrical network is applied to the feed network of the differential antenna. At the same time, single-ended Ports 2, 6, 3, and 7 will, respectively, output quadrature signals with equal amplitudes and phases of 0°, 180°, 90°, and 270°. This property can also be applied to subharmonic mixers or subharmonic modulators.

To implement the fully integrated eight-port differential coupler circuit depicted, it is essential to first establish the constraints that the eight-port network needs to satisfy. Assuming that the eight-port network is symmetric in both the up and down directions, its S-parameter matrix can be expressed as follows:

$$S_8 = \begin{bmatrix} a & b & c & d & e & f & g & h \\ b & a & d & c & f & e & h & g \\ c & d & a & b & g & h & e & f \\ d & c & b & a & h & g & f & e \\ e & f & g & h & p & q & r & s \\ f & e & h & g & q & p & s & r \\ g & h & e & f & r & s & p & q \\ h & g & f & e & s & r & q & p \end{bmatrix} \tag{4.25}$$

The S-parameter matrix of its mixed mode can be obtained by

$$S_{\text{mixed}} = \begin{bmatrix} S^{dd}_{(4\times4)} & S^{dc}_{(4\times4)} \\ S^{cd}_{(4\times4)} & S^{cc}_{(4\times4)} \end{bmatrix} = MS_8M^{-1} \tag{4.26}$$

where

$$M = \frac{1}{\sqrt{2}} \begin{bmatrix} 1 & 0 & 0 & 0 & -1 & 0 & 0 & 0 \\ 0 & 1 & 0 & 0 & 0 & -1 & 0 & 0 \\ 0 & 0 & 1 & 0 & 0 & 0 & -1 & 0 \\ 0 & 0 & 0 & 1 & 0 & 0 & 1 & -1 \\ 1 & 0 & 0 & 0 & 1 & 0 & 0 & 0 \\ 0 & 1 & 0 & 0 & 0 & 1 & 0 & 0 \\ 0 & 0 & 1 & 0 & 0 & 0 & 1 & 0 \\ 0 & 0 & 0 & 1 & 0 & 0 & 0 & 1 \end{bmatrix} \tag{4.27}$$

In differential mode and common mode signals, the S-parameter matrices $S^{dd}_{(4\times4)}$ and $S^{cc}_{(4\times4)}$ represent their characteristics, respectively. However, the S-parameter matrices $S^{cd}_{(4\times4)}$ and $S^{dc}_{(4\times4)}$ represent the conversion between differential mode signal and common mode signal, that is, the characteristics of cross mode. Substituting (4.25) and (4.27) into (4.26), we have

$$S^{dd}_{(4\times4)} = \begin{bmatrix} a/2-e+p/2 & b/2-f+q/2 & c/2-g+r/2 & d/2-h+s/2 \\ b/2-f+q/2 & a/2-e+p/2 & d/2-h+s/2 & c/2-g+r/2 \\ c/2-g+r/2 & d/2-h+s/2 & a/2-e+p/2 & b/2-f+q/2 \\ d/2-h+s/2 & c/2-g+r/2 & b/2-f+q/2 & a/2-e+p/2 \end{bmatrix} \tag{4.28a}$$

$$S^{cc}_{(4\times4)} = \begin{bmatrix} a/2+e+p/2 & b/2+f+q/2 & c/2+g+r/2 & d/2+h+s/2 \\ b/2+f+q/2 & a/2+e+p/2 & d/2+h+s/2 & c/2+g+r/2 \\ c/2+g+r/2 & d/2+h+s/2 & a/2+e+p/2 & b/2+f+q/2 \\ d/2+h+s/2 & c/2+g+r/2 & b/2+f+q/2 & a/2+e+p/2 \end{bmatrix} \tag{4.28b}$$

$$S_{(4\times4)}^{cd} = S_{(4\times4)}^{dc\ T} = \begin{bmatrix} a/2-p/2 & b/2-q/2 & c/2-r/2 & d/2-s/2 \\ b/2-q/2 & a/2-p/2 & d/2-s/2 & c/2-r/2 \\ c/2-r/2 & d/2-s/2 & a/2-p/2 & b/2-q/2 \\ d/2-s/2 & c/2-r/2 & b/2-q/2 & a/2-p/2 \end{bmatrix} \tag{4.28c}$$

For a differential coupler, the following constraints should be satisfied.

When the differential Port A feeds the signal, there is no energy reflection, that is, $S_{AA}^{dd} = a/2 - e + p/2 = 0$. Differential Port D is isolated, then get $S_{AD}^{dd} = d/2 - h + s/2 = 0$. For 3-dB equal power output, get $\left|S_{AB}^{dd}\right| = \left|S_{Ac}^{dd}\right| = 1/\sqrt{2}$, that is, $|b/2 - f + q/2| = |c/2 - g + r/2| = 1/\sqrt{2}$. If the phase difference between differential Port B and differential Port C is 90°, $b/2 - f + q/2 = \pm j \cdot (c/2 - g + r/2)$ can be obtained.

In addition, common-mode signals need to be suppressed, that is, no common-mode signals are transmitted from Port A to Port B, Port C, and Port D, and $\left|S_{AA}^{cc}\right| = |a/2 + e + p/2| = 1$, $S_{AB}^{cc} = b/2 + f + q/2 = 0$, $S_{AC}^{cc} = c/2 + g + r/2 = 0$, $S_{AD}^{cc} = d/2 + h + s/2 = 0$.

At the same time, it is necessary to suppress the cross-mode signal, that is, there is no conversion between the differential-mode signal and the common-mode signal, and it is necessary to satisfy $S_{(4\times4)}^{cd} = S_{(4\times4)}^{dc} = 0$, that $a = p$, $b = q$, $c = r$, $d = s$.

Combining the above three constraints, (4.25) can be simplified as:

$$S_8 = \begin{bmatrix} a & b & c & 0 & a & -b & -c & 0 \\ b & a & 0 & c & -b & a & 0 & -c \\ c & 0 & a & b & -c & 0 & a & -b \\ 0 & c & b & a & 0 & -c & -b & a \\ a & -b & -c & 0 & a & b & c & 0 \\ -b & a & 0 & -c & b & a & 0 & c \\ -c & 0 & a & -b & c & 0 & a & b \\ 0 & -c & -b & a & 0 & c & b & a \end{bmatrix} \tag{4.29}$$

where $|a| = 1/2$, $|b| = |c| = \sqrt{2}/4$, $b = \pm j \cdot c$. Since the eight-port network is symmetrical horizontally and vertically, it to be analyzed using the method of odd and even modes.

Then, using the conversion relationship between the admittance (Y) matrix and the S-parameter matrix, the Y-parameters corresponding to each mode of the two-port circuit can be obtained, respectively:

$$Y_{11}^{ee} = Y_{55}^{ee} = \frac{1 - 4ac - 4ab}{Z_0(2a+1)(2b+2c+1)} \tag{4.30a}$$

$$Y_{15}^{ee} = Y_{51}^{ee} = \frac{-2a + 2b + 2c}{Z_0(2a+1)(2b+2c+1)} \tag{4.30b}$$

$$Y_{11}^{eo} = Y_{55}^{eo} = \frac{-4ab - 4ac - 1}{Z_0(2a+1)(2b+2c-1)} \tag{4.30c}$$

$$Y_{15}^{eo} = Y_{51}^{eo} = \frac{2a + 2b + 2c}{Z_0(2a+1)(2b+2c-1)} \tag{4.30d}$$

$$Y_{11}^{oe} = Y_{55}^{oe} = \frac{-4ab + 4ac + 1}{Z_0(2a+1)(2b-2c+1)} \tag{4.30e}$$

$$Y_{15}^{oe} = Y_{51}^{oe} = \frac{-2a + 2b - 2c}{Z_0(2a+1)(2b-2c+1)} \tag{4.30f}$$

$$Y_{11}^{oo} = Y_{55}^{oo} = \frac{4ab - 4ac + 1}{Z_0(2a+1)(-2b+2c+1)} \tag{4.30g}$$

$$Y_{15}^{oo} = Y_{51}^{oo} = \frac{-2a - 2b + 2c}{Z_0(2a+1)(-2b+2c+1)} \tag{4.30h}$$

To achieve 3-dB quadrature output of the differential mode signal, while suppressing the common-mode and cross-mode signals in the eight-port differential coupler with symmetrical structure, the conditions that need to be satisfied are shown in (4.30), where $|a| = 1/2$, $|b| = |c| = S_{AB}^{dd}$ $b = \pm j \cdot c$. It should be noted that, for the situation where the nonconstant-amplitude quadrature output of the differential mode signal is required, the constraint conditions can be adjusted accordingly, so that another set of constraint formulas can be obtained, which will not be discussed here.

This design proposes a transformer-based differential coupler circuit, as shown in Figure 4.19, consisting of 2 pairs of transformers and 8 pairs of capacitors. The coupling coefficient of the transformer is represented by k_1 or k_2, and each transformer includes a pair of inductances whose inductance values are equal, represented by L_1 or L_2, respectively. Connect the same end of each pair of inductors with a capacitor C_2 or C_3, at the same time, connect a capacitor C_1 or C_4 in parallel with both ends of each inductor. There are eight single-ended ports, among which Port 1 and Port 5, Port 2 and Port 6, Port 3 and Port 7, and Port 4 and Port 8 are combined in pairs to form four differential ports, namely differential Port A, differential Port B, differential Port C, and differential Port D.

It can be seen that the differential coupler is based on a transformer structure and includes other lumped parameter elements such as capacitors, rather than relying on a transmission line structure that is dependent on wavelength. This approach offers the advantage of a smaller circuit area, particularly at lower frequency bands where its benefits become even more pronounced.

Using the T-type equivalent circuit of the transformer, and analyzing the odd and even modes of the differential coupler, four simplified two-port circuits can be obtained, as shown in Figure 4.20. Assuming that the central angular frequency is ω_0 the Y parameters of the four two-port networks can be obtained as follows:

$$Y_{11}^{ee} = Y_{55}^{ee} = jw_0(C_2 + C_3) \tag{4.31a}$$

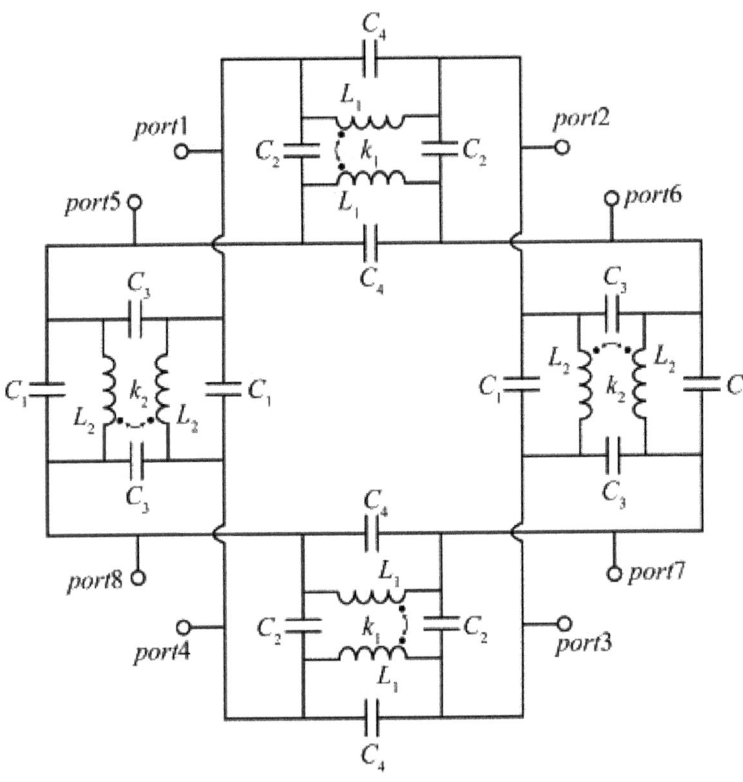

Figure 4.19 Schematic diagram of the proposed transformer-based differential coupler [10].

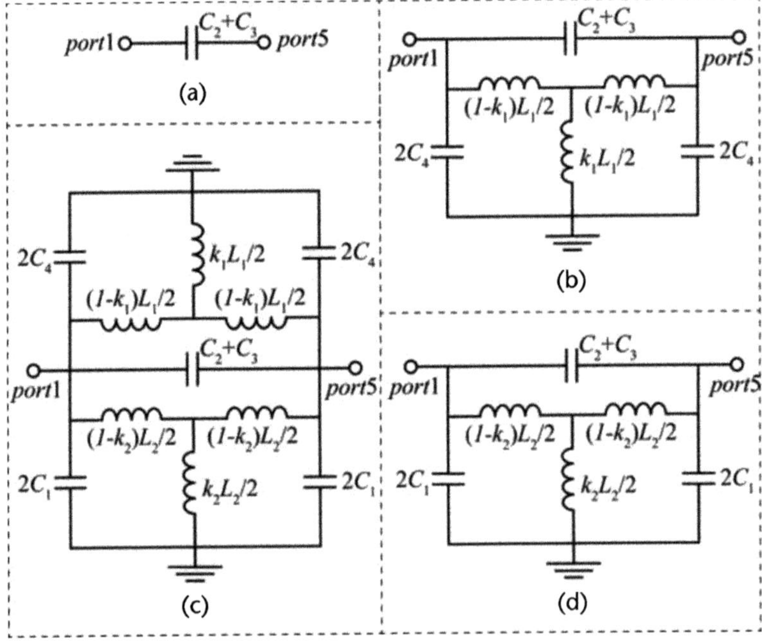

Figure 4.20 The four simplified two-port circuits of [10]: (a) even-even, (b) even-odd, (c) odd-even, and (d) odd-odd modes.

$$Y_{15}^{ee} = Y_{51}^{ee} = -jw_0\left(C_2 + C_3\right) \tag{4.31b}$$

$$Y_{11}^{eo} = Y_{55}^{eo} = jw_0\left(C_2 + C_3 + 2C_4\right) + \frac{j2}{wL_1\left(k_1^2 - 1\right)} \tag{4.31c}$$

$$Y_{15}^{eo} = Y_{51}^{eo} = -jw_0\left(C_2 + C_3\right) - \frac{j2k_1}{wL_1\left(k_1^2 - 1\right)} \tag{4.31d}$$

$$Y_{11}^{oe} = Y_{55}^{oe} = jw_0\left(C_2 + C_3 + 2C_1\right) + \frac{j2}{w_0L_2\left(k_2^2 - 1\right)} \tag{4.31e}$$

$$Y_{15}^{oe} = Y_{51}^{oe} = -jw_0\left(C_2 + C_3\right) - \frac{j2k_2}{w_0L_2\left(k_2^2 - 1\right)} \tag{4.31f}$$

$$\begin{aligned} Y_{11}^{oo} &= Y_{55}^{oo} \\ &= jw_0\left(C_2 + C_3 + 2C_4 + 2C_1\right) + \frac{j2}{w_0L_1\left(k_1^2 - 1\right)} + \frac{j2}{w_0L_2\left(k_2^2 - 1\right)} \end{aligned} \tag{4.31g}$$

$$Y_{15}^{oo} = Y_{51}^{oo} = -jw_0\left(C_2 + C_3\right) - \frac{j2k_1}{w_0L_1\left(k_1^2 - 1\right)} - \frac{j2k_2}{w_0L_2\left(k_2^2 - 1\right)} \tag{4.31h}$$

Therefore, based on the differential coupler structure in Figure 4.19, achieving 3-dB quadrature differential-mode signal operation, as well as common-mode and cross-mode suppression, can be achieved by making a one-to-one correspondence between (4.31) and (4.30). From the (4.31a) and (4.31b), it can be seen that $Y_{11}^{\omega} = -Y_{15}^{\alpha e}$, from (4.30a) and (4.30b), we can derive $1 - 4ac - 4ab = -(-2a + 2b + 2c)$, which simplifies to $(1 - 2a)(2b + 2c + 1) = 0$. Considering the constraints before $|a| = 1/2$, $|b| = |c| = \sqrt{2}/4$, $b = \pm j \cdot c$ and $(2b + 2c + 1) \neq 0$, it can be $a = 1/2$. Substituting this value into (4.30) and simplifying, we can finally get the desired solution:

$$L_1 = \frac{jk_1Z_0\left(8bc - 1\right)}{2\omega_0\left(b + c\right)\left(1 - k_1^2\right)} \tag{4.32a}$$

$$L_2 = \frac{k_2Z_0\left(8b^2 + 4b + 1\right)}{j2\omega_0 c\left(1 - k_2^2\right)} \tag{4.32b}$$

$$C_1 = \frac{1}{\omega_0^2L_2\left(k_2 + 1\right)} = \frac{j2c\left(1 - k_2\right)}{\omega_0k_2Z_0\left(8b^2 + 4b + 1\right)} \tag{4.32c}$$

$$C_4 = \frac{1}{\omega_0^2 L_1 (k_1 + 1)} = \frac{2(b + c)(1 - k_1)}{jk_1\omega_0 Z_0 (8bc - 1)} \tag{4.32d}$$

$$C_2 + C_3 = \frac{j(b + c - 1/2)}{\omega_0 Z_0 (2b + 2c + 1)} \tag{4.32e}$$

Among them, $|b| = |c| = \sqrt{2}/4,\ b = \pm j \cdot c$. It should be noted that there are eight variables k_1, k_2, L_1, L_2, C_1, C_2, C_3, and C_4 in the above five equations, so there are multiple solutions. However, (4.32e) reveals that C_2 and C_3 can be combined arbitrarily, as long as the sum of the capacitance values of the two remains constant. This enables improved design flexibility without altering the frequency response of the circuit. In addition, when k_1 and k_2 are determined, L_1, L_2, C_1, C_4, and $C_2 + C_3$ can be calculated by (4.32).

It is apparent from the previous section that for the proposed transformer-based differential coupler, satisfying (4.32) is sufficient to achieve the desired response at the center frequency. The values of the coupling coefficients k_1 and k_2 of the transformer are directly related to the frequency response at the center frequency and will not be affected. Next, the influence of the coupling coefficients k_1 and k_2 on the frequency response within a certain bandwidth will be discussed. First, define the angular frequency as $w = n\omega_0$, the range of n is 0.8 to 1.2, which represents 40% of the frequency bandwidth. Let $a = 1/2$, $b = -j\sqrt{2}/4$, and $c = \sqrt{2}/4$. Substituting (4.32) into (4.31), and using the conversion relationship from Y matrix to S matrix to obtain the corresponding simplified two-port S-parameters, so that the complete eight-port S matrix and parameters can be obtained, and then the S-parameters of its differential mode, common-mode and cross-mode can be obtained by (4.26). The relevant S-parameter curve can be obtained by programming the formula through MATLAB software. When the coupling coefficient k_1 or k_2 of the transformer is kept constant, change the value of k_1 or k_2 to observe the frequency response. It has been found that when the coupling coefficient k_1 and k_2 of the two transformers are equal, the response in the frequency band is more flat and symmetrical. Therefore, in the following analysis, let $k = k_1 = k_2$ directly.

Figure 4.21 illustrates the relationship between the frequency response of the differential coupler and k, where the arrows indicate the direction in which k increases. For this differential coupler, the cross-mode suppression calculated by the formula is ideal throughout the whole frequency band, and thus it is not presented here. It can be seen from Figure 4.21 that when k increases, within a certain bandwidth range, the differential mode response and common mode rejection gradually become better. In other words, increasing k leads to an increase in the relative bandwidth of the differential coupler.

To better illustrate the influence of the coupling coefficient k on the overall performance of a transformer, the relationship between various performance indicators and k are also analyzed. In this case, $n = 0.9$, meaning that each performance index is examined at a frequency 10% away from the center frequency. As k increases from 0.3 to 0.9, the return loss (S_{AA}^{dd}) and isolation (S_{AB}^{dd}) of the differential-mode signal both improve gradually. Additionally, the common-mode rejection (S_{AB}^{cc},

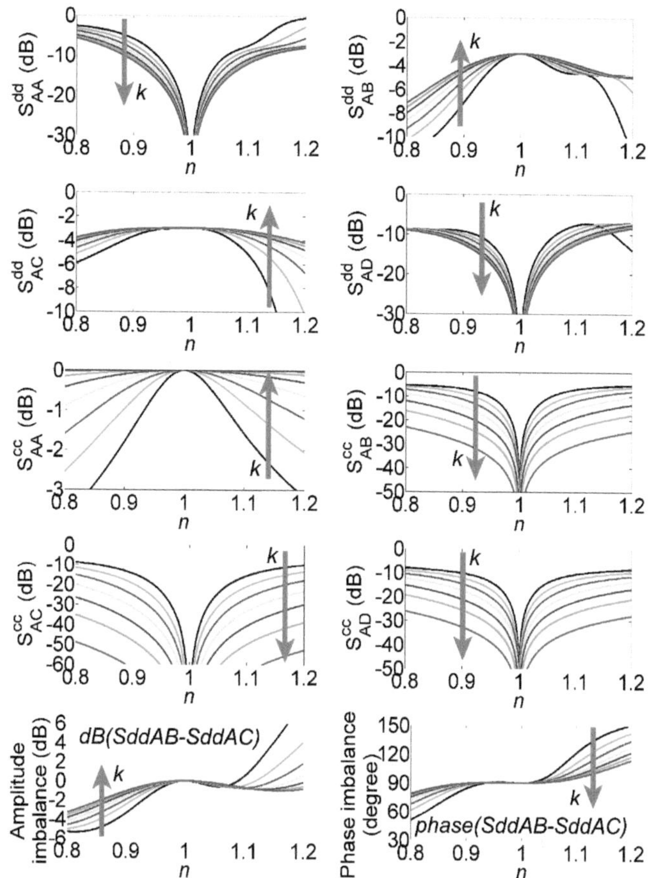

Figure 4.21 The frequency response versus k. $k = k_1 = k_2 = 0.3$ to 0.9. The direction of the arrow indicates the increase of k [10].

S^{cc}_{AC}, S^{cc}_{AD}) is gradually enhanced, and the amplitude unbalance of the differential mode signal ($S^{dd}_{AB} - S^{dd}_{AC}$) also improves gradually.

It is important to note that regardless of the value of k, as long as the parameter relationship in (4.32) is satisfied, the response at the center frequency point is ideal, and k will only affect the performance outside the center frequency point. The bandwidth of the differential-mode signal and the common-mode rejection will also change with k, where the bandwidth of the differential mode signal is represented by |phase (S^{dd}_{AB}) – phase (S^{dd}_{AC})| – 90° < 10°, $S^{dd}_{AA} < -10$ dB, $S^{dd}_{AD} < -10$ dB, $\left|S^{dd}_{AB} - S^{dd}_{AC}\right| < 1$ dB to define, common-mode suppression bandwidth is defined by $S^{cc}_{AA} > -0.2$ dB, $S^{cc}_{AB} < -20$ dB, $S^{cc}_{AC} < -20$ dB, $S^{cc}_{AD} < -20$ dB. As k gradually increases to above 0.8, the common-mode rejection bandwidth shows significant improvement, while the bandwidth improvement of the differential mode signal gradually slows down. When k gradually approaches the theoretical maximum value of 1, the bandwidth of the differential mode signal is about 25%. Additionally, it can be observed from (4.32) that when the coupling coefficient of the transformer increases, the inductance values L_1 and L_2 also need to increase accordingly, which means that the ground circuit area and loss will also increase. Therefore, there is a trade-off between circuit area and loss and frequency bandwidth.

Additionally, to provide a quick design guide, Tables 4.2 and 4.3 list multiple sets of parameter values for 3-dB differential couplers corresponding to different frequencies and coupling coefficients. It can be observed from Table 4.3 that as the coupling coefficient (k) increases, the value of $C_2 + C_3$ remains unchanged, the values of L_1 and L_2 both increase, and the values of C_1 and C_4 decrease.

It will be based on the differential coupler topology shown in Figure 4.20, which is designed and implemented on the SISL platform using the same dielectric substrate material and thickness as before. The planar view of the design is presented in Figure 4.22, where the main circuit is implemented on metal layer 5 and metal layer 6. As shown in Figure 4.22, two pairs of transformer structures, named

Table 4.2 Set of Parameters with Different Frequency

Frequency(GHz)	L_1 *(nH)*	L_2 *(nH)*	C_1 *(pF)*	C_4 *(pF)*	$C_2 + C_3$ *(pF)*
0.5	15	21.2	3.18	4.5	7.68
1	7.5	10.6	1.59	2.25	3.84
2	3.75	5.3	0.795	0.113	1.92
3	2.5	3.53	0.53	0.75	1.28
4	1.87	2.65	0.398	0.563	0.96
5	1.5	2.12	0.318	0.45	0.77
6	1.25	1.76	0.265	0.375	0.64
7	1.07	1.52	0.227	0.322	0.55
8	0.94	1.33	0.199	0.281	0.48
9	0.83	1.18	0.177	0.25	0.43
10	0.75	1.06	0.159	0.225	0.38

$k = k_1 = k_2 = 0.5$.
The bandwidth = 15.05% (differential-mode) and 7.2% (common-mode).
Source: [10].

Table 4.3 Set of Parameters with Different *k*

k (Coupling Coefficient)	L_1 *(nH)*	L_2 *(nH)*	C_1 *(pF)*	C_4 *(pF)*	$C_2 + C_3$ *(pF)*	*Bandwidth Defined by Differential-Mode Response*	*Bandwidth Defined by Common-Mode Response*
0.3	3.71	5.25	3.71	5.25	3.84	10.2%	3.16%
0.4	5.36	7.58	2.39	3.38	3.84	12.8%	4.83%
0.5	7.5	10.6	1.56	2.25	3.84	15.05%	7.16%
0.6	10.6	14.9	1.06	1.5	3.84	17.04%	10.8%
0.7	15.4	21.8	0.68	0.96	3.84	19.04%	16.8%
0.8	25	35.36	0.4	0.56	3.84	20.63%	28.6%
0.9	53.3	75.39	0.177	0.25	3.84	22.12%	64.6%

Frequency = 1 GHz.

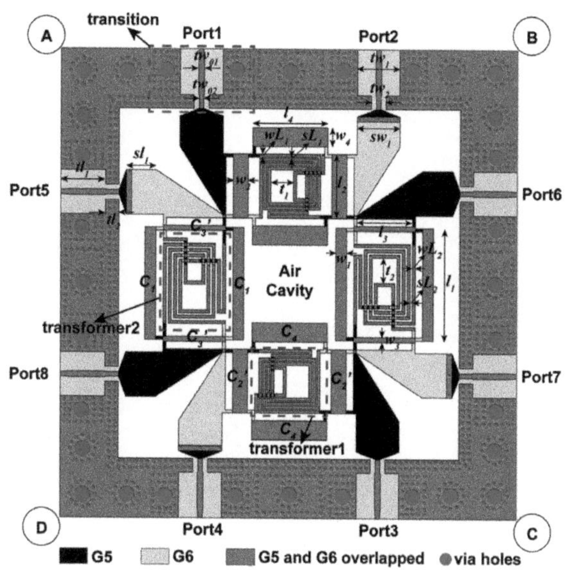

(a)

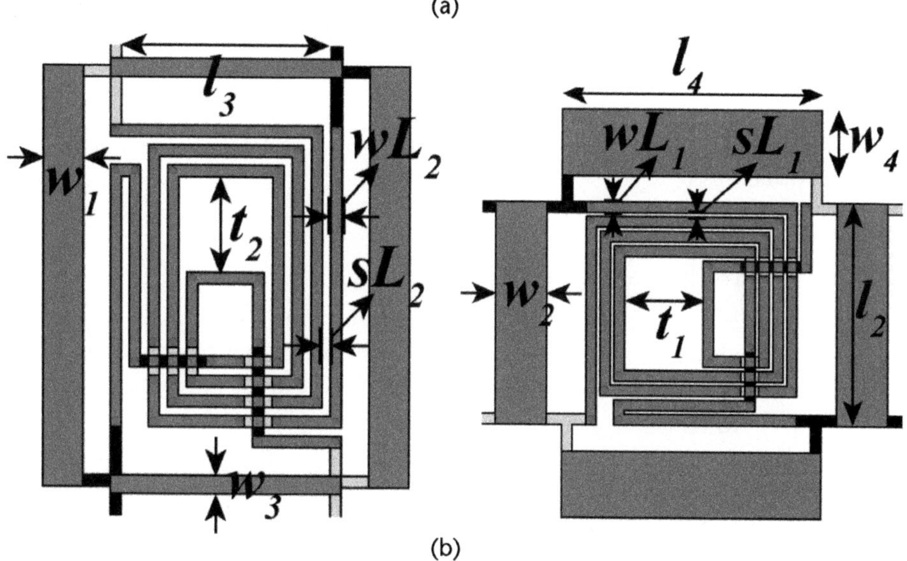

(b)

Figure 4.22 Circuit topology of the proposed transformer-based differential coupler implemented based on SISL technology [10]: (a) complete view, and (b) zoomed view of the transformers.

transformer 1 and transformer 2, are employed, and all inductors use a double-layer wiring structure to reduce circuit loss. Additionally, the eight pairs of capacitors are designed as plate capacitors to facilitate rapid modeling. For signal interconnection between different metal layers, namely metal layer 5 and metal layer 6, metalized through-holes are used for interconnection.

To better explain the design method and steps for the SISL differential coupler structure shown in Figure 4.22, a design flow chart has been provided in Figure 4.23.

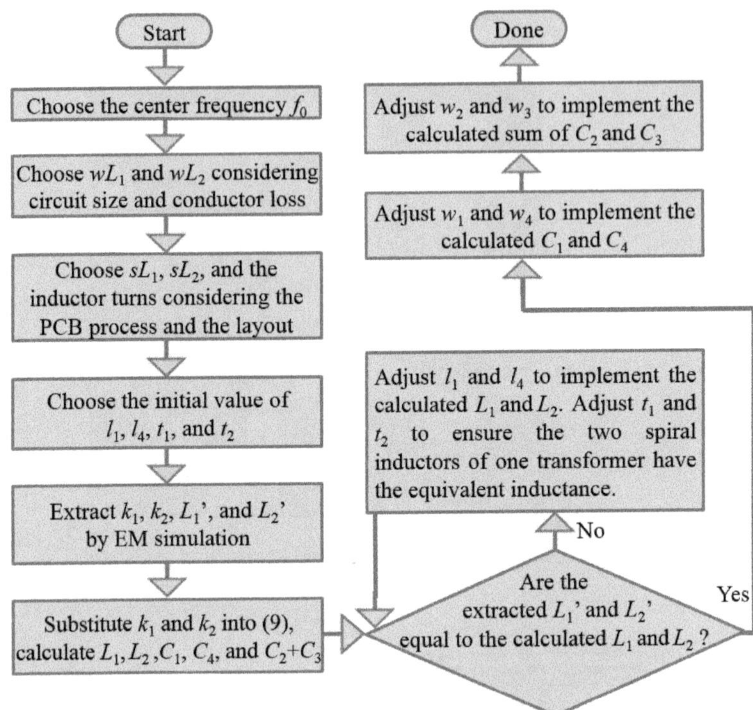

Figure 4.23 Design flow chart of the transformer-based differential coupler implemented on SISL platform [10].

Step 1: Choose the center frequency f_0.

Step 2: Choose the width of the inductor line (i.e., w_{L1} or w_{L2}). For the spiral inductor in Figure 4.22, a narrower line width results in a larger inductance value, which reduces the circuit area. However, there is a trade-off to consider since it also increases conductor loss.

Step 3: Choose the appropriate values of s_{L1} and s_{L2} and the number of turns of the inductor according to the PCB process capability used by the SISL and the consideration of the circuit layout. According to the previous analysis, when a wider frequency bandwidth is required, a larger coupling coefficient k is required. When the gaps s_{L1} and s_{L2} between the spiral inductors become smaller, the coupling coefficient will increase to expand the bandwidth, but the influence of PCB processing accuracy should also be considered.

Step 4: Choose initial values for l_1, l_4, t_1, and t_2. By adjusting the length of L_1 or L_2 in Figure 4.22, the inductance value of L_1 or L_2 can be adjusted. In addition, since the layout of the two spiral inductors forming a transformer is not exactly the same, two adjustment variables t_1 and t_2 are added for fine-tuning the inductance value.

Step 5: Use the electromagnetic simulation software and use (4.33) to extract the coupling coefficients k_1, k_2 and inductance values L_1', L_2' of the two transformers.

Step 6: Substitute k_1 and k_2 into (4.32), and calculate the values of L_1, L_2, C_1, C_4 and the sum of C_2 and C_3.

Step 7: Compare the extracted L'_1 and L'_2 with the calculated L_1 and L_2. If they are not equal, proceed to step 8. If they are the same, proceed to step 9.

Step 8: Adjust the value of l_1 and l_4 to match its inductance value to be equal to the calculated L_1 and L_2, respectively. Make sure to adjust t_1 and t_2 simultaneously to ensure that the two spiral inductors of the same transformer have equal inductance values. Once completed, proceed to step 9.

Step 9: Adjust w_1 and w_4 and extract the capacitance value that matches the calculated C_1 and C_4. The capacitance value can be extracted using the two-port π-type equivalent circuit method.

Step 10: Adjust w_2 and w_3 and extract the capacitance value $C'_2 + C'_3$ to make it slightly smaller than the calculated value of $C_2 + C_3$. Here, due to the parasitic mutual capacitance between the two spiral inductors of the same transformer, this part of the capacitance can replace part of the original $C_2 + C_3$ value, so the actual capacitance should be smaller than the calculated value.

$$L = \frac{\mathrm{Im}(Z_{11})}{2\pi f} \tag{4.33a}$$

$$L' = \frac{\mathrm{Im}(Z_{22})}{2\pi f} \tag{4.33b}$$

$$M = \frac{\mathrm{Im}(Z_{12})}{2\pi f} \tag{4.33c}$$

$$k = \frac{M}{\sqrt{LL'}} \tag{4.33d}$$

In this design, the center frequency is chosen as 0.5 GHz due to the small parasitic effects at low frequencies and the ease of layout, which facilitates the rapid design and verification of the proposed differential coupler circuit. The values of w_{L1} or w_{L2} are selected as 0.4 mm, the values of s_{L1} and s_{L2} are 0.15 mm and 0.3 mm, respectively, with 2.5 turns of the spiral inductor. Since the two transformers are not exactly the same in size and structure, and the influence factors of the layout need to be considered, the coupling coefficients k_1 and k_2 of the two transformers in this design are 0.56 and 0.5, respectively. Although they are not exactly equal, it also ensures the excellent characteristics of the circuit. When $f_0 = 0.5$ GHz, $\omega = 2\pi f_0$, $b = -j\sqrt{2}/4$, and $c = \sqrt{2}/4$, $k_1 = 0.56$, $k_2 = 0.5$, according to (4.32), it can be calculated that $L_1 = 18.36$ nH, $L_2 = 21.22$ nH, $C_1 = 3.18$ pF, $C_4 = 3.54$ pF, and $C_2 + C_3 = 7.68$ pF.

Based on the design steps described before, it is necessary to perform electromagnetic simulations using software to optimize the necessary parameters. The final values obtained, as shown in Figure 4.22, are as follows: $l_1 = 14.8$ mm, $l_2 = l_3 = 8.4$ mm, $l_4 = 10.15$ mm, $w_1 = 1.45$ mm, $w_2 = 2$ mm, $w_3 = 0.65$ mm, $w_4 = 2.5$ mm, $w_{L1} = w_{L2} = 0.4$ mm, $s_{L1}= 0.15$ mm, $s_{L2} = 0.3$ mm, $t_1 = 3.07$ mm, $t_2 = 3.37$ mm, $tw_{01}= 0.78$m, $tw_{02} = 0.6$ mm, $tw_1 = 5.8$ mm, $tw_2 = 2$ mm, $tl_1 = 6$ mm, $tl_2 = 1.8$ mm, $sl_1 = 4$ mm, and $sw_1 = 5.85$ mm.

The simulation results of the SISL differential coupler are in good agreement with the test results. Within the relative frequency bandwidth of 13.4%, the tested differential-mode signal amplitude imbalance is ±0.6 dB, and the phase imbalance is 90° ± 3°. Additionally, the tested differential mode return loss (S^{dd}_{AA}) and isolation (S^{dd}_{AD}) are both better than 13 dB. The tested S^{dd}_{AB} and S^{dd}_{AC} are −3.9 dB and −4.2 dB, respectively, indicating a loss of about 1 dB. Regarding common-mode rejection, within the relative bandwidth of 40% from 0.4 GHz to 0.6 GHz, the tested S^{cc}_{AB} and S^{cc}_{AD} are both better than −10 dB, while the S^{cc}_{AC} is better than −20 dB, with a maximum common mode rejection of 30 dB. Within the 40% frequency bandwidth range, the tested cross-mode rejection is better than 30 dB. The processed object with a circuit area is 0.08 $\lambda_g \times 0.08\ \lambda_g$, where λ_0 is the waveguide wavelength.

4.2.4 Branch-Line Coupler on the SISL Platform

A branch line coupler is a four-port microwave device that consists of two parallel main transmission lines and two branch lines. Power enters from one input port and is then divided equally between the two output ports, with the fourth port being an isolated port. Since the length of the transmission line and the branches is constant, the phase difference between the two output ports is 90°, which is commonly used for power distribution and synthesis. Because of these features, branch-line couplers play an important role in many RF and microwave circuits. This section will introduce three types of branch line couplers: DSISL branch line coupler, DGS branch line coupler, and slow-wave branch-line coupler.

4.2.4.1 DSISL Branch-Line Coupler

The SISL platform has advantages such as small size, low loss, low cost, high reliability, high integration, and self-packaging. In [11], double-sided interconnected strip lines (DSISL) have been introduced to further reduce loss.

The structure of DSISL is shown in Figure 4.24(a), two parallel strips on each side of Substrate 1 connected by connecting via or plated through-hole (PTH). Compared with the electric field distribution in the traditional suspended stepped-impedance line structure, more electric energy is distributed in the air in the DSISL structure, which therefore has a lower effective dielectric constant.

A branch-line coupler has been realized on the SISL platform using the DSISL structure. The second and fourth layers of this structure are hollowed out in a specific rectangular shape, and all five layers of the substrate are bonded together to form the basic cavity of the suspended line. The circuit structure on metal layers

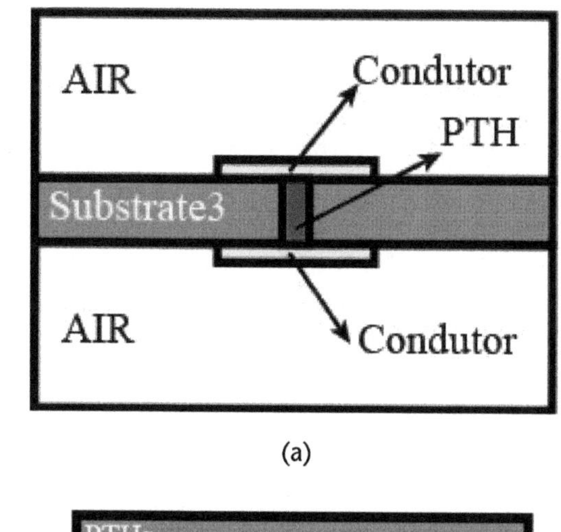

(a)

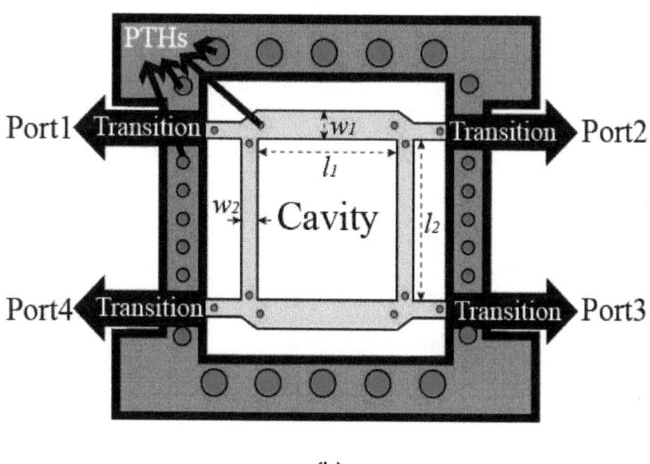

(b)

Figure 4.24 (a) Cross-sectional view of proposed DSISL structure [11], and (b) the top view of the branch-line coupler based on SISL and DSISL [11].

G_5 and G_6, located in the coupler, is shown in Figure 4.24(b). Plated through-holes around the coupler and the top and bottom ground layers provide electromagnetic shielding. When the signal is fed into Port 2, Ports 3 and 1 receive equal power, while Port 4 is the isolated port.

The measurement results of the proposed SISL coupler agree well with the simulation ones. The losses of transitions and the SMA connectors have been removed from the final measured results. At a center frequency of 4.6 GHz, the measured return loss and isolation performance are better than 28 dB. Over the entire operating bandwidth from 4.2 GHz to 5 GHz, the measured S_{21} is −3.16 dB with an amplitude imbalance of ±0.28 dB, and the S_{31} is −3.49 dB with an amplitude imbalance of ±0.036 dB. The phase difference between the two output ports is 89.4° ± 0.9°. The loss of its coupler is calculated using (4.1), which is lower than that of the prior art. DSISL branch couplers based on SISL technology achieve very low losses compared to existing technologies.

$$\text{Loss} = 1 - \left|S_{11}\right|^2 - \left|S_{21}\right|^2 - \left|S_{31}\right|^2 - \left|S_{41}\right|^2 - \left|S_{41}\right|^2 \quad (4.34)$$

4.2.4.2 Branch-Line Coupler Using DGS

In [12], this novel describes a branch line coupler that features harmonic suppression by using defect grounding structures (DGS). The DGS coupler is based on the SISL structure design, Substrate 3 is the main part of the SISL circuit, and the metal pattern of the design is etched on G_6, that is, the GDS structure. As can be seen from Figure 4.25(a), due to the existence of G_9, the external ground structure of SISL is still complete, so the internal DGS can be packaged inside the SISL multilayer board, which reduces radiation loss. At the same time, it can generate a certain stopband beyond the operating frequency.

Figure 4.25(b) shows a planar view of the proposed SISL branch coupler, wherein the pattern ground structure is applied to the branch coupler and placed on the corresponding G_6 under the T-junction. Its equivalent circuit can be represented as three LC circuits connected in parallel in the form of a T-junction, as shown in Figure 4.25(c). The L_1, L_2, and L_3 lines can be represented as two-port π equivalent networks, respectively. Therefore, under lossless conditions, the complete equivalent circuit of the coupler can be obtained as shown in Figure 4.25(d).

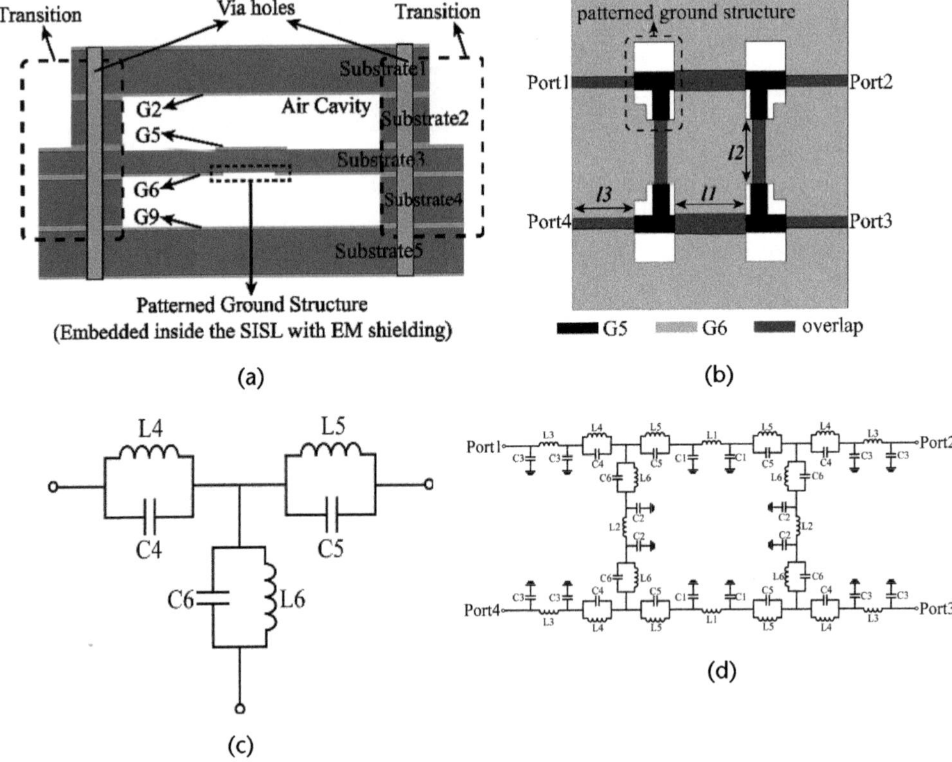

Figure 4.25 (a) Cross-sectional view of the proposed SISL coupler [12], (b) planar view of the proposed wideband branch-line coupler with stop band [12], (c) the patterned ground structure of equivalent circuit, and (d) equivalent circuits of the proposed wideband branch-line coupler with harmonics suppression [12].

It uses the even-mode/odd-mode analysis method to simplify the equivalent circuit in Figure 4.25 into four single-ended equivalent circuits for analysis, and after converting to the equivalent circuit, the S-parameter can be given by

$$S_{11} = \frac{\left(\Gamma_{ee} + \Gamma_{eo} + \Gamma_{oe} + \Gamma_{oo}\right)}{4} \tag{4.35a}$$

$$S_{21} = \frac{\left(\Gamma_{ee} - \Gamma_{eo} + \Gamma_{oe} - \Gamma_{oo}\right)}{4} \tag{4.35b}$$

$$S_{31} = \frac{\left(\Gamma_{ee} - \Gamma_{eo} - \Gamma_{oe} + \Gamma_{oo}\right)}{4} \tag{4.35c}$$

$$S_{41} = \frac{\left(\Gamma_{ee} + \Gamma_{eo} - \Gamma_{oe} - \Gamma_{oo}\right)}{4} \tag{4.35d}$$

In this design, the equivalent circuit is primarily used to explain the principle of operation of the coupler. The detailed dimensions of the coupler are optimized by using electromagnetic simulation software, and by correctly optimizing the size of the DGS, stopbands beyond the coupler's operating frequency band can be realized. According to electromagnetic simulation optimization, the electrical length of the transmission line is about one-twelfth of the guided wave wavelength. The specific size parameters of the coupler are shown in Figure 4.26. A number of via holes are designed around the cavity for electromagnetic shielding, and in order to avoid parasitic resonance in the passband, the position of the vias around the cavity is appropriately adjusted.

The circuit size of the coupler is 48 mm × 51 mm. In the frequency range of 2.12 ~ 3.46 GHz, (i.e., a relative bandwidth range of 48%), the phase imbalance of

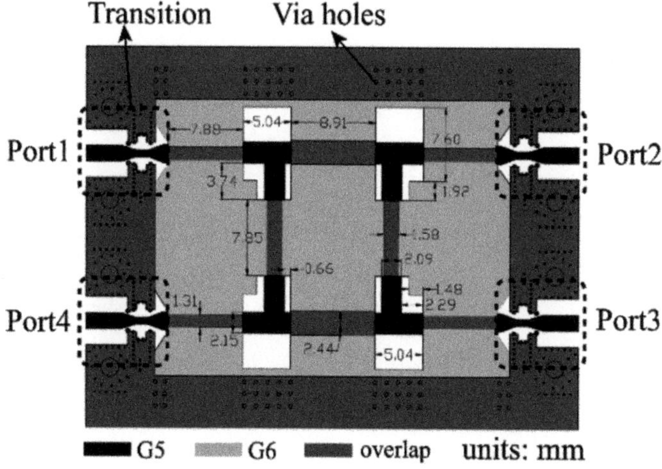

Figure 4.26 Planar view and dimensions of the SISL coupler [12].

the coupler is measured to be 89.6° ± 1.5°, the return loss and isolation are better than 14.5 dB. The stopband ranges from 4.25 GHz to 8.57 GHz, with stopband rejection exceeding 15 dB and a maximum of 40 dB.

4.2.4.3 Slow-Wave Branch Line Coupler

Slow-wave structures are widely for circuit miniaturization. Traditional slow-wave circuits generally require metal enclosures for necessary packaging. In recent years, research on slow-wave structures based on SISL technology with self-packaging platforms has also made progress.

In [13], a new SISL slow-wave structure is composed of meandering high- and low-impedance lines connected in series. The space for two adjacent low-impedance lines is small enough to generate more mutual capacitance, which will contribute to better slow-wave performance. The width and length of the low-impedance line are denoted by W and L, respectively, while the width of the high-impedance line and the distance between adjacent lines are represented by S. The specific structure is illustrated in Figure 4.27(a).

Figure 4.27(b) shows the equivalent circuit of a periodic unit cell of a slow-wave structure. The series inductance produced by the high-impedance line is denoted as L_1, the shunt capacitance to ground produced by the low-impedance line is denoted as C_1, and the mutual capacitance between adjacent lines due to the narrow

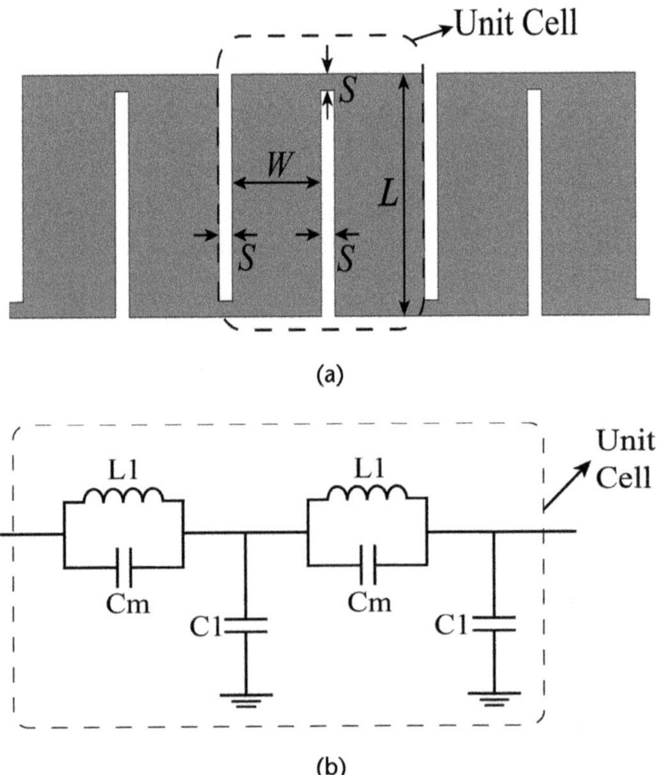

Figure 4.27 (a) Configuration of the proposed slow-wave structure [13], and (b) the equivalent circuit model of the unit cell [13].

edge-coupling effect is denoted as C_m. Thus, the total inductance value L_t, capacitance value C_t, characteristic impedance Z_C, phase constant β_1 of the slow-wave structure, and slow-wave coefficient SWF of the slow-wave unit can be given as

$$L_t = \frac{2L_1}{\left(1 - \omega^2 L_1 C_m\right)} \tag{4.36a}$$

$$C_t = 2C_1 \tag{4.36b}$$

$$Z_c = \sqrt{\frac{L_t}{C_t}} = \sqrt{\frac{2L_1}{\left(1 - \omega^2 L_1 C_m\right) C_1}} \tag{4.36c}$$

$$\beta_s = \frac{\omega\sqrt{L_t C_t}}{2(W + S)} = \frac{\omega}{W + S}\sqrt{\frac{2L_1 C_{g1}}{1 - \omega^2 L_1 C_m}} \tag{4.36d}$$

$$\mathrm{SWF} = \frac{\beta_1}{\beta_0} \tag{4.36e}$$

According to the formula analysis and electromagnetic simulation, the characteristic impedance (Z_C) and SWF versus S, W, and L with different numbers of cells can be obtained. As shown in Figure 4.28, when L and S remain unchanged and W increases, both the characteristic impedance Z_C and the slow-wave factor SWF decrease. When W and S remain unchanged and L increases, the characteristic impedance Z_C decreases, and the slow-wave factor SWF increases. When w_1 and l_1 remain unchanged and s_1 increases, the slow-wave factor SWF shows a decreasing trend overall. This also indicates that the narrow edge-coupling effect can increase the slow-wave effect. In contrast, traditional meander lines have a larger spacing, resulting in a lower slow-wave factor. At the same time, for this transmission-line structure, the slow-wave response and impedance characteristics do not change significantly with different unit cell numbers, indicating that its slow-wave characteristics are relatively stable for different electrical lengths.

A branch-line coupler using the meander slow-wave structure was implemented on the SISL platform. The second and fourth layers are hollowed out with rectangular openwork, which will form two air cavities. G_1 through G_{10} are metal layers. G_2 and G_9 are suspension circuits. The main circuit is designed on the G_5 layer. Figure 4.28(d) shows the circuit topology of the proposed compact SISL branch-line coupler. The electrical lengths of the four branches are all 90°, and the overall structure is symmetrical. The characteristic impedance values of the two types of transmission lines are 50Ω and 35Ω, respectively. Four periodic unit cells were used for the quarter-wavelength 50Ω transmission line, and two periodic unit cells were used for the quarter-wavelength 35Ω transmission line. The transition consists of three parts, including CBCPW, stripline, and suspended line. The holes around the air cavity are used to achieve electrical connections between the different metal

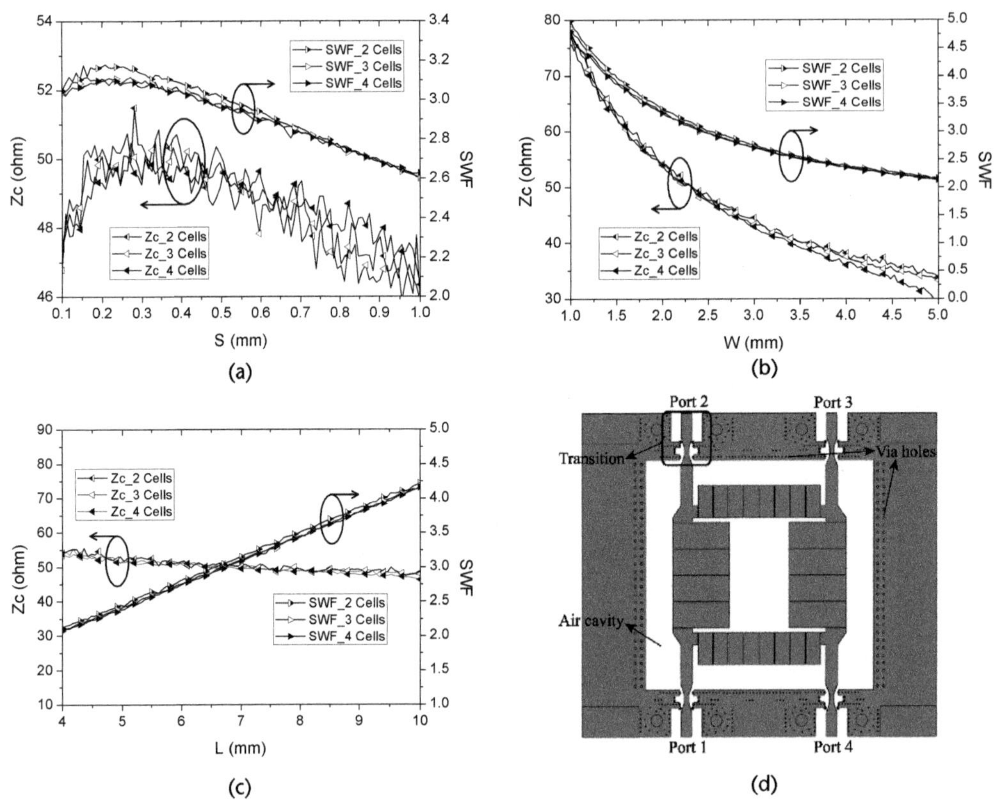

Figure 4.28 The characteristic impedance and slow-wave factor versus: (a) *W* at 1.5 GHz for *L* = 7 mm and *S* = 0.254 mm [13]; (b) *W* at 1.5 GHz for *L* = 7 mm and *S* = 0.254 mm [13]; (c) *L* at 1.5 GHz for *W* = 2.31 mm and *S* = 0.254 mm [13]; and (d) circuit topology of the proposed compact SISL branch-line coupler [13].

layers G_1 to G_{10}. In addition, these vias, together with the ground layers of G_2 and G_9, will constitute an electromagnetic shielding environment.

Based on the selected structural parameters, the designed slow-wave branch-line coupler was simulated and measured on the three-layer SISL platform. The simulation and measured results are in good agreement, and at the center frequency of 1.5 GHz, the measured return loss and isolation are both better than 20 dB. In the frequency range of 1.38 GHz to 1.7 GHz, the measured phase difference is 90° ± 1°. Moreover, the implemented SISL slow-wave branch-line coupler has a circuit area of only 35% of that of a traditional branch-line coupler, achieving circuit miniaturization. Its self-packaging, compact size, low cost, and good EMC characteristics will give it a significant advantage in miniaturization applications.

4.2.5 Rat-Race Coupler on the SISL Platform

4.2.5.1 Slow-Wave Rat-Race Coupler

A rat-race coupler, also known as a ring coupler, is in the shape of a circle. In [14], the rat coupler uses a periodic slow-wave DSISL structure.

As shown in Figure 4.29(a), the attenuation constant and the phase constants in the DSISL structure is smaller than the attenuation constant of SISL, indicating that there is a smaller loss. The cross-sectional view of the electric field distribution of DSISL and SISL shows that the electric field density distributed in the dielectric substrate 3 of DSISL is smaller than that of SISL, indicating that the double-layer wiring not only reduces the conductor loss but also the dielectric loss of the intermediate substrate.

Combining the low-loss DSISL structure and the high-low impedance cascaded meander slow-wave transmission line structure, this section proposes a new DSISL slow-wave transmission line structure, as shown in Figure 4.29(b). Unlike the traditional uniform meandering line, the proposed slow-wave DSISL is formed by cascading high and low-impedance sections with very small gaps, which helps to achieve the desired slow-wave effect. The two layers of transmission lines are on G_5 and G_6, the shape and size of the two layers are the same, and the upper and lower transmission lines are connected with metalized through-holes at both ends of the transmission line, thereby reducing the loss of the circuit.

Figure 4.30(a) shows a simplified electric field distribution in the DSISL slow-wave structure, in addition to the narrow-edge coupling effect on the same layer, there is a cross-coupling between different layers, which can increase the mutual capacitance value, thereby further improving the slow-wave coefficient. The equivalent circuit diagram is shown in Figure 4.30(b), where the series inductance value generated by the high impedance line is represented by L_2, the capacitance to ground generated by the low impedance line is represented by C_2, the mutual capacitance between the same metal layer is represented by C_{m1}, and the cross-mutual capacitance

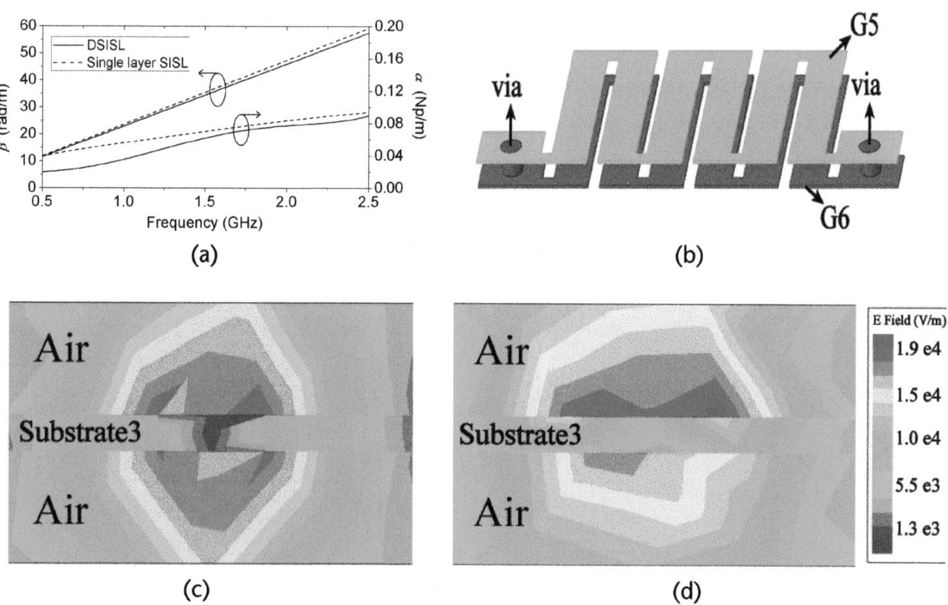

Figure 4.29 (a) Phase constant and attenuation constant of the DSISL and the single-layer SISL [14], (b) electric field distribution of single-layer SISL [14], electric field distribution of the (c) DSISL [14], and (d) single-layer SISL [14].

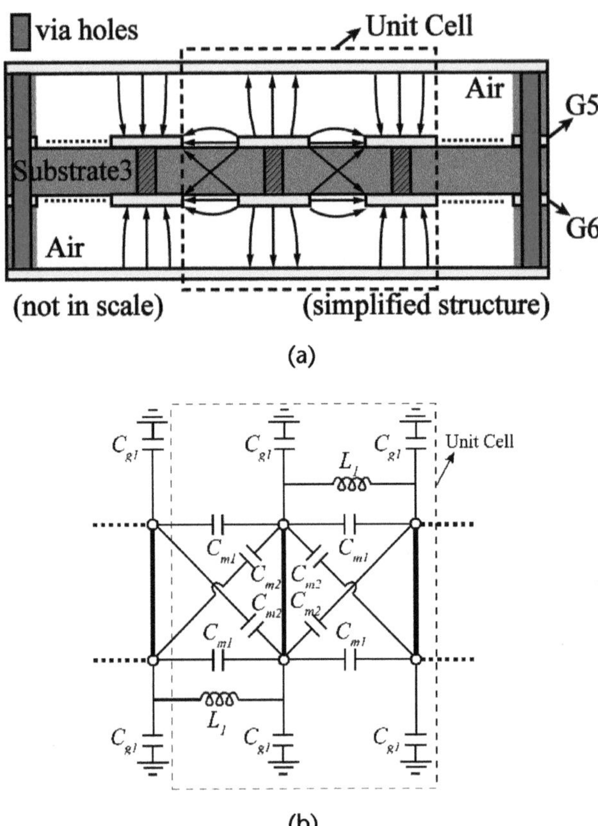

Figure 4.30 (a) Electric field distribution of the proposed slow-wave DSISL [14], and (b) equivalent circuit of the proposed slow-wave DSISL structure [14].

between different metal layers is represented by C_{m2}. The characteristic impedance Z_C and slow-wave coefficient SWF of the DSISL slow-wave transmission line can be given by

$$C_t = 4C_{g1} \tag{4.37a}$$

$$L_t = \frac{2L_1}{1 - 2\omega^2 L_1 \left(C_{m1} + C_{m2}\right)} \tag{4.37b}$$

$$Z_c = \sqrt{\frac{L_t}{C_t}} = \sqrt{\frac{2L_1}{2\left(1 - 2\omega^2 L_1 \left(C_{m1} + C_{m2}\right)\right)C_{g1}}} \tag{4.37c}$$

$$\beta_s = \frac{\omega\sqrt{L_t C_t}}{2(W + S)} = \frac{1}{W + S}\sqrt{\frac{2L_1 C_{g1}}{\frac{1}{\omega^2} - 2L_1\left(C_{m1} + C_{m2}\right)}} \tag{4.37d}$$

$$\mathrm{SWF} = \frac{\beta_s}{\beta_0} \tag{4.37e}$$

where C_t and L_t represent the total capacitance and inductance of the cell, respectively. β_S is the wavenumber of the proposed slow-wave DSISL, and β_0 is the free-space wavenumber and ω is the angular frequency.

With the help of electromagnetic simulation software of high-frequency structure simulator (HFSS), Z_C, and SWF can also be extracted using the above formula. Taking the frequency of 1.5 GHz as an example, the characteristic impedance (Z_C) and SWF versus S, W, and L, respectively, are plotted in Figure 4.31. To improve the mutual capacitance value and produce a better slow-wave effect, the line spacing S should be as small as possible. SWF and Z_C of the proposed slow wave DSISL barely

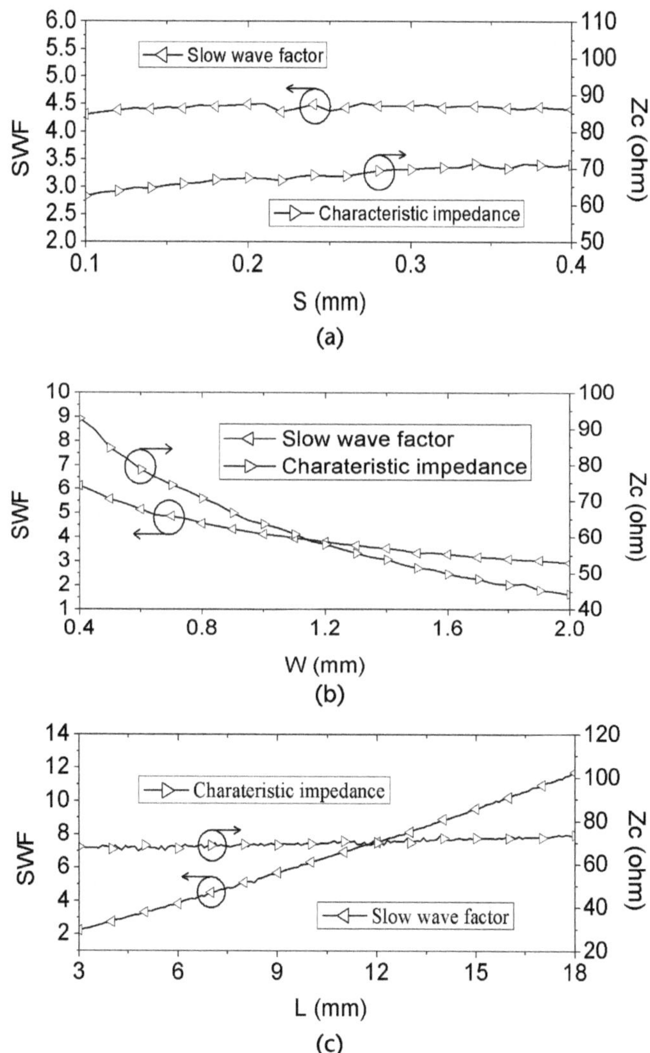

Figure 4.31 *SWF* and characteristic impedance versus: (a) S at 1.5 GHz for $W = 0.85$ mm and $L = 7$ mm [14]; (b) W at 1.5 GHz for $L = 7$ mm and $S = 0.25$ mm [14]; and (c) W at 1.5 GHz for $L = 7$ mm and $S = 0.25$ mm [14].

change when S varies from 0.1 to 0.4 mm, as shown in Figure 4.31(a). Z_C is mainly determined by the W factor. When W widens, Z_C decreases and SWF changes little, as shown in Figure 4.31(b). As shown in Figure 4.31(c), SWF is mainly determined by L. As L becomes longer, SWF increases significantly, while Z_C remains almost unchanged. Therefore, Z_C and SWF of the proposed slow-wave DSISL can be controlled separately. The extracted results plotted in Figure 4.31 can also validate the calculations in (4.37a)–(4.37e).

A conventional rat-race coupler is made up of three 70.7Ω $\lambda/4$ sections and one 70.7Ω $3\lambda/4$ section. Figure 4.32 shows the simulated S-parameters and group delay of the 70.7Ω slow-wave DSISL with different electric length. The frequency where S_{21} rapid decline corresponds to the cutoff frequency. A larger L indicates a higher SWF, but as SWF gets higher, the cutoff frequency decreases. For the $\lambda/4$ case with $L = 4$ mm, the cutoff frequency is about 20 GHz, much higher than the $3\lambda/4$ case with $L = 16.5$ mm. For a slow-wave DSISL and $L = 4$ mm with three cells at 70.7Ω, the simulation group delay is 1.5 ns at 0.07 GHz. The longer the electrical length of the slow wave DSISL, the larger the group delay.

The organizational structure based on this chapter is shown in Figure 4.32(c). In order to make full use of the layout space for further miniaturization, 6 cycle units were selected for both quarter-wavelength and three quarter-wavelength slow-wave transmission lines in this design, and the L_2 value of the three quarter-wavelength slow-wave transmission line was increased so that the slow-wave line extended inward, thus making full use of the internal space of the ring coupler. When the sum port (Port 1) excites, Ports 2 and 3 will gain equal power in phase, and Port 4 is isolated. When the differential port (Port 4) is excited, Ports 2 and 3 will receive equal power, and Port 1 is isolated. The CPW to DSISL conversion on the four ports is used to connect the DSISL signal to an external SMA connector for measurement.

The proposed slow wave DSISL rat-race coupler has been fabricated using the standard PCB process. Simulation results and measurement results show good agreement. From the 1.31 to 1.71-GHz frequency range (i.e., 26.5% bandwidth), return loss and isolation are better than 19 dB. For the sum port and the difference port excitations, the output phase error is within ±63° of 5.1 to 43.1 GHz, and the amplitude imbalance is ±1.7 dB in the 0.5 to 1.34-GHz range. The area of the circuit is only 9.5% of the area of the conventional ring coupler, which greatly reduces the size of the original circuit and realizes the miniaturization of the circuit.

4.2.5.2 180° Coupler

The 180° hybrid coupler is one of the most important passive components. In [15], based on the SISL platform, a low-loss 180° hybrid ring coupler with double-sided SISL and patterned substrate is proposed.

The entire structure of the proposed SISL 180° hybrid coupler with double-sided stripline and patterned substrate consists of five layers of substrate and ten layers of metal. The 5 substrates are stacked on top of each other, and the hollow of Substrate 2 and Substrate 4 is formed into a circle to form the upper and lower cavities of the suspended line. The 180° hybrid coupler still uses the DSISL structure, as shown in Figure 4.33(a). With the help of through-holes and upper and lower metal layers,

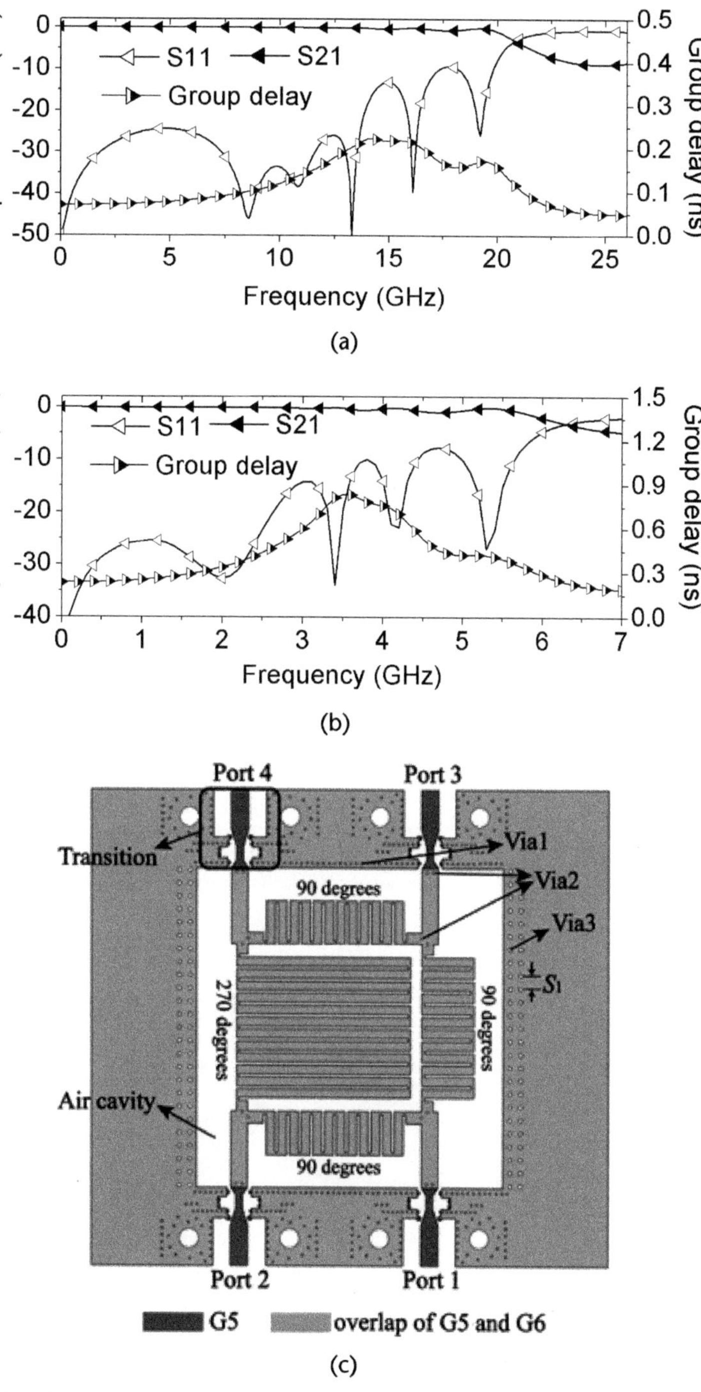

Figure 4.32 Simulated S-parameters and group delay of the 70.7Ω slow-wave DSISL with six unit cells for: (a) S = 0.254 mm, W = 0.85 mm, L = 4 mm (λ/4) [14]; (b) S = 0.254 mm, W = 0.85 mm, L = 16.5 mm (3λ/4) [14]; and (c) topology of the proposed slow-wave DSISL rat-race coupler [14].

the electromagnetic field will be mostly confined within the cavity. In addition to connecting the double-sided metal layers M_5 and M_6 through-hole, the redundant part of the suspended substrate where the circuit is not located is cut off, and the dielectric loss is reduced step by step. At the same time, Substrate 3 is also hollowed out according to the shape of the circuit, as shown in Figure 4.33(b).

The structure of the coupler is different from the hybrid ring formed by the conventional annular microstrip line, and the structure of the rectangular groove in the center of the circular patch is adopted, which improves the coupling level. The relationship between the change in coupling coefficient and the slot size of L is shown in Figure 4.34(a). As the length of the rectangular groove increases, the amplitude imbalance gradually increases. Its frequency is determined by the physical size of the circuit, which is mainly affected by the radius of the circle represented by R and L. Figure 4.34(b) shows the center frequency with R. When determining the initial values of R and L, the parameters of the w_1 and b_1 can be adjusted to achieve better impedance matching.

The proposed SISL 180° coupler has been fabricated by the PCB manufacturing process and measured. The measured return loss is better than 15 dB and the

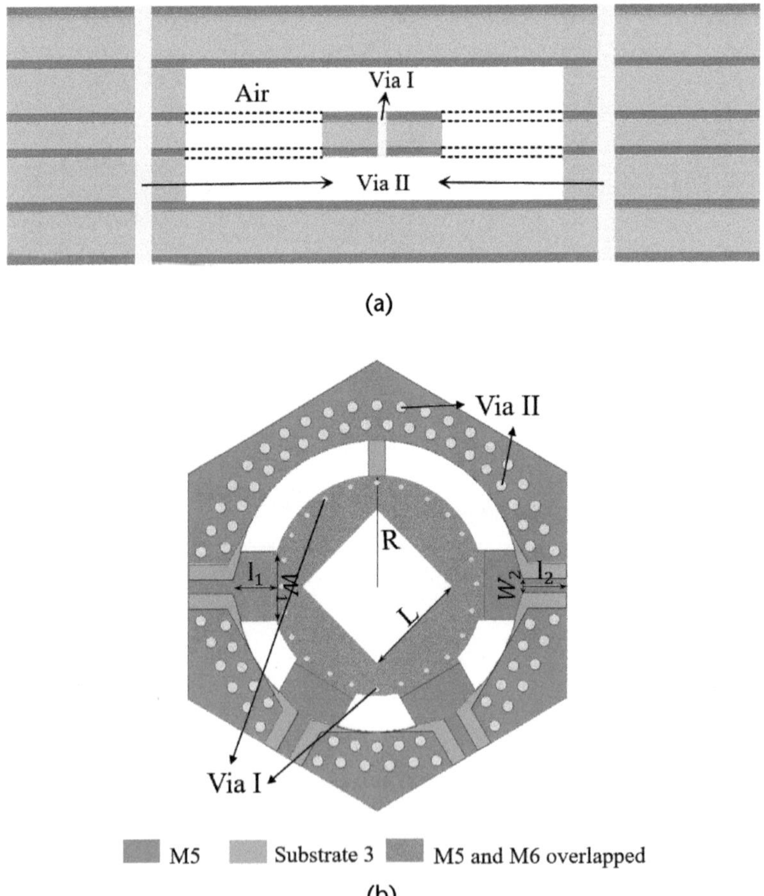

Figure 4.33 (a) Cross section view of DSISL with patterned substrate [15]. (b) Planar view of core circuit on M5 and M6, of the proposed low loss 180° coupler based on double-sided SISL with patterned substrate [15].

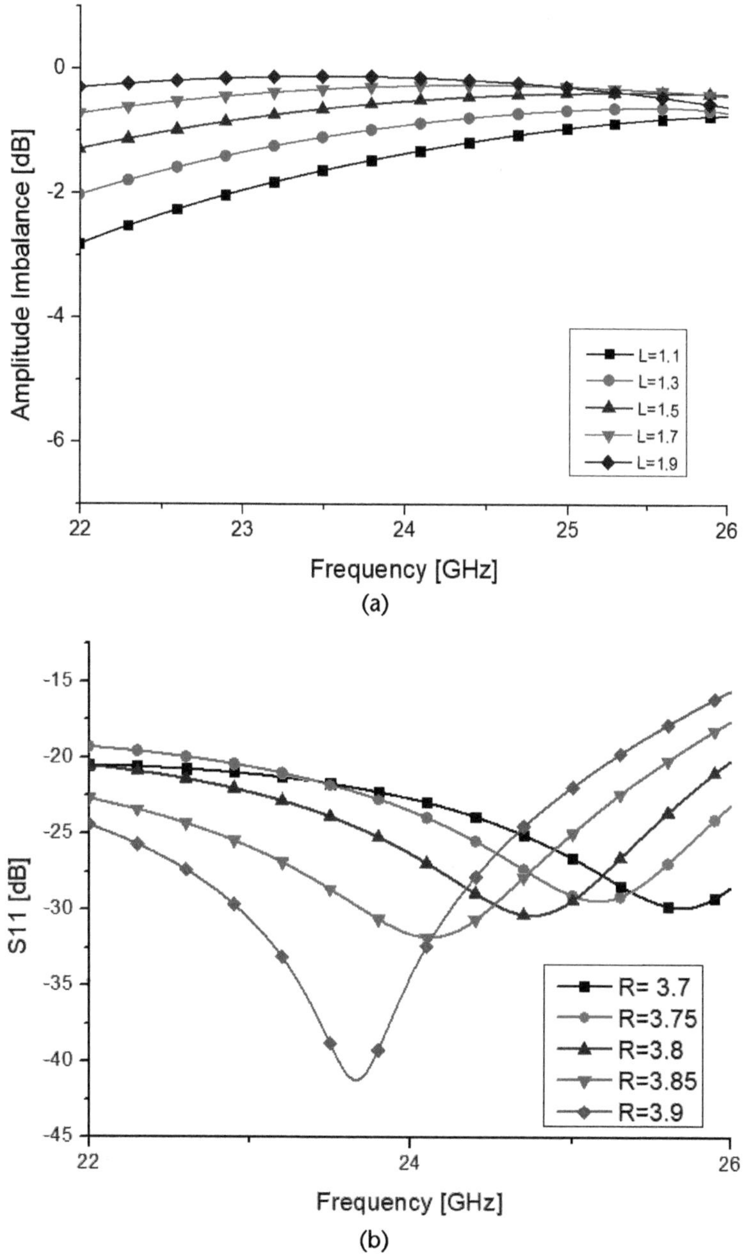

Figure 4.34 Simulated results of: (a) amplitude imbalance responses of the 180° coupler for different side length of the rectangle slot *L* [15], and (b) frequency responses of the 180° coupler for different *R* [15].

isolation is better than 22 dB from 22 GHz to 26 GHz. At a center frequency of 24 GHz, the measured insertion loss is only about 0.38 dB. In the range of 22 GHz to 25.5 GHz, the measured amplitude imbalance is less than ±0.5 dB, and the phase difference between the two outputs is 180° ± 5° and 0° ± 3.5° when operating in noninverting and inverting, respectively. The 180° coupler of the double-sided SISL can easily control the coupling coefficient through a rectangular groove and has the advantages of low loss and self-packaging, which has broad application prospects.

4.2.6 Lange Coupler on SISL Platform

The Lange coupler [16] was proposed by Lange in 1969. Conventional parallel coupled line couplers are difficult to achieve strong coupling [17]. The Lange coupler uses several parallel lines to increase the overall coupling degree, making it easier to achieve a 3-dB coupling ratio.

As shown in Figure 4.35 [17], Lange couplers can be divided into folded couplers and unfolded couplers. The folded Lange coupler has the coupled port and the through port on the same side, making it particularly suitable to achieve a compact circuit layout. Lange couplers are commonly found in on-chip circuit designs but are difficult to implement on PCBs. Because achieving a tight 3-dB coupling requires very narrow line widths and spacings, often exceeding the range allowed by PCB manufacturing precision. In addition, the traditional methods of connecting the spaced conductors using bonding wires or air bridges pose high demands on

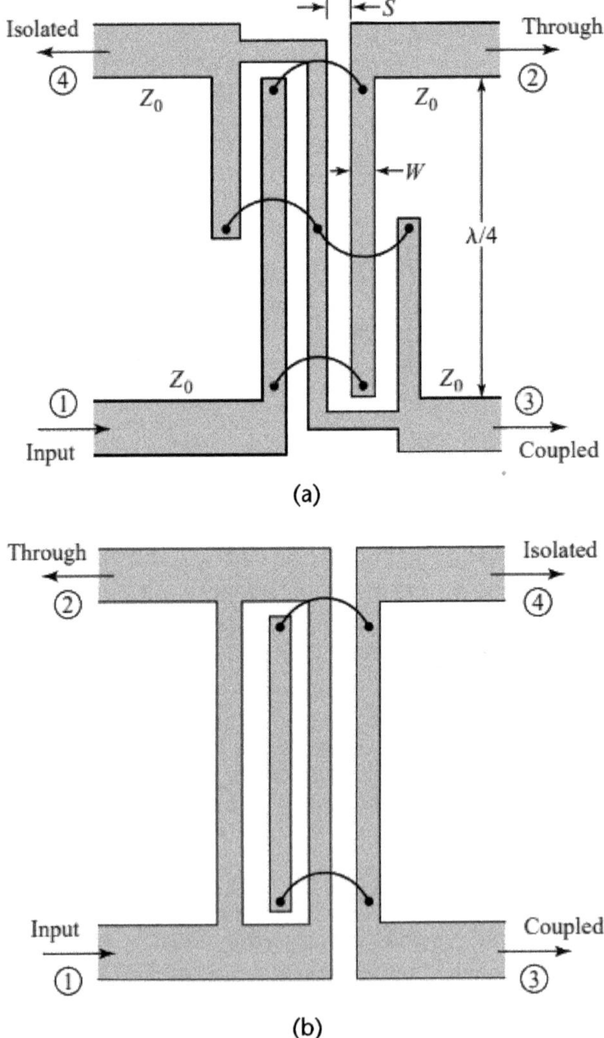

Figure 4.35 The Lange coupler [17]: (a) the folded Lange coupler, and (b) the unfolded Lange coupler.

the processing technology and are difficult to reproduce. Therefore, scholars have proposed some methods to solve this problem [18, 19]. In [18], it combines CPW with microstrip lines to increase the coupling line width and spacing and adjusts the position of the CPW ground plane to regulate the Lange coupler's coupling line width and spacing. Reference [19] used DGS structures at the coupling locations to relax the requirements for coupling line width and spacing, but the use of bonding wires cannot be avoided. Consequently, an SISL multilayer structure is adopted, using via bridges to avoid the use of bonding wires and air bridges, and selecting an appropriate SISL cavity height to increase the Lange coupler's coupling line width and spacing, making it suitable for PCB processing, as detailed in the following analysis.

Figure 4.36 shows a 3D view of the proposed Lange coupler based on the SISL structure [20]. As depicted in Figure 4.36, the SISL is composed of five stacked substrates (Sub1, Sub2, Sub3, Sub4, and Sub5), with metal layers covering the top and bottom surfaces of each substrate. Sub1, Sub2, Sub4, and Sub5 choose FR4 material, with characteristics parameters: $\varepsilon_r = 4.4$, $\tan\delta = 0.027$, and thicknesses of 0.6 mm, h_0, h_0, and 0.6 mm, respectively. Sub3 chooses Rogers5880 ($\varepsilon_r = 2.2$, $\tan\delta = 0.0009$) with a thickness of 0.254 mm.

The main circuit design of the Lange coupler is on the top and bottom metal layers G_5 and G_6 of Sub3, while Sub2 and Sub4 are hollowed out to form air cavities, and Sub1 and Sub5 act as covers. The designed plate through via holes, together with G_2 and G_9 metal layers build an electromagnetic shielding box. The core circuit of the Lange coupler is embedded inside the constructed electromagnetic shielding box. It can help reduce electromagnetic radiation and achieve good electromagnetic shielding.

In addition, two spaced conductive leads are connected by metal bridges constructed through metal strips and via holes, thereby avoiding the use of bonding wires and air bridges, as shown in Figure 4.36. The folded Lange coupler essentially belongs to a four-wire crossover coupling. As known from the literature [17], the even-mode impedance Z_{4e}, odd-mode impedance Z_{4o}, and coupling coefficient k of the Lange coupler satisfy the following relationship:

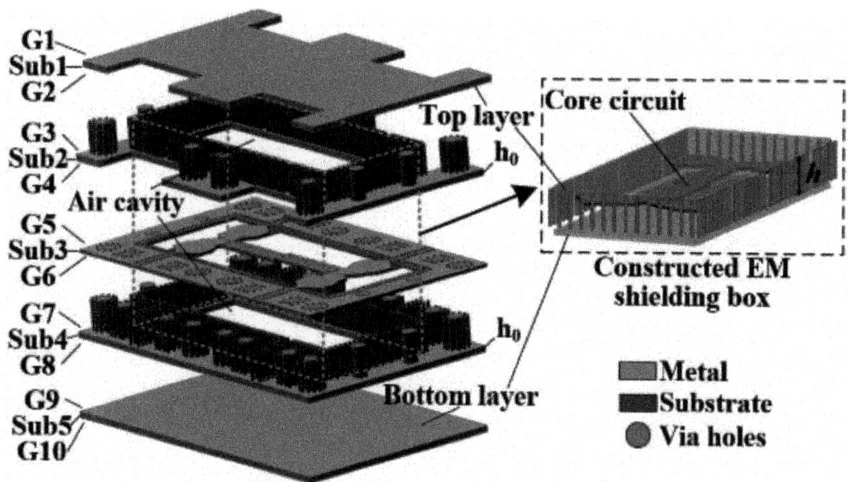

Figure 4.36 The 3D view of the SISL Lange coupler [20].

$$k = \frac{(Z_{4e} - Z_{4o})}{(Z_{4e} + Z_{4o})} \tag{4.38a}$$

$$Z_{4e} = \frac{Z_{2e}(Z_{2o} + Z_{2e})}{(3Z_{2o} + Z_{2e})} \tag{4.38b}$$

$$Z_{4o} = \frac{Z_{2o}(Z_{2o} + Z_{2e})}{(3Z_{2e} + Z_{2o})} \tag{4.38c}$$

where Z_{2e} and Z_{2o} are the even-mode and odd-mode impedances of the planar coupled-line structure.

As the thickness of the metal layer of the adopted PCB is small compared to the substrate thickness, it can be neglected. Furthermore, except for the part that provides support and metal coverage, the remaining suspended substrate layer Sub3 is entirely cut out. Thus, the SISL structure satisfies the assumption made by Cohn in deriving the impedance formula for air stripline parallel coupled lines (i.e., Z_{2e} and Z_{2o}) satisfy the impedance formula of parallel coupled striplines proposed by Cohn [21], as follows:

$$\begin{aligned} Z_{2e} &= \frac{\left(\dfrac{94.15}{\sqrt{\varepsilon_r}}\right)}{\left\{\dfrac{w}{h} + \dfrac{\ln 2}{\pi} + \ln\left[1 + \tanh\left(\dfrac{\pi}{2}\dfrac{s}{h}\right)\right]/\pi\right\}} \\ Z_{2o} &= \frac{\left(\dfrac{94.15}{\sqrt{\varepsilon_r}}\right)}{\left\{\dfrac{w}{h} + \dfrac{\ln 2}{\pi} + \ln\left[1 + \coth\left(\dfrac{\pi}{2}\dfrac{s}{h}\right)\right]/\pi\right\}} \end{aligned} \tag{4.39}$$

Here, w is the coupling line width, s is the coupling line spacing, h is the thickness of the filling substrate, and, in this case, h equals 0.254 mm + $2h_0$, and ε_r is the relative dielectric constant of the filling substrate.

For the SISL structure, the filled substrate is air, so $\varepsilon_r = 1$. When $\varepsilon_r = 1$, assume that s is 0.1 mm, the relationship between the substrate thickness h and the coupling coefficient k is given in Figure 4.37(a). It can be seen that k increases when h increases, and k decreases when w increases. The relationship between substrate thickness h and the even-mode/odd-mode impedance (Z_{4e}, Z_{4o}) is plotted in Figure 4.37(b). As seen from Figure 4.37(b), when h increases, the value of Z_{4e} increases, and the value of Z_{4o} decreases. When w increases, both the values of Z_{4e} and Z_{4o} decrease. So the coupling coefficient k and even-mode/odd-mode impedance (Z_{4e}, Z_{4o}) can be adjusted by changing the substrate thickness h. When $\varepsilon_r = 1$, $k = 0.707$, the relationship between substrate thickness h and coupled line gap s is shown in

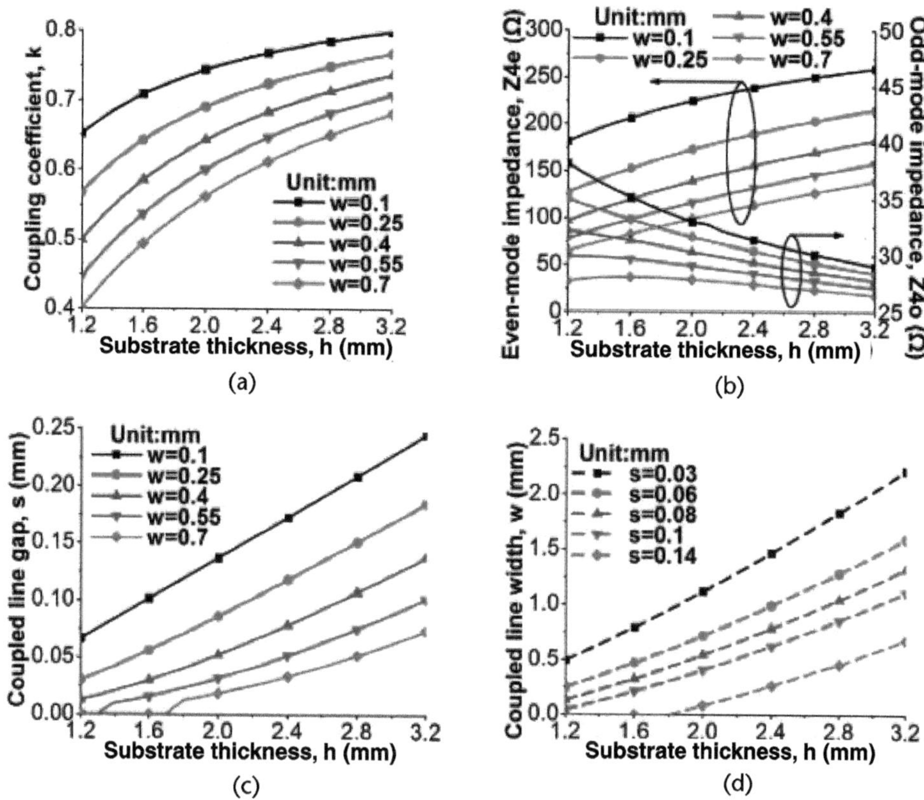

Figure 4.37 (a) Coupling coefficient k as a function of h when $\varepsilon_r = 1$, $s = 0.1$ mm [20]; (b) even-mode/odd-mode impedance (Z_{4e}, Z_{4o}) versus h when $\varepsilon_r = 1$, $s = 0.1$ mm [20]; (c) for a fixed coupling coefficient $k = 0.707$, the coupled-line gap s versus h when $\varepsilon_r = 1$ [20]; and (d) for a fixed coupling coefficient $k = 0.707$, the coupled-line width w versus h when $\varepsilon_r = 1$ [20].

Figure 4.37(c). It can be seen that s increases when h increases, and s decreases when w increases. The relationship between the substrate thickness h and the coupled line width w is plotted in Figure 4.37(d). It can be seen that w increases when h increases, and w decreases when s increases. Therefore, by properly selecting a larger substrate thickness h, the coupled line width w and coupled line spacing s can be increased for ease of PCB fabrication.

According to the relationship curves in Figure 4.37, when the gap is larger than 0.1 mm, the standard PCB process can be used for fabrication. Therefore, considering the available FR4 thickness, we choose $h \approx 2.7$ mm, that is, $h_0 = 1.2$ mm. According to Figure 4.37, when the desired coupling coefficient k is 0.707 (i.e., 3-dB coupling), the coupled line width w should be 0.4 mm, and the coupled line gap s should be 0.1 mm. As compared with the traditional Lange coupler with line width 0.11 mm and line gap 0.08 mm in [16], the proposed SISL Lange coupler increased the coupled line width and spacing. Then we can see from Figure 4.37(b) that $Z_{4e} = 166.65\Omega$ and $Z_{4o} = 28.69\Omega$.

The minimum radius of the via holes implemented on PCB is 0.25 mm; considering the fabrication limitation, we chose $w = 0.55$ mm. The SISL Lange coupler is optimized with the assistance of electromagnetic simulation software. To improve the ports matching, three chamfered lines are added at the input and output, respectively. The final parameter values are as follows: $w_1 = 2.2$, $w_2 = 3.8$, $w_3 = 5.8$, $d_1 = 1.7$, $d_2 = 2.2$, $d_3 = 1.5$, $d_4 = 0.8$, $d_5 = 1.95$, $d_6 = 1.3$, $d_7 = d_9 = 1.85$, $d_8 = 3.15$, and $d_{10} = 22.98$ (units: mm).

As Figure 4.36 shows, the designed Lange coupler is embedded inside the air cavity of the multilayer SISL. The electric field is mostly distributed inside the air rather than in the substrate, and the dielectric loss is dramatically reduced. To further reduce the dielectric loss, the suspended Sub3 is cut out in pattern. The simulated loss comparison of the SISL Lange coupler with and without patterned substrate is shown in Figure 4.38(a). It can be seen that the SISL Lange coupler with patterned substrate has a loss reduction of about 2% compared with that without patterned substrate.

To measure the SISL Lange coupler, four transitions (as marked in Figure 4.36) from the SISL to conducted-back coplanar waveguide (CBCPW) are used to connect the SISL Lange coupler to SMA connectors, and the loss of transitions will be calibrated.

The proposed 3-dB SISL Lange coupler is fabricated and measured. Figure 4.38(b) gives the photograph, the measured and simulated S-parameters of the patterned SISL Lange coupler. The measured return loss and isolation are both better

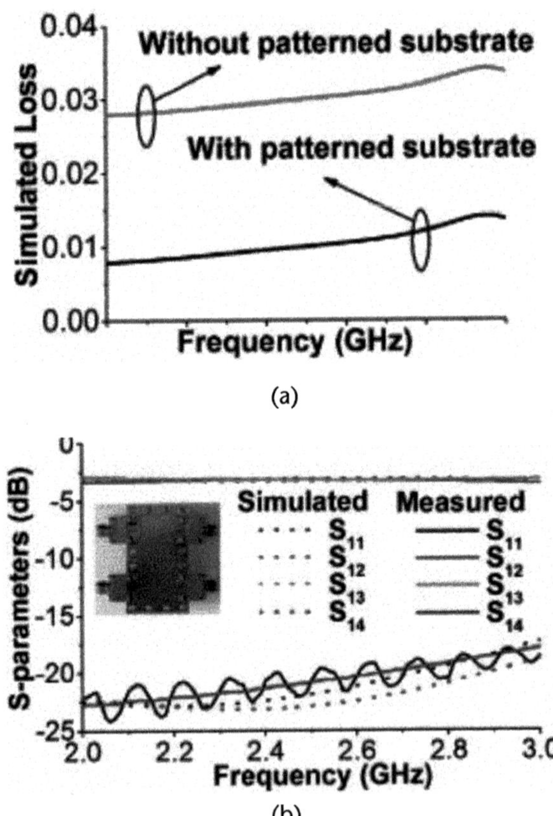

Figure 4.38 (a) The simulated loss of the Lange coupler with and without patterned substrate [20], and (b) simulated and measured S-parameters of the proposed 3-dB SISL Lange coupler [20].

than 17.8 dB from 2 GHz to 3 GHz. The measured S_{21} and S_{31} are 3.15 ± 0.41 dB and −3.16 ± 0.35 dB, respectively. The measured phase difference between the through port and the coupled port is 91° ± 1°.

The SISL Lange coupler has superior performance compared with other published couplers based on coupled lines. Compared to the couplers in [22–24], its size is comparable to [22], and it has a smaller coupling-line spacing than [24]. Above all, benefiting from the low radiation loss of the SISL and the substrate removal method, the SISL-based Lange coupler has the lowest loss. By appropriately increasing the height of the SISL cavity, the designed Lange coupler achieves the largest and easy-to-fabricate coupling line width for PCB processing. Moreover, the SISL Lange coupler has the closest 3-dB tight coupling and is the only one with a self-packaging feature.

Based on the multilayer SISL structure, a 3-dB tightly coupled Lange coupler is designed in this section, which solves the problem that it is difficult to achieve 3-dB tight coupling on the PCB boards due to the limitation of processing accuracy.

4.2.7 Patch Coupler on the SISL Platform

4.2.7.1 A Self-Packaged Patch Coupler

Patch elements have been widely studied and applied in patch coupler designs [25–31] due to their low conductor loss and ease of fabrication. Reference [25] presented a design procedure for a 3-dB orthogonal patch coupler, which obtains the circuit structure dimensions associated with its operating frequency by using an external two-port matching network. A dual-band patch coupler structure with an arbitrary power distribution ratio is reported in [28], which generates two operating frequency bands by loading separated ring resonators and obtains different power distribution ratios by adjusting the circuit structure dimensions, and an equivalent circuit is provided to explain and illustrate the working mechanism of the circuit. However, patch structures usually have a larger size. A rectangular patch coupler reported in [29] reduces the circuit size by etching patterns on the ground plane, achieving a maximum reduction of 27.7% compared to traditional rectangular patch couplers.

Nonetheless, patch structures tend to radiate energy outward, resulting in significant radiation loss. In [31], a cross-slot is loaded in the middle of the square patch, and a substrate with a higher dielectric constant is used to simultaneously reduce the circuit size and radiation loss. However, in practical applications, the available substrate's dielectric constant is usually limited and generally not high. Additionally, an extra metal enclosure is required to further reduce the circuit's radiation loss, which is not conducive to miniaturization and light weighting of the circuit.

Based on SISL platform, the metalized via holes surrounding the cavity and upper and lower metal ground layers provide an electromagnetic shielding environment to reduce the radiation loss of the internal circuit.

In [32], a 3-dB patch coupler is used as an example to illustrate the low-loss design method for SISL patch element circuits. The SISL patch coupler consists of 5 substrate layers. Substrate 2 and Substrate 4 are hollowed out to form air cavities, and the third substrate layer is also partially removed to reduce dielectric loss.

As shown in Figure 4.39, the patch coupler adopts a square patch structure, and the cross-slot structure designed in the middle can reduce the circuit area [15].

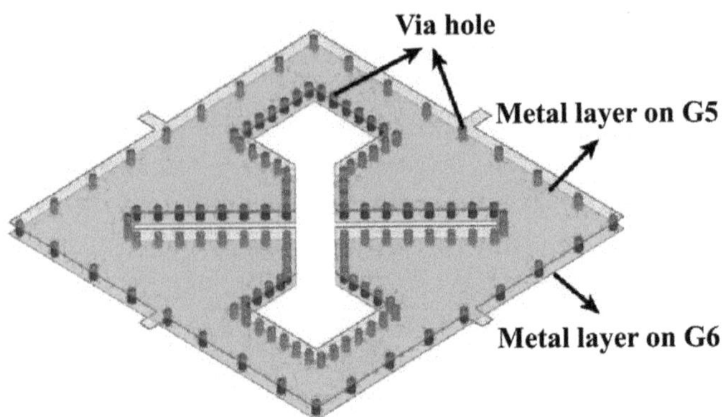

Figure 4.39 Double-layer metal interconnect patch coupler structure.

When a signal is fed from Port 1, Ports 2 and 3 output signals, while Port 4 is in isolation. The working frequency of the patch coupler is mainly determined by its side length. Adjusting the two asymmetric cross-slot structures can achieve a certain power distribution ratio (i.e., the coupling degree of the coupler).

Circuit loss generally includes three parts: dielectric loss, conductor loss, and radiation loss. The following will detail how to reduce loss in these three aspects.

1. As shown in Figure 4.39, the patch shapes on metal layers 5 and 6 are identical. Metalized via holes are designed at the edge of the patch, connecting the metal of layers 5 and 6, thereby helping to reduce conductor loss.
2. As shown in Figure 4.40, for the cross-shaped substrate slot excavated at the patch center, the slot width is slightly larger, making it easier to remove the substrate in this area using existing PCB processes. For common mechanical cutting processes, the slot width should be greater than 0.6 mm. While removing as much excess substrate as possible from the noncircuit part of Substrate 3, it is necessary to ensure the required physical connections and

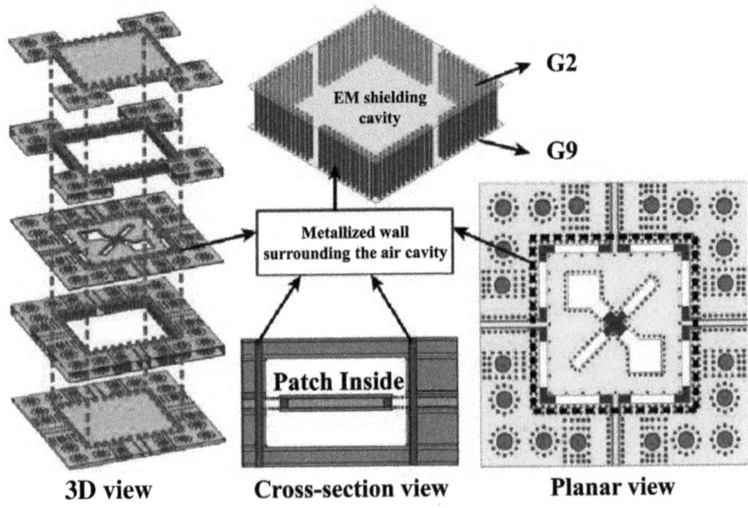

Figure 4.40 Electromagnetic shielding environment of SISL substrate-removed patch coupler.

maintain the substrate's mechanical strength. Furthermore, as shown in Figure 4.39, when metalized via holes connect the two layers of circuits along the patch periphery, the electromagnetic field distribution in the internal substrate region also decreases, thereby reducing dielectric loss to some extent.

3. As shown in Figure 4.40, numerous metalized via holes are placed around the cavity, forming an approximate metal wall. These via holes, together with ground layers of metal layers 2 and 9, provide electromagnetic shielding characteristics, causing almost no radiation from the internal circuit, and thus resulting in low radiation loss.

In addition, the SISL transition structure used in this circuit is different from that in [33], as shown in Figure 4.40. The opening width w_3 of the suspended line circuit within the cavity connected to the external transition is designed to be relatively small, which helps to provide better electromagnetic shielding characteristics. Moreover, the gap s of the CPW at the port is much smaller than the design in [33], which can better suppress parasitic higher-order modes, making the transition structure applicable to higher-frequency bands.

Sub1, Sub2, Sub4, and Sub5 in this patch coupler choose FR4 with thicknesses of 0.6 mm, 2 mm, 2 mm, and 0.6 mm, respectively. Sub3 chooses Rogers 5880 with thicknesses of 0.254 mm. Since this patch coupler uses a cross-slot structure, its field distribution is relatively complex, so electromagnetic simulation software is required for simulation and optimization. In this design, the center frequency is set to 6 GHz, and the final dimensional parameters are as follows:

a = 4.02 mm, b = 5.4 mm, c = 1.95 mm, e = 17.2 mm, w_1 = 1.56 mm, w_2 = 1.04 mm, w_3 = 1.5 mm, w_4 = 3.56 mm, w_5 = 3.04 mm, w_6 = 1 mm, w_{01} = 0.63 mm, w_{02} = 0.6 mm, s = 0.18 mm, l_1 = 6 mm, l_2 = 1.5 mm, sl_1 = 5.6 mm, sl_2 = 4 mm.

Figure 4.41 presents the simulation and measured results, which are in good agreement. The loss of the connectors and transition structures have been subtracted from the final loss. At the center frequency of 6 GHz, the measured S_{21} and S_{31} are −3.095 dB and −3.218 dB, respectively, with an insertion loss of about 0.15 dB. From 4.8 GHz to 6.8 GHz, the measured return loss and isolation are both better than 10 dB, with maximum values better than 30 dB. The measured phase difference is 90° ± 1° in the frequency range of 5.5 GHz to 6.6 GHz, and the measured amplitude imbalance is less than 0.6 dB. The overall size of the fabricated patch coupler is 19.2 × 19.2 mm.

Through experimental measuring, the low-loss characteristics of the proposed SISL substrate-removed patch coupler are verified. Three comparative examples are as follows:

- *Example 1:* The patch coupler is not packaged and uses a single-layer metal layer.
- *Example 2:* The coupler is packaged based on the SISL multilayer platform and uses a single-layer metal layer.
- *Example 3:* The coupler is packaged based on the SISL multilayer platform, and patch structures are designed on metal layers 5 and 6, which are connected using metalized via holes.

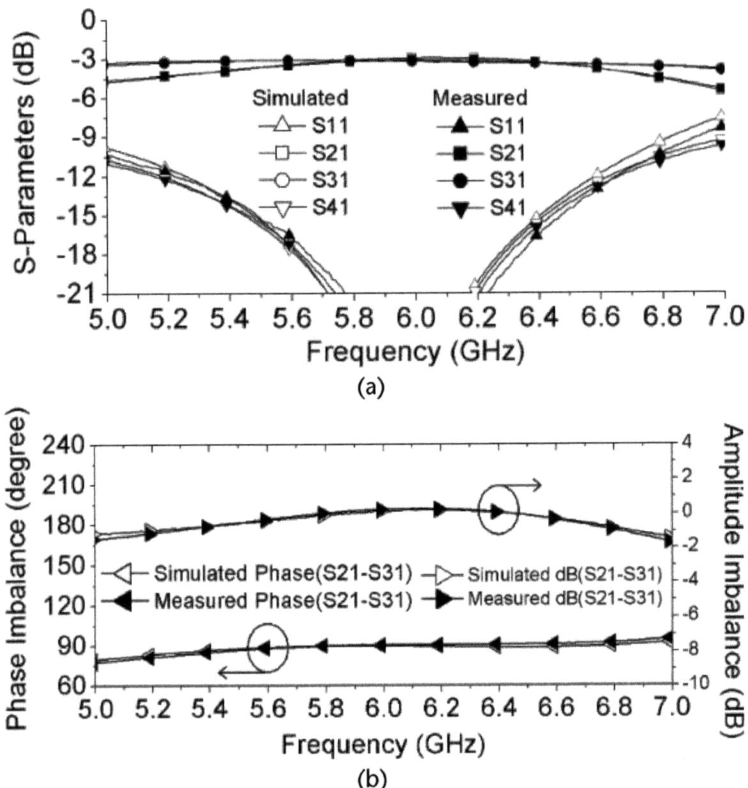

Figure 4.41 Simulated and measured results of the proposed SISL patch coupler with low loss [32]: (a) scattering parameters, and (b) phase imbalance and amplitude imbalance.

After measuring and using (4.40) to calculate the loss percentage, the following conclusions are obtained. First, compared to Example 1, the measured loss of Example 2 decreased by 76%, mainly due to reduced radiation loss. Second, compared to Example 2, the measured loss of Example 3 decreased by 20%, mainly due to reduced conductor loss and dielectric loss. Finally, compared to Example 3, the substrate-removed SISL patch coupler has a measured loss reduction of 35%, primarily due to reduced dielectric loss. Therefore, by packaging the patch elements in the SISL cavity, using metalized via holes to connect the double-layer patches, and performing substrate removal on Substrate 3, the total loss compared to Example 1 was reduced by 87.5%. Thus, the SISL patch coupler using this method exhibits extremely low-loss characteristics. The loss is expressed as a percentage in (4.40). It can be calculated that this design has a low loss of 0.032%.

$$\text{Loss} = 1 - |S_{11}|^2 - |S_{21}|^2 - |S_{31}|^2 - |S_{41}|^2 \tag{4.40}$$

4.2.7.2 A Low-Loss Patch Coupler with Arbitrary Coupling Coefficient

A structure of an SISL patch coupler with arbitrary coupling degrees is also proposed, using the same multilayer materials and thicknesses as the 3-dB patch coupler. This patch coupler also adopts the substrate removal method and employs double-layer patches interconnected to reduce loss, as shown in Figure 4.42. It employs an asymmetric square patch structure, with two side lengths of a and $a - 2d$, where a is the longer outer perimeter side length, and $a - 2d$ is the shorter inner perimeter side length. In this design, Port 1 is the input port, Port 2 is the coupling port, Port 3 is the through port, and Port 4 is the isolation port. By controlling the difference of the side length between the inner and outer perimeters, the coupling coefficient of the patch coupler can be adjusted (i.e., the proportion of energy coupled from Port 1 to Port 2). In addition, four identical slot structures are set at the center of the square patch, characterized by dimensions w_2 and w_3. These slot structures help the patch coupler achieve better port matching, specifically manifested as return loss and isolation.

First, we defined the ratio of the a to $(a - 2d)$ as

$$r_1 = \frac{a}{a - 2d} \tag{4.41}$$

The ratio of the slot area to the total square area is denoted by

$$r_2 = \frac{S_{slot}}{S_{square}} = \frac{4\left(w_2^2 - \left(w_2 - w_3\right)^2\right)}{a^2} \tag{4.42}$$

where S_{slot} is the area of the four slots and it is denoted by w_2 and w_3. S_{square} is the area of a square with side length of a.

Due to the internal slot structure in the SISL patch coupler, the electric field distribution is relatively complex. Therefore, an equivalent circuit is not provided here. Instead, a design guidance method is presented in conjunction with electromagnetic simulation.

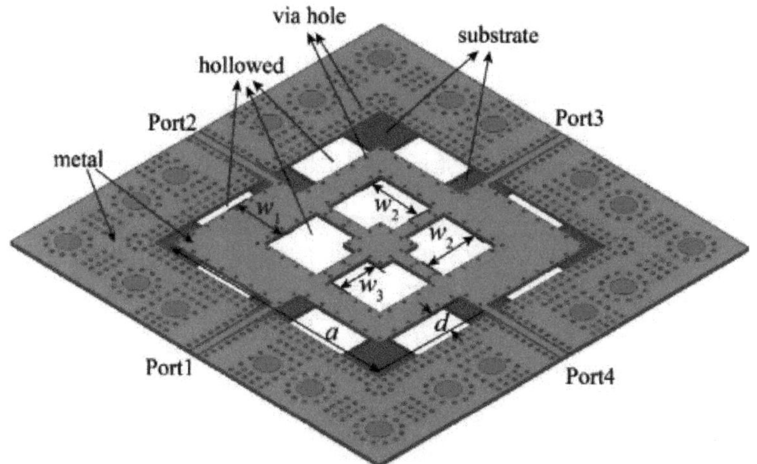

Figure 4.42 The 3D view of Substrate 3 of the proposed SISL patch coupler with arbitrary coupling coefficient [34].

Ultimately, the design guidance steps for the SISL patch coupler are summarized as follows:

1. Select the center frequency. In this design and analysis, the center frequency is chosen as 6 GHz.
2. Determine the value of a. The value of a is approximately half the guided wavelength at the center frequency.
3. Determine the required coupling coefficient.
4. Determine the values of r_1 and r_2 based on the required coupling coefficient.
5. Calculate d and S_{slot} using (4.41) and (4.42).
6. Select the value of w_1. Since the variation of w_1 has little impact on the overall performance when a and d are determined, an arbitrary initial value can be selected for w_1.
7. Select the values of w_2 and w_3 based on S_{slot}.
8. Utilize electromagnetic simulation software and adjust the parameters of a, d, w_1, w_2, and w_3, and finally do the optimization.

According to the above design steps, arbitrary coupling coefficients between 3 dB and 10 dB can be achieved. Table 4.4 lists the parameter values corresponding to different coupling coefficients. It can be seen that when the coupling coefficient changes from 3 dB to 10 dB, the values of a and w_1 increase slightly, while the values of d, w_2, and w_3 decrease.

As validation, couplers with coupling coefficients of 3 dB, 6 dB, and 8 dB were fabricated and measured. At the center frequency of 6 GHz, for the 3-dB, 6-dB, and 8-dB couplers, the measured S_{21} and S_{31} are −3.21 dB and −3.43 dB, −5.96 dB and −1.59 dB, and −7.97 dB and −1 dB, respectively. The measured return loss and isolation at the center frequency are better than 25 dB.

The strongly coupled couplers have a wider bandwidth. The phase characteristics of all three couplers at the center frequency are close to the ideal 90° phase difference.

4.3 Magic-T on the SISL Platform

The Magic-T (Magic Tee) [35–45] is a four-port passive device that can be applied to power synthesis and distribution, balanced mixers, and power amplifiers, among

Table 4.4 Parameters of the SISL Patch Coupler for Arbitrary Coefficients

C (dB)	*a (mm)*	*d (mm)*	w_1 *(mm)*	w_2 *(mm)*	w_3 *(mm)*
3	21.4	1.1	2.03	8.11	6.77
4	21.7	0.85	2.05	8	5.98
5	22.4	0.7	2.28	7.72	4.46
6	23.1	0.61	2.9	6.85	2.32
7	23.15	0.45	3.04	6.83	1.85
8	23.2	0.35	3.05	6.73	1.28
9	23.33	0.27	3.57	6.55	0.73
10	23.36	0.18	3.86	6.37	0.55

other applications. Ring coupler [46] is a special form of Magic-T, but its frequency bandwidth is usually narrow. Magic-T circuits based on slotline-to-microstrip or slotline-to-CPW transition structures [38–44] have a wider frequency bandwidth and, compared to waveguide magic-T circuits [47], offer advantages such as smaller size and lighter weight. However, due to the larger radiation loss typically associated with slotline and its transition structures, the realized Magic-T circuits exhibit greater loss. In practical engineering applications, slotline circuits also require additional devices to support their circuit boards, leading to certain packaging challenges.

As shown in Figure 4.43(a), we first consider a TE_{10} mode incident at Port 1. The resulting E_y field lines, as shown in Figure 4.43(b), are odd-symmetric around

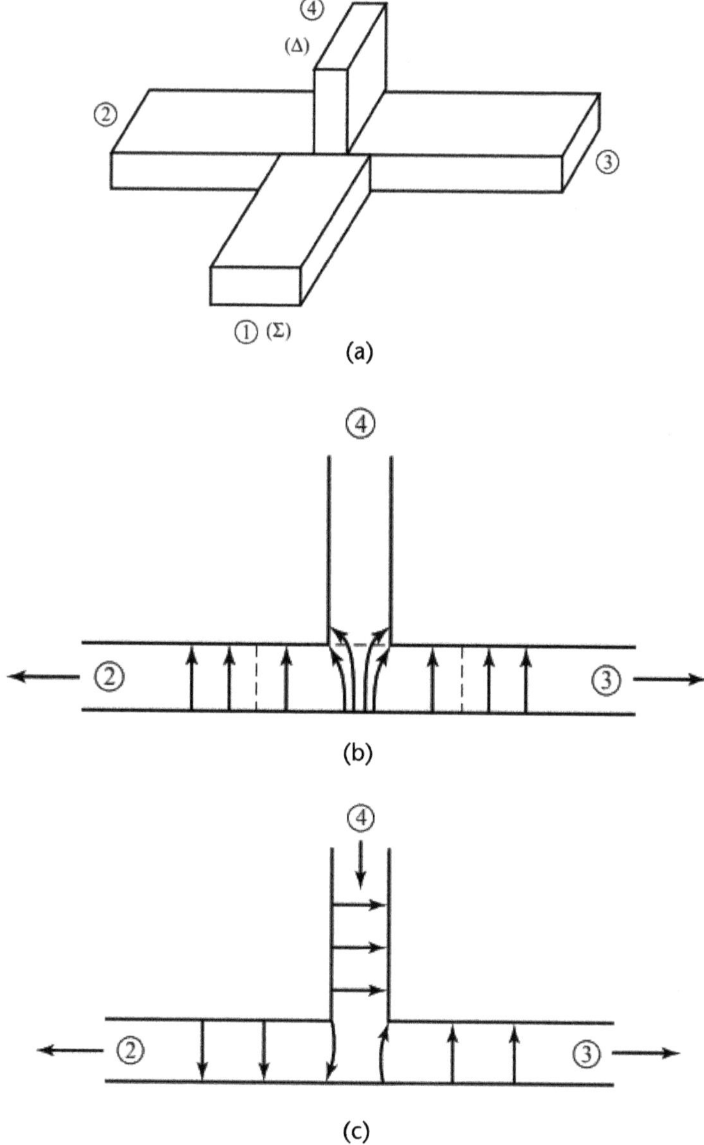

Figure 4.43 (a) A waveguide hybrid junction or magic-T [17], (b) electric field lines for incident wave at Port 1, and (c) electric field lines for incident wave at Port 4.

waveguide 4. Since the field lines of the TE_{10} mode in waveguide 4 are even-symmetric, there is no coupling between Ports 1 and 4. However, there is the same coupling for Ports 2 and 3, resulting in an in-phase, equal-split power division.

When the TE_{10} mode is input from Port 4, the field lines are shown in Figure 4.43(c). Similarly, due to symmetry (or reciprocity), there is no coupling between Ports 1 and 4. Ports 2 and 3 are excited equally by the input wave but have a 180° phase difference. In practice, tuning posts or irises are often used for matching; these matching components must be placed symmetrically to maintain the proper operation of the hybrid network.

To address the loss and packaging issues of Magic-T, a new SISL Magic-T structure is proposed. The main circuit components are located on metal layers 5 and 6, with the microstrip circuit structures on metal layer 5 and the slotline structures on metal layer 6. The outer perimeter of metal layer 6 remains connected to the external signal ground of the SISL. Metal layers 2 and 9 serve as the external ground for the overall circuit, creating an ideal electromagnetic shielding environment together with the metalized via holes around the cavity. This confines all the electromagnetic energy of the internal slotline circuit within the multilayer SISL, significantly reducing its radiation loss.

The planar view is shown in Figure 4.44(a), with a total of four ports [48]. Port 1 is the differential port (E-port). When a signal is fed from Port 1, it passes through the microstrip-to-slotline transition structure and then outputs equal amplitude, out-of-phase signals from Ports 2 and 3, while Port 4 is in isolation. When a signal is fed from Port 4 (H-port), it passes through the microstrip power divider structure and then outputs equal amplitude, in-phase signals from Ports 2 and 3, while Port 1 is in isolation. As can be seen from Figure 4.44(a), except for the transition structure used for measuring, the entire Magic-T circuit is designed within the cavity of the SISL, enabling self-packaging of the circuit.

Figure 4.44(b) shows the equivalent circuit of the SISL Magic-T, where the two microstrip-to-slotline transition sections are equivalent to n:1 and 1:1 transformer circuits, respectively. The microstrip feedline connected to Port 1, with a width of w_3, is represented by a microstrip line (MLN), with electrical length and impedance denoted as θ_1 and Z_1, respectively. A square microstrip open circuit with a side length of r_1 is formed across the slotline and is represented by a microstrip open circuit (MOC). A square slotline short circuit with a side length of r_2 is represented by a slotline short circuit (SSC1), while a rectangular slotline short circuit with lengths l_1 and l_2 is represented by SSC2. The uniform slotline transmission line with width w_s and length l_s is represented by SLN (slotline), with electrical length and impedance denoted as θ_{S1} and Z_{S1}, respectively. The electrical length and impedance of the microstrip connected to Ports 2 and 3 and leading to the middle slot transmission line are denoted as θ_2 and Z_2, while the electrical length and impedance of the microstrip line connected to port 4 are denoted as θ_3 and Z_3, respectively.

When designing the microstrip-to-slotline transition structure in this circuit, the initial parameters can be obtained by referring to the method in [49]. However, since the electric field distribution of the Magic-T circuit is relatively complex, and there are multiple solutions for the parameter values of the equivalent circuit structure in Figure 4.44(b), electromagnetic simulation is still relied upon to obtain the final size parameters. The focus of this section is to propose a method using the SISL

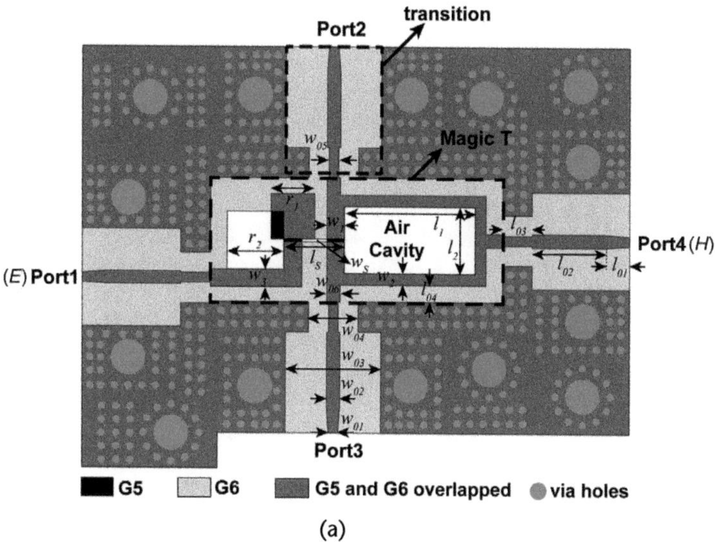

(a)

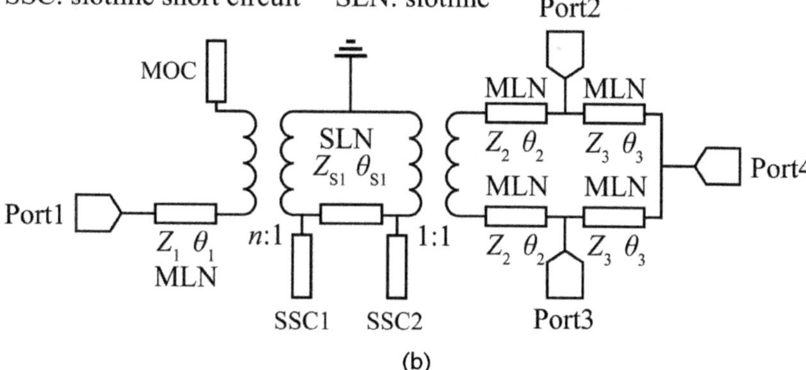

(b)

Figure 4.44 The proposed SISL Magic-T circuit [48]: (a) planar view of the SISL Magic-T, and (b) equivalent circuit of the SISL Magic-T.

platform to reduce slotline radiation loss, with the Magic-T circuit as an example for illustration and verification. For the equivalent circuit in Figure 4.44(b), a set of solutions is given: $Z_1 = 40\Omega$, $\theta_1 = 63°$, $Z_{S1} = 95\Omega$, $\theta_{S1} = 36°$, $Z_2 = 42\Omega$, $\theta_2 = 23°$, $Z_3 = 56\Omega$, $\theta_3 = 116°$, and $n = 0.73$.

Figure 4.45 shows the schematic diagrams of the electric field and current distributions when Ports 1 and 4 are excited, respectively.

1. When Port 1 is excited, as shown in Figure 4.45(a), the signal is fed along the microstrip line on the metal layer 5 and then couples to the slotline. Due to the inherent characteristics of energy transmission in the slotline, the electric field direction of the slot mode points from one metal side to the other. Next, the signal energy in the slotline couples to the microstrip structure located at the BB′ plane. Since the two branches of the microstrip lines at this location form a cascaded structure, when the signal energy converts from slot mode to microstrip mode, the AA′ plane can be regarded as a virtual microstrip

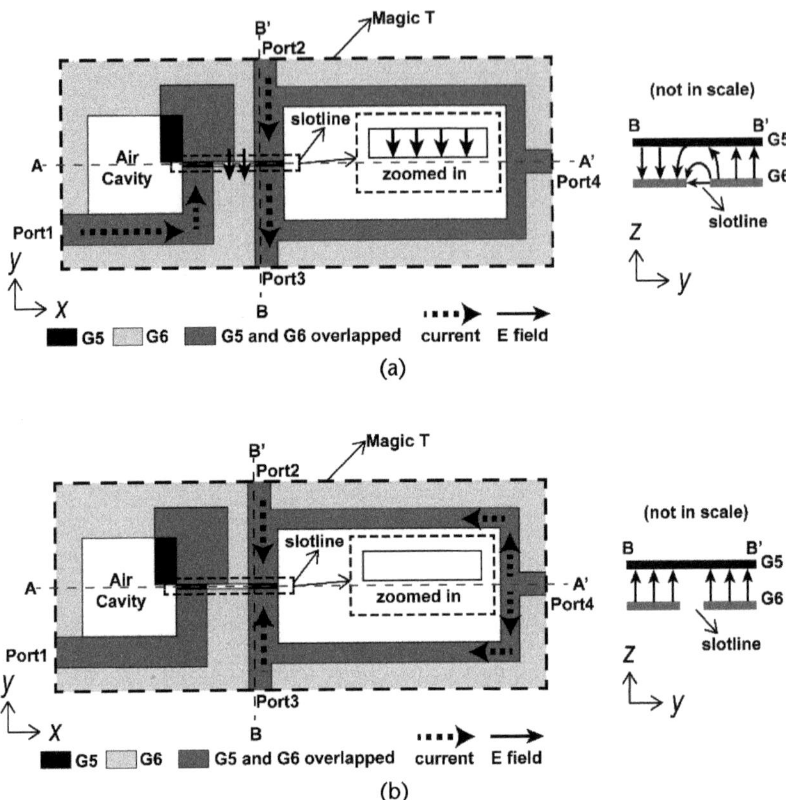

Figure 4.45 Electric field and current distribution of the proposed SISL Magic-T [48]: (a) out-of-phase case, and (b) in-phase case.

short-circuit plane. The electric field directions on both sides of the microstrip line are opposite; thus, Ports 2 and 3 will output signals with a 180° phase difference. The signals from the two microstrip branches connected to Port 4 have equal amplitude and opposite phase, which cancel each other out at Port 4, making it isolated.

2. When Port 4 is excited, as shown in Figure 4.45(b), the AA′ plane can now be regarded as a virtual microstrip open-circuit plane. The signal is fed along the microstrip line on metal layer 5 and flows with the same phase towards the symmetrical microstrip branches, resulting in Ports 2 and 3 outputting signals with a 0° phase difference. For the slotline structure at the BB′ plane, since the electric field directions on both sides are opposite, the slot mode will not be excited, making Port 1 isolated.

The proposed magic-T structure is ultimately designed and implemented based on the SISL platform. The multilayer board substrate materials and thicknesses used are the same as those in Section 4.2.7. In the design process, in order to reduce the overall circuit area, the gap between the metalized vias around the cavity and the internal Magic-T circuit is relatively small, as shown in Figure 4.44(a). This approximate electric wall will have a slight impact on the characteristics of the internal circuit, such as the characteristic impedance of the transmission line. Therefore, the influence of relevant boundary effects should be considered, and the overall circuit should be electromagnetically simulated and optimized. The final dimension parameters are as follows:

w_{01} = 0.58 mm, w_{02} = 0.78 mm, w_{03} = 5.8 mm, w_{04} = 3 mm, w_{05} = 0.6 mm, w_{06} = 0.78 mm, l_{01} = 1.5 mm, l_{02} = 4.5 mm, l_{03} = 1.8 mm, l_{04} = 1 mm, w_1 = 1.03 mm, w_2 = 0.69 mm, w_3 = 1.09 mm, r_1 = 2.67 mm, r_2 = 3.43 mm, l_1 = 8.01 mm, l_2 = 3.99 mm, l_S = 3.67 mm, and w_S = 0.16 mm.

The simulated and measured results of the SISL Magic-T are shown in Figure 4.46, and the simulation results are in good agreement with the measured results. In the frequency range of 5 GHz to 9 GHz, which is a 57% relative bandwidth, the return loss of all ports measured is better than 10 dB, the isolation between Ports 1 and 4 is greater than 35 dB, and the amplitude imbalance is within ±0.14 dB. The minimum insertion loss measured is 3.1 dB, making the structure very close to the

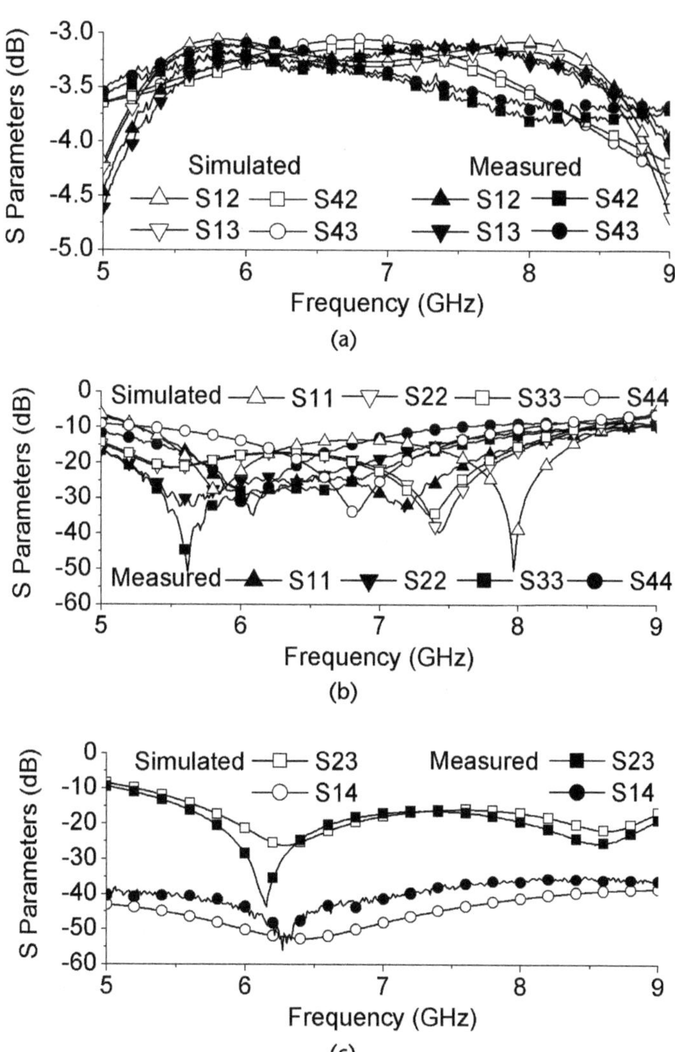

Figure 4.46 Simulated and measured insertion loss of the proposed self-packaged SISL Magic-T with compact size [48]: (a) insertion loss when Ports 1 and 4 are individually excited; (b) return loss for each port; and (c) port isolation.

ideal 3-dB power division. The loss here does not include the transition structure and connector losses. It should be noted that the transition structure and connectors are used only for measuring convenience. When multiple circuit modules are integrated on the same SISL platform, the internal modules can be directly connected without the need for additional transition structures.

The phase error of the SISL Magic-T is ±0.8° throughout the entire frequency band. The core circuit area of the SISL Magic-T is 0.18 $\lambda_g \times$ 0.43 λ_g, where λ_g represents the guided wavelength. Compared to the designs in the existing literature, the proposed SISL magic-T has the advantages of low loss, miniaturization, and self-packaging.

Acknowledgments

We would like to thank Shupeng Zhang, Shutao He, Chen Lei, Jingxin Zhao, and Haozhen Huang for their valuable support in this chapter.

References

[1] Feng, T., K. Ma, and Y. Wang, "A Miniaturized Bandpass Filtering Power Divider Using Quasi-Lumped Elements," *IEEE Transactions on Circuits Systems II*, Vol. 69, No. 1, January 2022, pp. 70–74.

[2] Feng, T., K. Ma, and Y. Wang, "A Self-Packaged Power Divider with Compact Size and Low Loss," *IEEE Transactions on Circuits Systems II*, Vol. 67, No. 11, November 2020, pp. 2437–2441.

[3] Luo, J., et al., "A Miniaturized and Self-Packaged Gysel Power Divider with Embedded Metamaterials in SISL Platform," *IEICE Electron. Express*, Vol. 18, No. 12, June 2021, pp. 20210196–20210196.

[4] Feng, T., K. Ma, and Y. Wang, "A Dual-Band Coupled Line Power Divider Using SISL Technology," *IEEE Transactions on Circuits Systems II*, Vol. 68, No. 2, February 2021, pp. 657–661.

[5] Qin, Z., et al., "An FR4-Based Self-Packaged Full Ka-Band Low-Loss 1:4 Power Divider Using SISL to Air-Filled SIW T-Junction," *IEEE Transactions on Components, Packaging and Manufacturing Technology*, Vol. 12, No. 3, March 2022, pp. 587–590.

[6] Pozar, D. M., *Microwave Engineering*, 2nd ed., New York: Wiley, 2004.

[7] Wang, Y., K. Ma, and S. Mou, "A High Performance Tandem Coupler Using Substrate Integrated Suspended Line Technology," *IEEE Microwave Wireless Component Letters*, Vol. 26, No. 5, May 2016, pp. 328–330.

[8] Wang, Y., K. Ma, and M. Yu, "A Low-Cost Substrate Integrated Suspended Line Platform with Multiple Inner Boards and Its Applications in Coupled-Line Circuits," *IEEE Transactions on Components, Packaging and Manufacturing Technology*, Vol. 10, No. 12, 2020, pp. 2087–2098.

[9] Wang, Y., K. Ma, and S. Mou, "A Compact Self-Packaged Lumped-Element Coupler Using Substrate Integrated Suspended Line Technology," *2016 IEEE MTT-S International Microwave Symposium (IMS)*, San Francisco, CA, May 2016, pp. 1–3.

[10] Wang, Y., K. Ma, and S. Mou, "A Transformer-Based 3-dB Differential Coupler," *IEEE Transactions on Circuits Systems I*, Vol. 65, No. 7, July 2018, pp. 2151–2160.

[11] Wang, Y., et al., "A Low Loss Branch Line Coupler Based on Substrate Integrated Suspended Line (SISL) Technology and Double-Sided Interconnected Strip Line (DSISL)," *2015 IEEE Asia-Pacific Microwave Conference (APMC)*, Nanjing, China, December 2015, pp. 1–3.

[12] Wang, Y., and K. Ma, "A Self-Packaged SISL Coupler Using Patterned Ground Structure

with Stopband," *2019 IEEE MTT-S International Wireless Symposium (IWS)*, Guangzhou, China, May 2019, pp. 1–3.

[13] Wang, Y., K. Ma, and S. Mou, "A Compact Branch-Line Coupler Using Substrate Integrated Suspended Line Technology," *IEEE Microwave Wireless Component Letters*, Vol. 26, No. 2, February 2016, pp. 95–97.

[14] Wang, Y., et al., "A Slow-Wave Rat-Race Coupler Using Substrate Integrated Suspended Line Technology," *IEEE Transactions on Components, Packaging and Manufacturing Technology*, Vol. 7, No. 4, April 2017, pp. 630–636.

[15] Ma, Y., K. Ma, and Y. Wang, "A Novel 180° Coupler Based on Double-Sided Substrate Integrated Suspended Line Technology with Patterned Substrate," *2018 48th European Microwave Conference (EuMC)*, September 2018, pp. 223–226.

[16] Lange, J., "Interdigitated Stripline Quadrature Hybrid," *IEEE Transactions on Microwave Theory and Techniques*, Vol. 17, No. 12, 1969, pp. 1150–1151.

[17] Pozar, D. M., *Microwave Engineering*, 4th ed., New York: John Wiley & Sons, 2012, pp. 95–345.

[18] Chiu, J. C., C. M. Lin, Y. H. Wang, "A 3-dB Quadrature Coupler Suitable for PCB Circuit Design," *IEEE Transactions on Microwave Theory and Techniques*, Vol. 54, No. 9, 2006, pp. 3521–3525.

[19] Zhang, J., et al., "An Interdigitated Coupler with Defect Ground Structure," *MTT-S International Microwave Workshop Series on Advanced Materials and Processes for RF and THz Applications (IMWS-AMP)*, Suzhou, 2015.

[20] Feng, T., K. Ma, and Y. Wang, "A Self-Packaged SISL 3-dB Lange Coupler Suitable for PCB Fabrication," *2019 IEEE Asia-Pacific Microwave Conference (APMC)*, Singapore, Singapore, December 2019, pp. 1083–1085.

[21] Cohn, S. B., "Shielded Coupled-Strip Transmission Line," *IEEE Transactions on Microwave Theory and Techniques*, Vol. 3, No. 5, 1955, pp. 29–38.

[22] Yoon, H. J., and B. W. Min, "Two Section Wideband 90° Hybrid Coupler Using Parallel-Coupled Three-Line," *IEEE Microwave and Wireless Components Letters*, Vol. 27, No. 6, 2017, pp. 548–550.

[23] Tseng, C. H., and Y. T. Chen, "Design and Implementation of New 3-dB Quadrature Couplers Using PCB and Silicon-Based IPD Technologies," *IEEE Transactions on Components, Packaging and Manufacturing Technology*, Vol. 6, No. 5, 2016, pp. 675–682.

[24] Chang, W. S., C. H. Liang, and C. Y. Chang, "Slow-Wave Broadside-Coupled Microstrip Lines and Its Application to the Rat-Race Coupler," *IEEE Microwave and Wireless Components Letters*, Vol. 25, No. 6, 2015, pp. 361–363.

[25] Page, M. J., and S. R. Judah, "A Flexible Design Procedure for Microstrip Planar Disk 3 dB Quadrature Hybrids," *IEEE Transactions on Microwave Theory and Techniques*, Vol. 38, No. 11, 1990, pp. 1733–1736.

[26] Zheng, S. Y., "A Compact Patch Quadrature Coupler with Enhanced Bandwidth and Harmonic Suppression," *2015 IEEE 4th Asia-Pacific Conference on Antennas and Propagation*, Kuta, Indonesia, 2015, pp. 429–430.

[27] Yang, J. P., and S. C. Shi, "A Fast Approach to the Design of Compact Couplers," *2015 Asia-Pacific Microwave Conference*, Nanjing, China, 2015, pp. 1–4.

[28] Zheng, S. Y., et al., "Dual-Band Hybrid Coupler with Arbitrary Power Division Ratios over the Two Bands," *IEEE Transactions on Components, Packaging and Manufacturing Technology*, Vol. 4, No. 8, 2014, pp. 1347–1358.

[29] Zheng, S. Y., et al., "Size-Reduced Rectangular Patch Hybrid Coupler Using Patterned Ground Plane," *IEEE Transactions on Microwave Theory and Techniques*, Vol. 57, No. 1, 2009, pp. 180–188.

[30] Jing, X., and S. Sun, "Design of Impedance Transforming 90 Degree Patch Hybrid Couplers," *2014 Asia-Pacific Microwave Conference*, Sendai, Japan, 2014, pp. 25–27.

[31] Sun, S., and L. Zhu, "Miniaturised Patch Hybrid Couplers Using Asymmetrically Loaded

Cross Slots," *Antennas Propagation IET Microwaves*, Vol. 4, No. 6, 2010, pp. 1427–1433

[32] Wang, Y., K. Ma, and S. Mou, "A Low Loss and Self-Packaged Patch Coupler Based on SISL Platform," *2017 IEEE MTT-S International Microwave Symposium (IMS)*, Honolulu, HI, June 2017, pp. 192–195.

[33] Li, L., et al., "A Novel Transition from Substrate Integrated Suspended Line to Conductor Backed CPW," *IEEE Microwave and Wireless Components Letters*, Vol. 26, No. 6, 2016, pp. 389–391.

[34] Wang, Y., and K. Ma, "Low Loss SISL Patch-Based Coupler with Arbitrary Coupling Coefficient," *2019 IEEE 23rd Workshop on Signal and Power Integrity (SPI)*, Chambéry, France, June 2019, pp. 1–4.

[35] Laughlin, G., "A New Impedance-Matched Wide-Band Balun and Magic Tee," *IEEE Transactions on Microwave Theory and Techniques*, Vol. 24, No. 3, 1976, pp. 135–141.

[36] Aikawa, M., and H. Ogawa, "Double-Sided MICs and Their Applications," *IEEE Transactions on Microwave Theory and Techniques*, Vol. 37, No. 2, 1989, pp. 406–413.

[37] Tokumitsu, T., S. Hara, and M. Aikawa, "Very Small Ultra-Wide-Band MMIC Magic T and Applications to Combiners and Dividers," *IEEE Transactions on Microwave Theory and Techniques*, Vol. 37, No. 12, 1989, pp. 1985–1990.

[38] Aikawa, M., and H. Ogawa, "A New MIC Magic-T Using Coupled Slot Lines," *IEEE Transactions on Microwave Theory and Techniques*, Vol. 28, No. 6, 1980, pp. 523–528.

[39] Fan, L., et al., "Wide-Band Reduced-Size Uniplanar Magic-T, Hybrid-Ring, de Ronde's CPW-Slot Couplers," *IEEE Transactions on Microwave Theory and Techniques*, Vol. 43, No. 12, 1995, pp. 2749–2758.

[40] Kim, J. P., and W. S. Park, "Novel Configurations of Planar Multilayer Magic-T Using Microstrip-Slotline Transitions," *IEEE Transactions on Microwave Theory and Techniques*, Vol. 50, No. 7, 2002, pp. 1683–1688.

[41] U-yen, K., et al., "Slotline Stepped Circular Rings for Low-Loss Microstrip-to-Slotline Transitions," *IEEE Microwave and Wireless Components Letters*, Vol. 17, No. 2, 2007, pp. 100–102.

[42] Henin, B., and A. Abbosh, "Wideband Hybrid Using Three-Line Coupled Structure and Microstrip-Slot Transitions," *IEEE Microwave and Wireless Components Letters*, Vol. 23, No. 7, 2013, pp. 335–337.

[43] U-yen, K., et al., "A Broadband Planar Magic-T Using Microstrip-Slotline Transitions," *IEEE Transactions on Microwave Theory and Techniques*, Vol. 56, No. 1, 2008, pp. 172–177.

[44] Bialkowski, M. E., and Y. Wang, "Wideband Microstrip 180 Hybrid Utilizing Ground Slots," *IEEE Microwave and Wireless Components Letters*, Vol. 20, No. 6, 2010, pp. 495–497.

[45] Feng, W., Q. Xue, and W. Che, "Compact Planar Magic-T Based on the Double-Sided Parallel-Strip Line and the Slotline Coupling," *IEEE Transactions on Microwave Theory and Techniques*, Vol. 58, No. 11, 2010, pp. 2915–2923.

[46] Eccleston, K. W., and S. H. M. Ong, "Compact Planar Microstripline Branch-Line and Rat-Race Couplers," *IEEE Transactions on Microwave Theory and Techniques*, Vol. 51, No. 10, 2003, pp. 2119–2125.

[47] He, Y., et al., "A Ka-Band Waveguide Magic-T with Coplanar Arms Using Ridge-Waveguide Transition," *IEEE Microwave and Wireless Components Letters*, Vol. 27, No. 11, 2007, pp. 965–967.

[48] Wang, Y., K. Ma, and S. Mou, "A Low-Loss Self-Packaged Magic-T with Compact Size Using SISL Technology," *IEEE Microwave Wireless Components Letters*, Vol. 28, No. 1, January 2018, pp. 13–15.

[49] Shuppert, B., "Microstrip/Slotline Transitions: Modeling and Experimental Investigation," *IEEE Transactions on Microwave Theory and Techniques*, Vol. 36, No. 8, 1988, pp. 1272–1282.

CHAPTER 5

SISL Baluns, Phase Shifters, and Butler Matrices

This chapter describes the passive components including the balun, phase shifter, and Butler matrices that are widely employed in RF circuits and microwave circuits. With the proposal of the SISL platform that brings the strengths of low loss, weak dispersion, and self-packaging, these passive devices are also being studied for new structures and excellent performance. Hence, this chapter will describe the development of these passive components on the SISL platform.

5.1 Balun on the SISL Platform

Balun [1–3] is a three-port component that contains one unbalanced port and two balanced ports, which achieves single-ended to differential signal transformation.

In RF and microwave circuits, scattering matrices are generally for characterizing the performance of components. According to the microwave circuit network theory, a three-port network can be expressed as follows:

$$\begin{bmatrix} V_1^- \\ V_2^- \\ V_3^- \end{bmatrix} = \begin{bmatrix} S_{11} & S_{12} & S_{13} \\ S_{21} & S_{22} & S_{23} \\ S_{31} & S_{32} & S_{33} \end{bmatrix} \begin{bmatrix} V_1^+ \\ V_2^+ \\ V_3^+ \end{bmatrix} \tag{5.1}$$

where V_i^+ and V_i^- denote the amplitudes of the incident and reflected voltage waves at port i, respectively. The scattering matrix of a balun can be expressed as:

$$[S] = \begin{bmatrix} S_{11} & S_{21} & S_{31} \\ S_{21} & S_{22} & S_{32} \\ S_{31} & S_{32} & S_{33} \end{bmatrix} \tag{5.2}$$

In the case of an ideal balun, its scattering matrix should satisfy the following condition:

$$S_{21}(\text{dB}) = -10\lg\left|S_{21}\right|^2 = 3 \text{ dB} \tag{5.3}$$

$$S_{31}(\text{dB}) = -10\lg|S_{31}|^2 = 3\text{ dB} \tag{5.4}$$

$$S_{21}(\text{dB}) = S_{31}(\text{dB}) \tag{5.5}$$

$$S_{21}(\text{deg}) - S_{31}(\text{deg}) = 180° \tag{5.6}$$

In this section, we will illustrate two types of baluns: lumped balun and Marchand balun.

5.1.1 Lumped Balun

At lower operating frequencies, the lumped balun is widely used because of its simple circuit structure and small size. Based on the SISL platform, the lumped balun has also progressed in research.

The compensated interdigital capacitor (CIDC) [4] is a novel interdigital capacitor that consists of two layers of cross-connected metal G5 and G6, with via holes for connecting the same terminals distributing the two layers of metal, as shown in Figure 5.1. W_c, L_c, and S_c denote the length, width, intersection gap, and distance to the other end of the finger, respectively.

The electric field distributed on the compensated interdigital capacitor is shown in Figure 5.2. The capacitance generated by the vertical electric field of different layers of metal is indicated by C_V, and the capacitance induced by the horizontal electric field of the same layer of metal is represented by C_L. Thus, when the finger pair of compensated interdigital capacitors have n, the sum capacitance can be given by

$$C_T = nC_V + 2(n-1)C_L \tag{5.7}$$

By extracting the capacitance parameters of the CIDC model, it can be seen that the total capacitance is dependent on W_c, L_c, and S_c. The capacitance density increases as W_c, S_c decreases, while L_c has little effect on it. The self-resonant frequency decreases as W_c, L_c increases and S_c decreases. Thus, the expected capacitance can be attained by selecting the proper dimensional parameters. Then, for the same dimension, the three capacitance models of parallel plate capacitor, single-layer interdigital capacitor, and CIDC are compared as shown in Table 5.1, which reveals that CIDC has the advantage of large capacitance: smaller area with the same capacitance compared to parallel plate capacitor, and larger capacitance density compared to single-layer interdigital capacitor.

Due to the advantages of the large capacitance of this structure, a lumped balun is implemented on SISL platform using the CIDC structure as shown in Figure 5.3, which is based on the principle of a modified second-order lattice balun. The equivalent circuit diagram is shown in Figure 5.3(b), where the basic structure of the second-order lattice balun consists of two lowpass T-cells and two highpass T-cells, consisting of capacitors C_1 and C_2, and inductors L_1 and L_2, whose values are given by the following equations:

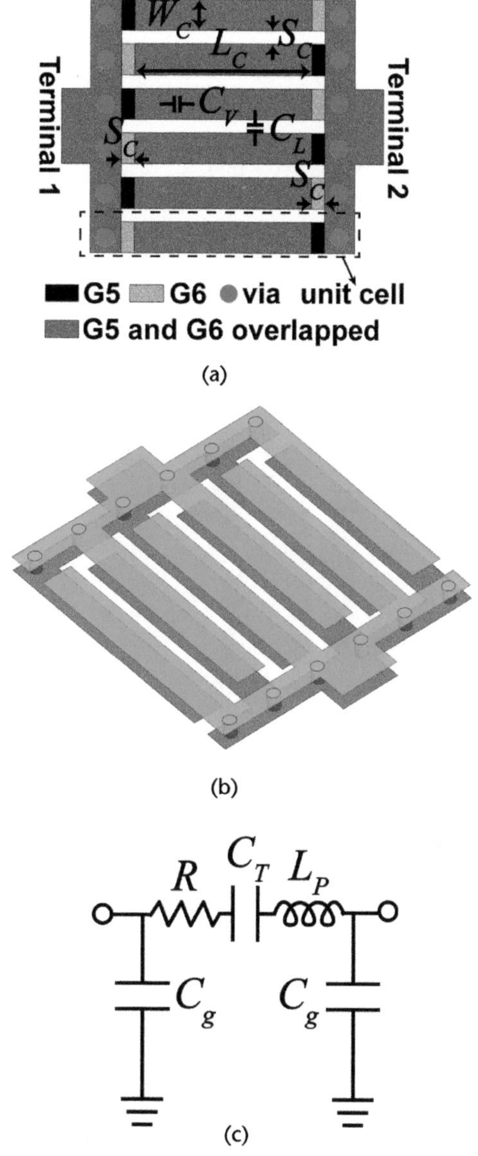

Figure 5.1 Structure of a compensated interdigital capacitor on the SISL platform [4]: (a) planar view, (b) 3D view, and (c) equivalent circuit.

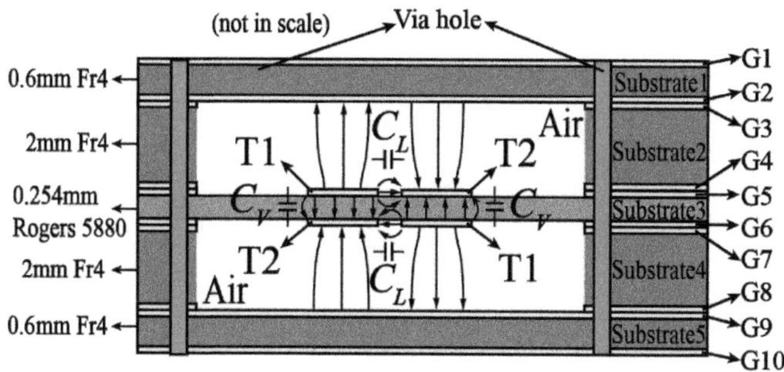

Figure 5.2 Electric field distribution of a compensated interdigital capacitor [4].

Table 5.1 Comparison of Capacitance Values of Three Capacitor Models

Reference	*Parallel-Plate Capacitor*	*Single-Layer Interdigital Capacitor*	*CIDC*
Capacitor	1.68 pF	1.85 pF	2.75 pF
Dimension	W_c = 0.3 mm, L_c = 0.4 mm, S_c = 0.1 mm, n = 10		

Source: [4].

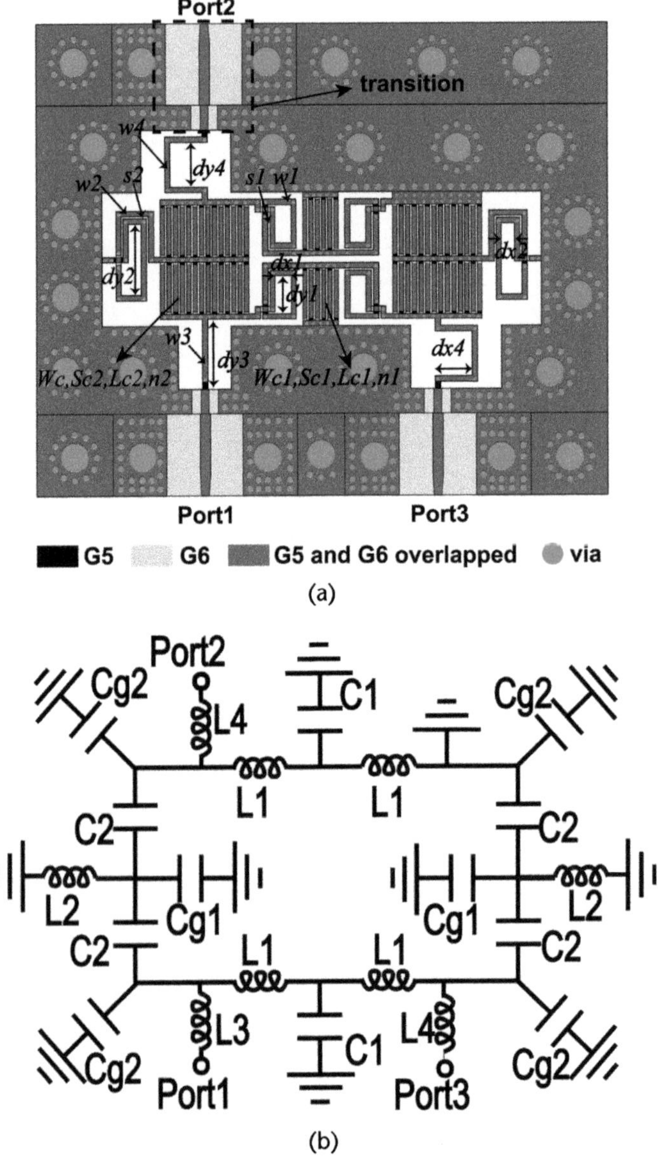

Figure 5.3 Structure of a lumped balun on the SISL platform [4]: (a) planar view, and (b) equivalent circuit.

$$C_1 = \frac{1}{mZ_0 \cdot n\omega_0} \tag{5.8}$$

$$L_1 = \frac{mZ_0}{n\omega_0} \tag{5.9}$$

$$C_2 = \frac{n}{mZ_0 \cdot \omega_0} \tag{5.10}$$

$$L_2 = \frac{mZ_0 \cdot n}{\omega_0} \tag{5.11}$$

where ω_0 is the center frequency, Z_0 is the port impedance, and the value of m, n in the formula is determined by the maximum tolerated amplitude imbalance.

In this design, capacitors C_1 and C_2 adopt CIDC structure, inductors L_1 and L_2 adopt double-sided interconnected spiral structure; meanwhile, inductors L_3 and L_4 are added to compensate for the parasitic capacitance to ground C_{g1} and C_{g2}.

Through practical measurements, as shown in Figure 5.4, the measured phase imbalance of the lumped balun is 181° ± 1.4° in the frequency from 0.7 GHz to 1.4 GHz, 66.7% of the fractional operating bandwidth, and the circuit size of the

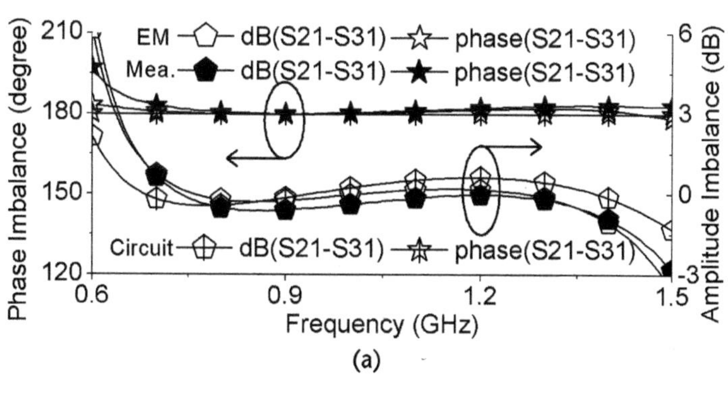

(a)

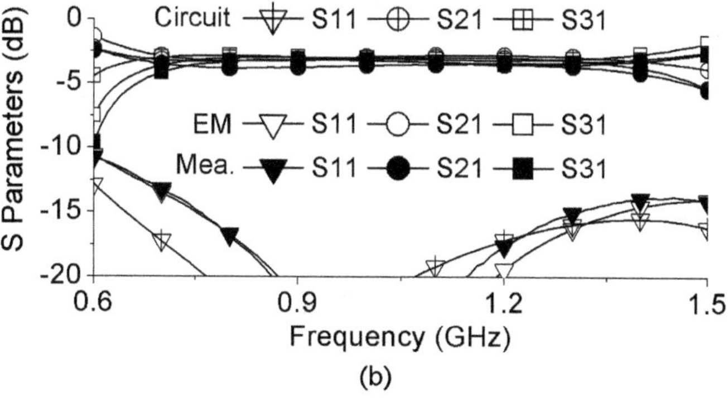

(b)

Figure 5.4 Simulated and measured results of the SISL lumped balun [4]. (a) Amplitude imbalance and phase imbalance. (b) S-parameters.

overall design is $0.06\lambda_g \times 0.11\lambda_g$. The balun exhibits good bandwidth and phase imbalance, as well as compactness, self-packaging, and low cost, and is believed to have broad application prospects.

5.1.2 Marchand Balun

The lumped balun introduced in the previous section is mainly suitable for lower operating frequency and narrowband design. This section will introduce the Marchand balun of the distributed balun, which has the advantages of good frequency characteristics and broadband, and it is a hot spot of current research. On the foundation of the SISL platform with multiple inner layers, the research of Marchand balun has also advanced.

A new structure of multiple inner board layers on SISL platform is proposed [5], which is made up of two inner boards, a hollow board for separation and support between the inner boards, two hollow boards forming a cavity, and two upper and lower cover boards. This structure has the advantage of effectively reducing the dielectric loss of the broadside coupled line structure circuit, especially for FR4 substrate with large loss angle tangent, which has a widespread prospect of application.

Based on the strengths of the multiple inner layers structure, a Marchand balun with a triple inner layer on the SISL platform is proposed, as shown in Figure 5.5. This balun adopts nine low-cost FR4 substrates, where TB3, TB5, and TB7 are triple inner layer boards with internal main circuit coupled line structure.

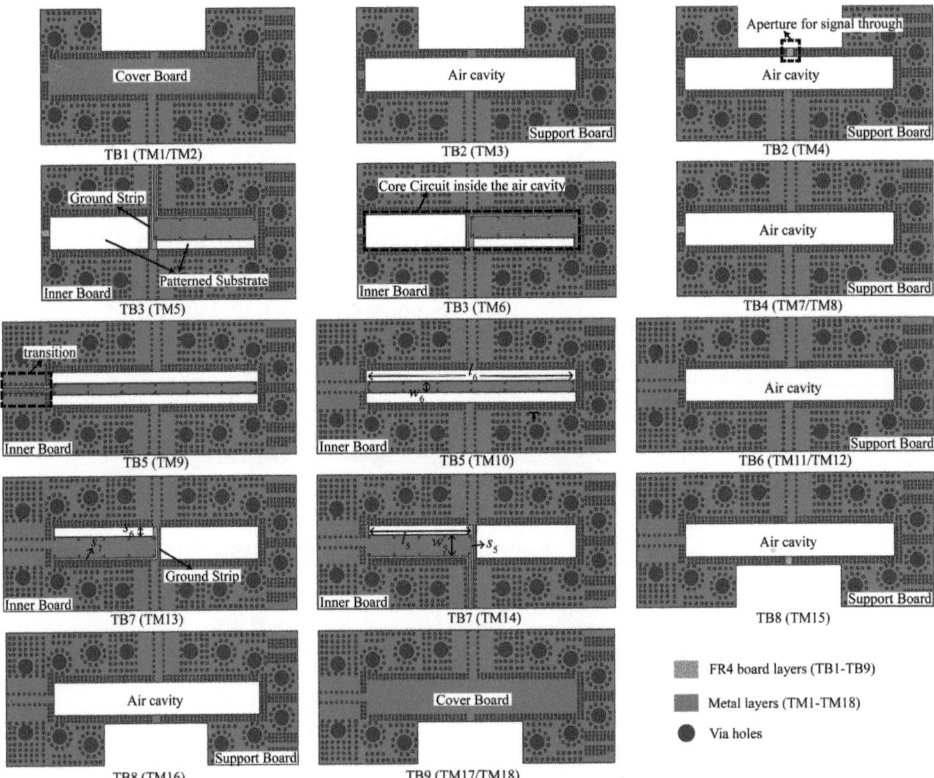

Figure 5.5 Structure of a Marchand balun with a triple inner layer on the SISL platform [5].

The designed Marchand balun circuit principle employs an asymmetric coupled line structure, which comprises two quarter-wavelength coupled transmission lines, as shown in Figure 5.6. The characteristic impedance of the two quarter-wavelength transmission lines connecting Port 1 is indicated by Z_1, and the characteristic impedance of the quarter-wavelength transmission lines connecting Ports 2 and 3 are indicated by Z_2. Both of them have different characteristic impedances, and the coupling coefficient of their constituent coupled lines is indicated by k_a.

The simulation shows that the bandwidth and in-band return loss of balun are related to the characteristic impedance Z_1 and Z_2, coupling coefficient k_a. It can be seen from Figure 5.7(a) that, while the balun bandwidth increases as the characteristic impedance Z_2 decreases, the in-band return loss performance becomes worse, and the coupling coefficient remains almost constant. Figure 5.7(b) shows the variation curve of balun return loss versus characteristic impedance Z_1. When measured with in-band return loss better than 25 dB, the growth of characteristic impedance Z_1 leads to an improvement in the 25-dB return loss bandwidth, which at the same time requires a corresponding increase in characteristic impedance Z_2 and coupling coefficient k_a. Therefore, we choose the appropriate value of the characteristic impedance and coupling coefficient, we can obtain the expected frequency response of the balun.

Based on the selection of the parameters of the asymmetric coupling line, the simulation and measurement results of the designed coupled line Marchand balun on SISL platform with triple inner layer are shown in Figure 5.8. The measured S_{21} and S_{31} are −3.23 dB and −3.4 dB, the insertion loss deviation is only 0.3 dB against the ideal case at the center frequency of 4.5 GHz, the return loss is better than 17 dB, and the phase imbalance is $178.62° \pm 0.9°$. The balun demonstrates low insertion loss at the center frequency, especially when using the inexpensive FR4 substrates with a high dielectric loss tangent, which provides low-cost, low-loss, and self-packing characteristics that will make it a great benefit for applications in complex feeding networks.

5.2 Phase Shifters on the SISL Platform

5.2.1 Introduction

The phase shifter [6, 7] is an important RF front-end device that controls and adjusts the phase of electromagnetic waves. It has important applications in phased arrays,

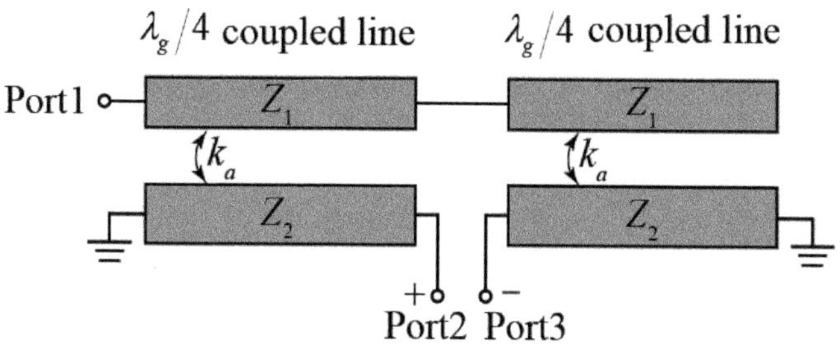

Figure 5.6 Structure of a Marchand balun with asymmetrical coupled lines [5].

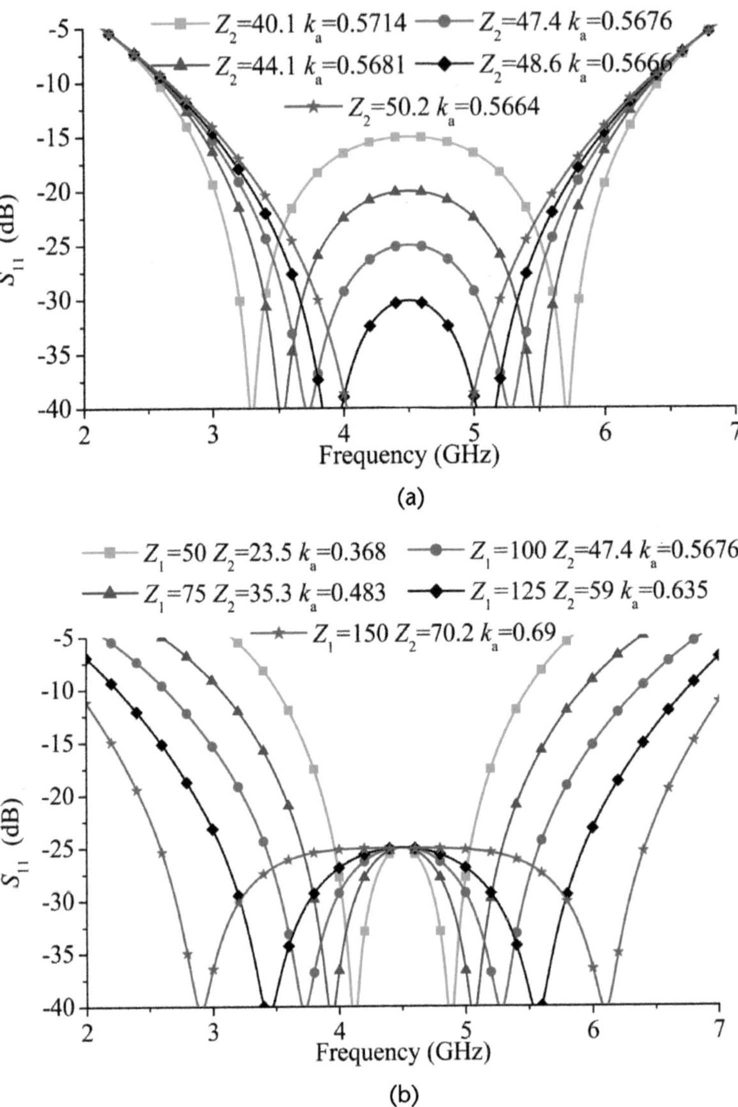

Figure 5.7 Simulation results of return loss variation with: (a) Z_2 and k_a (Z_1 = 100Ω) and (b) Z_1 [5].

modulators, beamforming and beam steering networks, antenna feed networks, and other system control circuits.

The phase shifter is a two-port device that is generally divided into two sections, namely, the main line and the reference line. Phase shift is a key parameter of the phase shifter; it indicates the difference Φ in phase between the main line and the reference line. Generally, the phase shift will be selected at 45°, 90°, 180°, and other fixed values. However, there are also tunable phase shifters, which can change the phase shift within a certain range by voltage or mechanically. In addition to the phase shift, researchers also focus on the operating frequency band, phase shift stability, and filtering function.

However, in beamforming networks or phased arrays, multiple phase shifters will be used instead of just one, which results in the total circuit loss of phase shifters

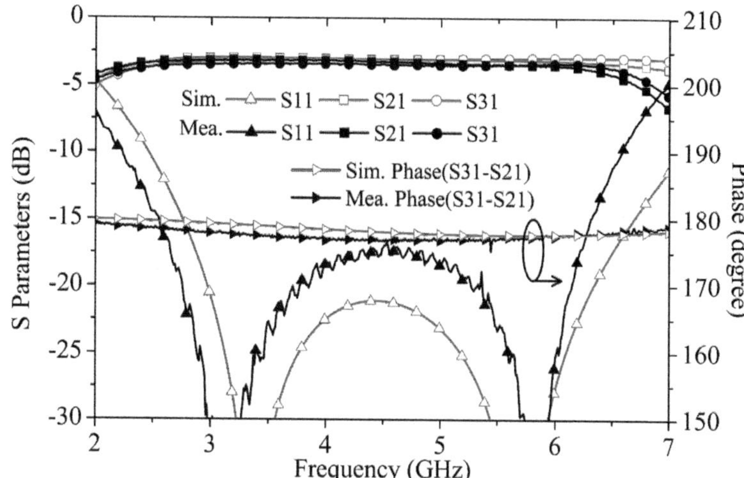

Figure 5.8 Simulation and measurement of the coupled-line Marchand balun on the SISL platform with a triple inner layer [5].

that cannot be ignored. In order to achieve better performance of the phase shifter, minimizing its loss is important. One effective way to achieve this is through the use of SISL, which offers several advantages such as low loss, high Q value, self-packaging, and low cost. In the next section, we will introduce a low-loss, patch-based phase shifter on the SISL platform.

5.2.2 SISL Patch-Based Phase Shifter

Unlike previous studies, this phase shifter's main path and reference path are both fabricated using a right-angled triangle patch, as shown in Figure 5.9. Besides, a slot is set on the hypotenuse to better adjust the performance of the proposed phase shifter. The main patch is grounded at the right-angle corner, while the reference patch is open at the right-angle corner.

The operational principle of this patch-based phase shifter is illustrated in Figure 5.10. The phase shift of the novel phase shifter can be written as

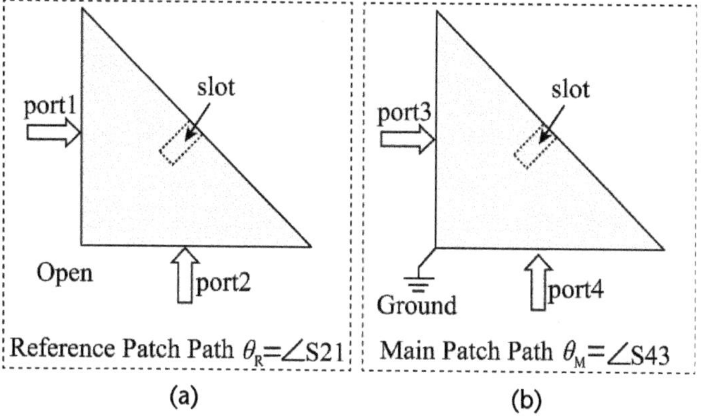

Figure 5.9 Basic topology of the novel patch-based phase shifter: (a) reference patch, and (b) main patch [8].

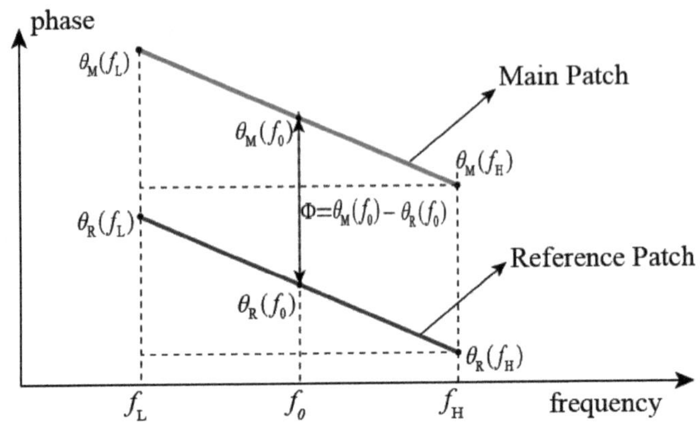

Figure 5.10 Operational principle of the novel patch phase shifter based on the SISL platform [8].

$$\Phi = \angle S_{43(\text{Main})} - \angle S_{21(\text{Ref})} = \theta_M(f_0) - \theta_R(f_0) \tag{5.12}$$

The phase shifter is based on a classic five-layer SISL structure, and its 3D view is depicted in Figure 5.11. A portion of Substrate 2 and Substrate 4 are excavated to form an air cavity. The novel patch phase shifter is placed on Substrate 3, which is 0.254-mm-thick Rogers 5880.

Figure 5.12 depicts the 2D view of the phase shifter's main patch and the reference patch. The phase shifter has the same pattern in G_5 and G_6, the only difference is that the feeding lines from phase shifter to the ports are set in G_5.

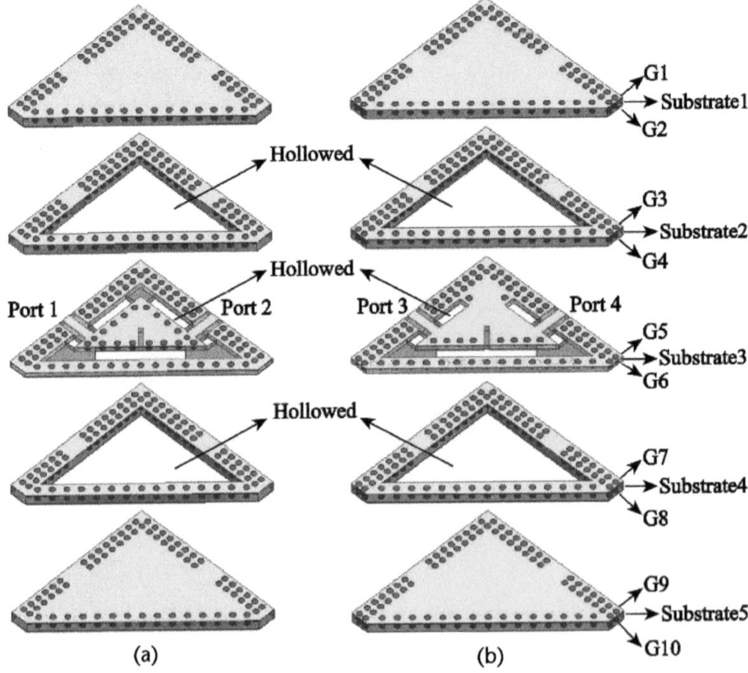

Figure 5.11 A 3D view of the novel low-loss SISL-based patch phase shifter: (a) reference patch, and (b) main patch [8].

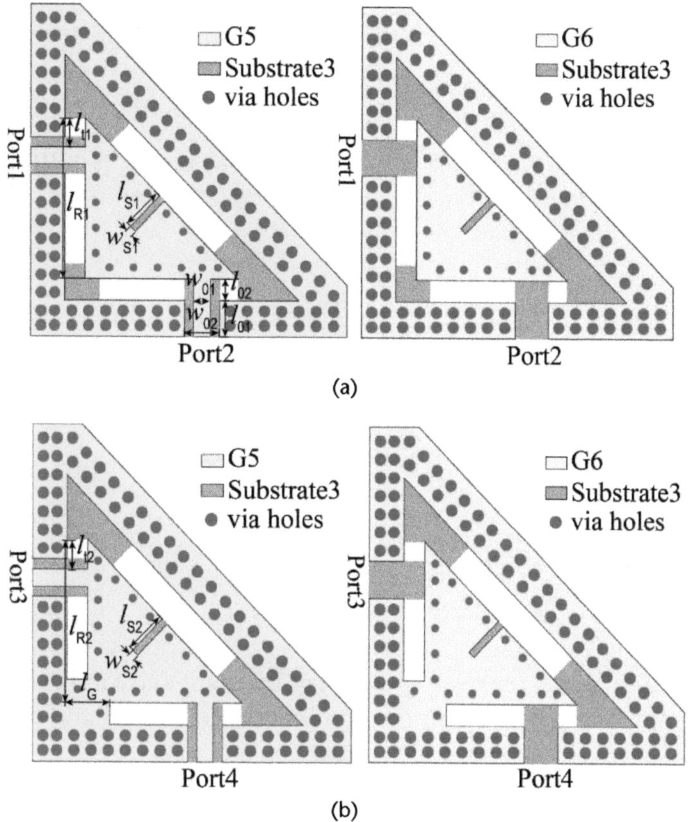

Figure 5.12 A 2D view of the SISL-based patch phase shifter: (a) reference patch, and (b) main patch (only shows Substrate 3 here) [8].

Of the phase shifter, the circuit loss is mainly composed of three aspects: conductor loss, radiation loss, and dielectric loss. First, we select patches with large surface areas to replace traditional uniform transmission lines with narrow trace widths, which can effectively reduce conductor losses. Furthermore, the larger size makes PCB fabrication more convenient. Moreover, the design uses two-layer patches that are connected by a series of metal via holes along the edge of the patch, and its conductor loss will be lower than that of a single-layer patch. Second, the large surface area also increases the contact area between the patch and the air, which leads to radiation loss increase. Therefore, we set up a series of via holes at every dielectric layer around a right-angled triangular cavity to reduce radiation loss. These vias and 10 metal layers will connect after pressing the five substrate layers together, which will produce excellent electromagnetic shielding properties and then the radiation loss decrease. Third, we design the core circuit at Substrate 3 and cut out the excess part of this layer to reduce dielectric loss. Moreover, the SISL platform contains air cavities in which the E-field is mainly distributed, which makes the circuit loss decrease ulteriorly.

For a right triangle patch without slots and feed, the TM-mode field [8] can be expressed using the following formula:

$$\left(\frac{\partial^2}{\partial x^2}+\frac{\partial^2}{\partial y^2}+k_{m,n}^2\right)E_z=0 \tag{5.13}$$

where E_z represents the component of the electric field in the z-direction, which is vertical to the patch plane. $K_{m,n}$ denotes the wavenumber [9], and it is written as the following formula:

$$k_{m,n}=\frac{\pi}{l_R}\cdot\sqrt{m^2+mn+2n^2} \tag{5.14}$$

where m and n reflect the resonance mode, and l_R represents the length of the right-angled side of triangular patch. Then the resonant frequency can be given by

$$f_R=\frac{ck_{m,n}}{\left(2\pi\sqrt{\varepsilon_r}\right)}\cdot\sqrt{m^2+mn+2n^2} \tag{5.15}$$

where ε_r indicates the relative dielectric constant and c indicates the velocity of light. The resonant frequency for the dominant mode ($m = 0$ and $n = 1$) can be written as

$$f_R=\frac{\sqrt{2}c}{\left(2l_R\sqrt{\varepsilon_r}\right)} \tag{5.16}$$

Afterward, side length l_R can be written as

$$l_R=\frac{\sqrt{2}c}{\left(2f_R\sqrt{\varepsilon_r}\right)}=\frac{\sqrt{2}}{2\lambda_g} \tag{5.17}$$

As can be seen from the above equation, we can change the resonant frequency of the proposed phase shifter by adjusting the side lengths l_{R1} and l_{R2} of the right triangle patch. The resonant frequency f_R increases with the decrease in the side length l_R. Furthermore, we set a slot on the hypotenuse side, and the operating frequency of the phase shifter is able to be changed by modifying the size of the slot.

As shown in Figure 5.13(a), if we make the slot length l_{S1} larger, S_{11} moves towards the low frequency. l_{t1} indicates the distance from the bottom corner to the feeding line. As shown in Figure 5.13(b), when increasing l_{t1}, S_{11} moves toward the high frequency.

Figures 5.14 and 5.15 depict the surface current distribution of the main patch and the reference patch under the fundamental mode. A significant change in surface current is observed near the slot, while the rest of the current distribution changes gently. The introduction of the slot increases the current path near it, resulting in a corresponding decrease in the resonant frequency. The slots and the two connected small triangular patches on both sides determine the operation of the fundamental

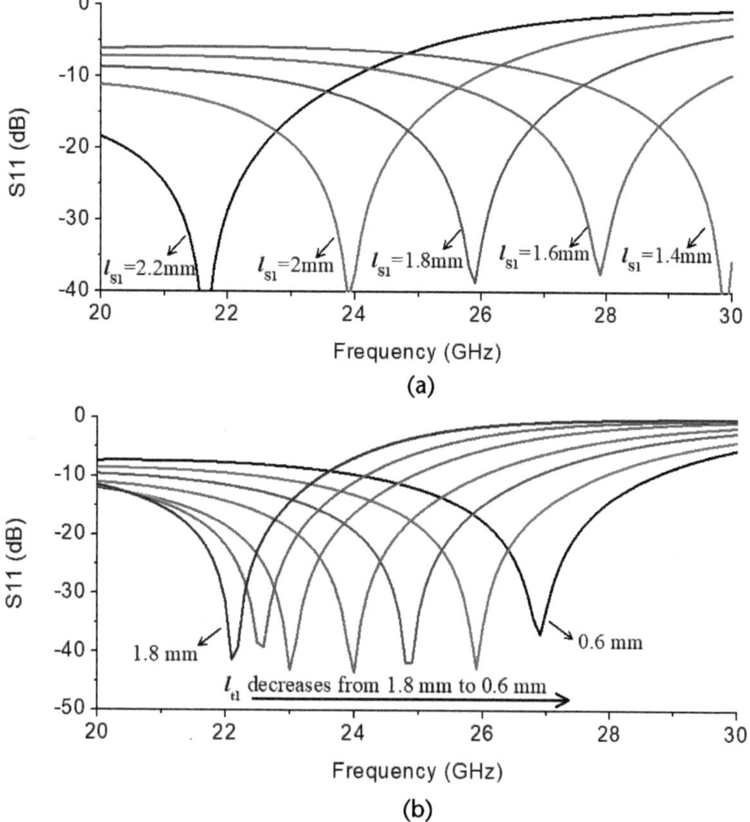

Figure 5.13 Simulated s_{11} of the reference patch: (a) l_{S1} varies from 1.4 to 2.2 mm (l_{t1} = 0.81 mm, l_R = 4.45 mm, and w_{S1} = 0.2 mm), and (b) l_{t1} varies from 0.6 to 1.8 mm (l_{t1} = 0.81 mm, l_R = 4.45 mm, and w_{S1} = 0.2 mm) [8].

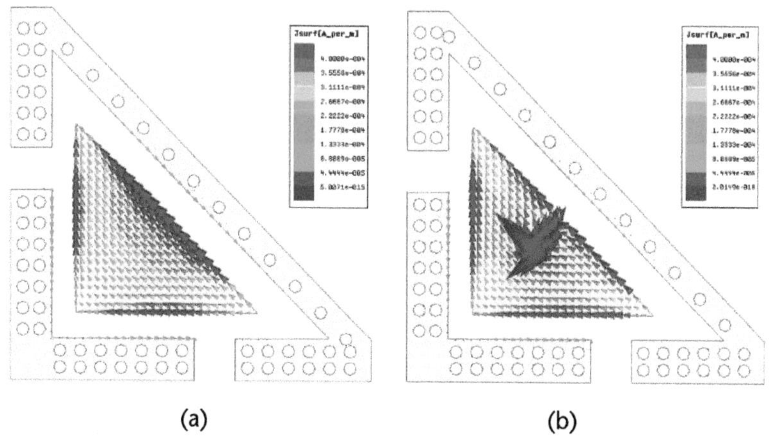

Figure 5.14 Surface current density of the reference patch under the fundamental mode. (a) l_{S1} = 0 mm, and (b) l_{S1} = 1 mm [8].

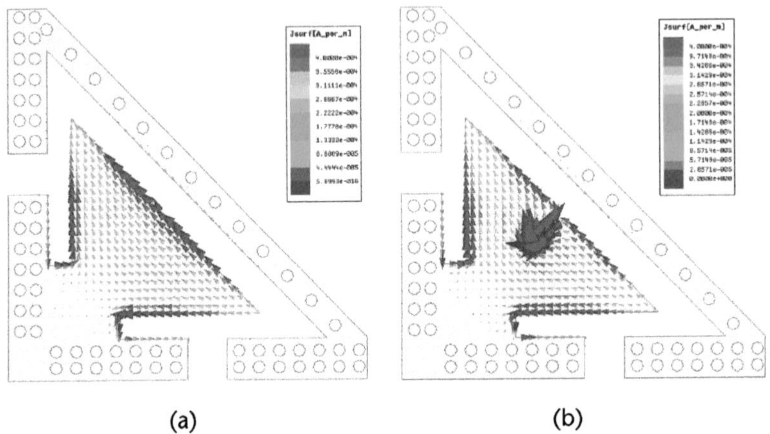

Figure 5.15 Surface current density of the main patch under the fundamental mode: (a) $l_{S2} = 0$ mm, and (b) $l_{S2} = 1$ mm [8].

mold. Figures 5.14(a) and 5.15(a) demonstrate that, after adding the grounded corner, the current density in the grounded part decreases rapidly.

To meet the design requirements, the phase shift at the center frequency should match the desired value, besides the phase slope of the main patch and reference patch required to be equal ($k_M = k_R$). Figure 5.16 shows the phase curve of the reference patch, and the slope can be expressed as

$$k_R = \frac{\left[\theta_R(f_H) - \theta_M(f_L)\right]}{\left(f_H - f_L\right)} \tag{5.18}$$

The phase properties of the main patch are able to be changed by adjusting the ground connection length l_G, length l_{S2}, and width w_{S2} of the slot. Figure 5.17 shows how the trend of phase shift Φ and phase slope change versus l_G, l_{S2}, and w_{S2}. As the value of l_G increases, the phase slope decreases while the phase shift Φ increases. As the value of l_{S2} or w_{S2} increases, both the phase shift Φ and the phase slope decrease.

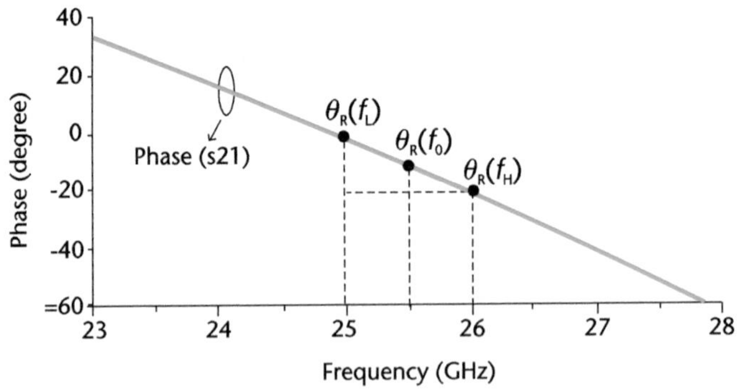

Figure 5.16 Simulated phase value of the SISL reference patch ($l_{t1} = 0.81$ mm, $l_{R1} = 4.45$ mm, $w_{S1} = 0.2$ mm, and $l_{S1} = 1.8$ mm) [8].

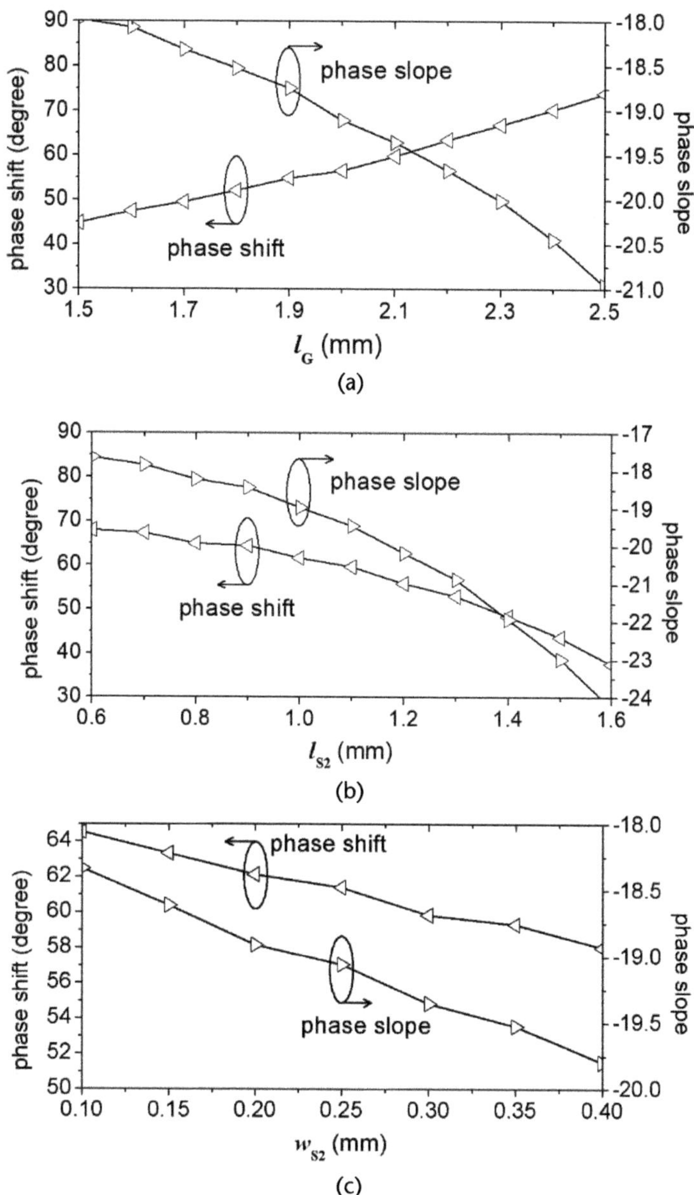

Figure 5.17 Simulation result of phase shift (Φ) at the working frequency and the phase slope of the main patch: (a) the value of l_G rises from 1.5 to 2.5 mm (l_{S2} = 1.08 mm, l_{R2} = 4.7 mm, and w_{S2} = 0.3 mm), (b) the value of l_{S2} rises from 1.5 to 2.5 mm (l_G = 2.1 mm, l_{R2} = 4.7 mm, and w_{S2} = 0.3 mm), and (c) the value of w_{S2} rises from 0.1 to 0.4 mm (l_G = 2.1 mm, l_{R2} = 4.7 mm, and l_{S2} = 1.1 mm) [8].

Figure 5.18 depicts the equivalent circuit structure of the reference patch and the main patch, which helps us to generate further comprehension of the working mechanism of the proposed phase shifter.

From Figure 5.18(a) we can get the transmission (ABCD) matrix of the reference patch,

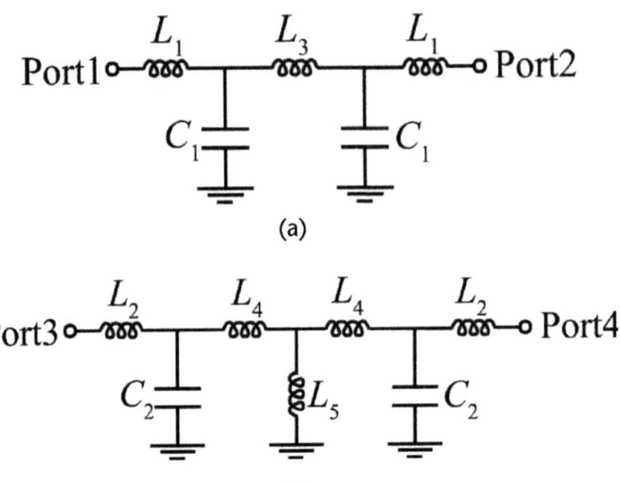

Figure 5.18 Equivalent circuit of: (a) the reference patch, and (b) the main patch [8].

$$
\begin{aligned}
M_{\text{Ref}} &= \begin{bmatrix} A_0 & B_0 \\ C_0 & D_0 \end{bmatrix} = M_1 M_2 M_3 M_2 M_1 \\
&= \begin{bmatrix} 1 & j\omega L_1 \\ 0 & 1 \end{bmatrix} \begin{bmatrix} 1 & 0 \\ j\omega C_1 & 1 \end{bmatrix} \\
&\times \begin{bmatrix} 1 & j\omega L_3 \\ 0 & 1 \end{bmatrix} \begin{bmatrix} 1 & 0 \\ j\omega C_1 & 1 \end{bmatrix} \begin{bmatrix} 1 & j\omega L_1 \\ 0 & 1 \end{bmatrix}
\end{aligned}
\tag{5.19}
$$

From Figure 5.18(b), we can get the transmission (ABCD) matrix of the main patch,

$$
\begin{aligned}
M_{\text{Main}} &= \begin{bmatrix} A_1 & B_1 \\ C_1 & D_1 \end{bmatrix} = M_4 M_5 M_6 M_7 M_6 M_5 M_4 \\
&= \begin{bmatrix} 1 & j\omega L_2 \\ 0 & 1 \end{bmatrix} \begin{bmatrix} 1 & 0 \\ j\omega C_2 & 1 \end{bmatrix} \begin{bmatrix} 1 & j\omega L_4 \\ 0 & 1 \end{bmatrix} \\
&\times \begin{bmatrix} 1 & 0 \\ 1/j\omega L_5 & 1 \end{bmatrix} \begin{bmatrix} 1 & j\omega L_4 \\ 0 & 1 \end{bmatrix} \\
&\times \begin{bmatrix} 1 & 0 \\ j\omega C_2 & 1 \end{bmatrix} \begin{bmatrix} 1 & j\omega L_2 \\ 0 & 1 \end{bmatrix}
\end{aligned}
\tag{5.20}
$$

In accordance with the conversion between the scattering matrix and ABCD matrices, the forward transmission coefficients for M_{Ref} and M_{Main} can be, respectively, written by

$$
S_{21(\text{Ref})} = \frac{2}{\left(A_0 + B_0/Z_0 + C_0 Z_0 + D_0\right)}
\tag{5.21}
$$

$$S_{43(\text{Main})} = \frac{2}{\left(A_1 + B_1/Z_1 + C_1 Z_0 + D_1\right)} \tag{5.22}$$

Then we can give the phase values of $S_{21(\text{Ref})}$ and $S_{43(\text{Main})}$, respectively

$$\angle S_{21(\text{Ref})} = \arctan\left[\frac{\text{imag}\left(S_{21(\text{Ref})}\right)}{\text{real}\left(S_{21(\text{Ref})}\right)}\right] \tag{5.23}$$

$$\angle S_{43(\text{main})} = \arctan\left[\frac{\text{imag}\left(S_{43(\text{main})}\right)}{\text{real}\left(S_{43(\text{main})}\right)}\right] \tag{5.24}$$

The corresponding phase slopes can be expressed as

$$PS_{\text{Ref}} = \frac{d\angle S_{21(\text{Ref})}}{df} \tag{5.25}$$

$$PS_{\text{Main}} = \frac{d\angle S_{43(\text{Main})}}{df} \tag{5.26}$$

For the main patch, the series of L_4 denote the slot and L_5 denote the connection line to ground. The phase value and phase slope of the simple transmission two-port network consisting of series L_4 or shunt L_5 components can be described by

$$\angle S_{43(L4)} = \arctan\left[\frac{-\omega L_4}{\left(2Z_0\right)}\right] \tag{5.27}$$

$$\angle S_{43(L5)} = \arctan\left[\frac{Z_0}{\left(2\omega L_5\right)}\right] \tag{5.28}$$

The corresponding phase slopes can be written by

$$PS_{L4} = \frac{-\pi Z_0}{\left(\pi^2 f^2 L_4 + Z_0^2/L_4\right)} \tag{5.29}$$

$$PS_{L5} = \frac{-4\pi Z_0}{\left(16\pi^2 f^2 L_5 + Z_0^2/L_5\right)} \tag{5.30}$$

where $\omega = 2\pi f$. From (5.27) to (5.30), as the value of L_4 increases, the phase value $\angle S_{43(L4)}$ decreases, and the phase slope decreases because the phase delay increases.

In terms of the correlation between the equivalent circuit model and the circuit layout, by increasing the length of the ground connection l_G, it is possible to decrease the shunt inductance L_5. Based on the analysis results above, the phase value is expected to rise while the phase slope will decrease. If we increase the value of the slot length l_{S2}, the series inductance L_4 in Figure 5.18(b) increases, and both the phase value and the phase slope are expected to decrease. In accordance with Figure 5.17, a rule can be found, when tuning for increased or decreased phase shifts, the changing in the phase slopes for l_G exhibit opposite trends to those of l_{S2} and w_{S2}. The design flow of the proposed SISL patch-based phase shifter should emphasize the tuning procedure of adjusting the dimension of the phase shifter to control its phase properties in detail.

Based on the design flowchart, the novel patch-based phase shifter utilizing the SISL platform was ultimately achieved, and, meanwhile, electromagnetic simulation software was employed to construct and optimize the model. Figure 5.19 shows the photographs of the fabricated SISL patch-based phase shifter, respectively containing the reference patch and the main patches for 45°, 60°, and 90°. Table 5.2 lists the parameter values of these patches.

Figure 5.20 shows the simulated and measured frequency responses and phase properties of the patch-based phase shifters. Well-agreed simulation and measurement

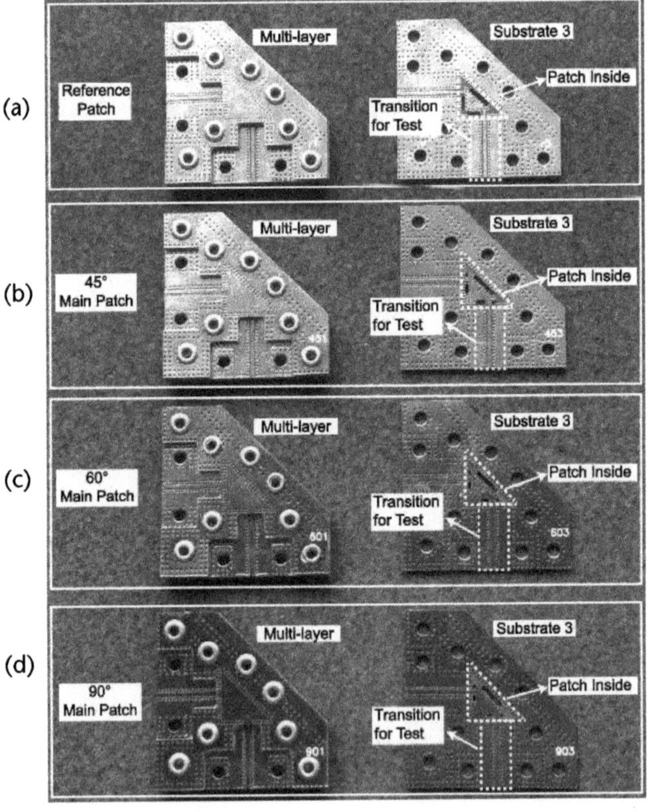

Figure 5.19 Photographs of the fabricated SISL patch-based phase shifters: (a) reference patch, (b) main patch for 45°, (c) main patch for 60°, and (d) main patch for 90° [8].

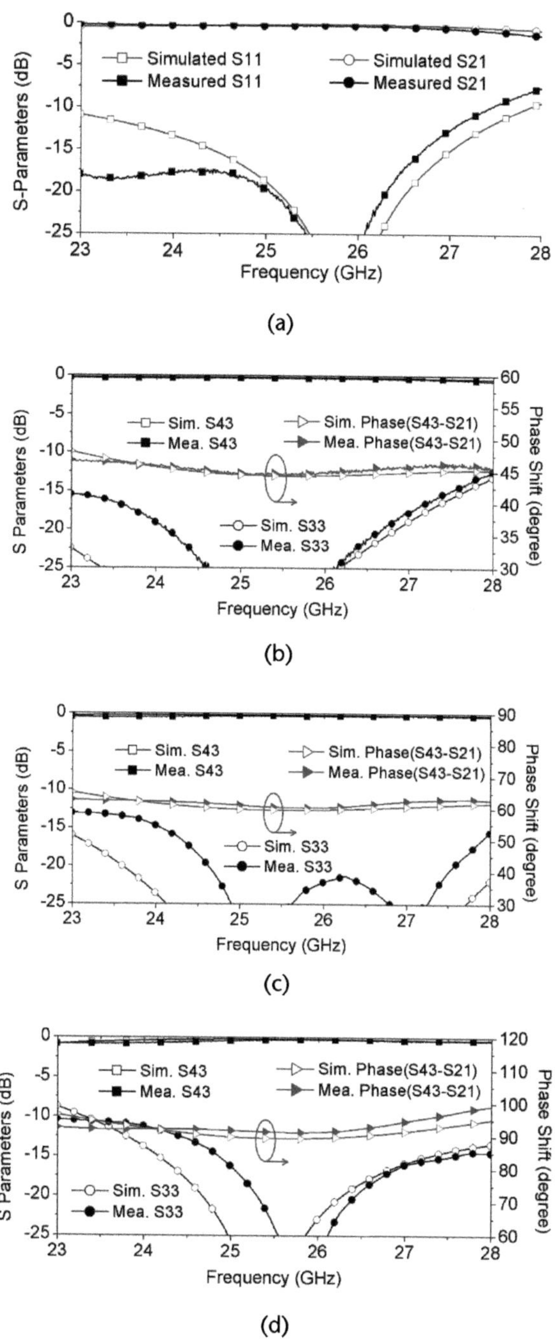

Figure 5.20 Simulated and measured results of the SISL patch-based phase shifters: (a) reference patch, (b) main patch for 45°, (c) main patch for 60°, and (d) main patch for 90° [8].

Table 5.2 Parameter Value of the SISL Phase Shifter

Reference Patch (Unit: mm)			
l_{R1}	4.45	l_{t1}	0.81
w_{S1}	0.2	l_{S1}	1.8
Main patch (unit: mm)			
Phase Shifters	*45°*	*60°*	*90°*
l_G	1.76	2.1	2.4
l_{S2}	1.2	1.08	0.6
w_{S2}	0.3	0.3	0.2
l_{R2}	4.74	4.71	4.6
l_{t2}	1.1	1.07	0.96

Source: [8].

results have been obtained. The obtained measurement results are well agreed with the simulation results.

5.3 Butler Matrices on the SISL Platform

5.3.1 Introduction

Butler matrices, which are well-known passive multiport networks with multiple input and output ports, have been widely used in contemporary microwave electronics. The differential phase of the various output signals relies on the input port selected by the $N \times N$ Butler matrix, which evenly splits the input signal into all output ports [10]. It provides a progressive phase difference signal between the output ports to perform a beamforming operation and achieve a multidirectional transmission of the signal. Due to their simple fabrication process, low cost, high precision, and stable and reliable performance, Butler matrices are one of the most common beamforming networks that are used in the information and communication field nowadays.

Figure 5.21 shows a 4 × 4 Butler matrix with the associated beams for different feeding input ports, which consist of four 90° couplers, two crossovers, two 45° phase shifters, and two 0° phase shifters. The crossover between Ports 6 and 7 causes a phase delay, which is made up for by the 0° phase shifters. When Ports 1, 2, 3, and 4 are excited separately, the phase differences between the adjacent output ports are –45°, +135°, –135°, and +45°.

Based on microstrip line, stripline, waveguide, coplanar waveguide (CPW), and substrate integrated waveguide (SIW) platforms, many high-performance Butler matrices have been designed. Since the Butler matrix is made up of multiple components, the size of the entire circuit is always big. As a result, reducing the size of the whole circuit is a significant consideration factor in the design of the Butler matrix in the low-frequency band. Nevertheless, when the frequency rises to a higher band, the circuit loss will become one of the most vital consideration factors. In order to reduce the circuit loss, a high-performance substrate with low loss is a good choice. However, the cost of the whole circuit will increase correspondingly.

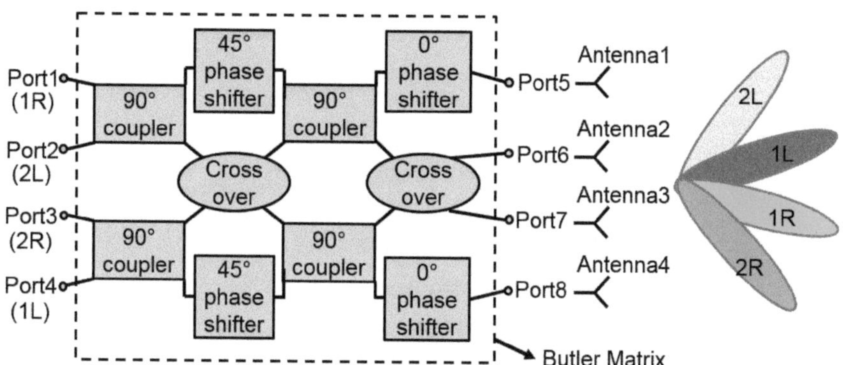

Figure 5.21 The circuit diagram and associated beams of the 4 × 4 Butler matrix [12].

SIW and waveguide structures are also widespread structures in the design of low-loss circuits. Nevertheless, system integration with other planar circuits is difficult on account of the large size and heavy weight of the traditional waveguide with metal housing.

On the basis of the SISL platform's low-loss, low-cost, and high-integration properties, it has been used in the design of many passive and active circuits. Unlike the traditional waveguide suspended-line circuits with metal housing, SISL can be fabricated by using standard PCB processes and avoid the use of mechanical housing for shielding [11]. When the frequency rises to a higher band, on account of the undesirable parasitic impact of the T- or cross-junctions and the coupling between adjacent lines, some types of transmission-line-based couplers and crossovers such as branch-line-based couplers and crossovers may not be appropriate [12]. Therefore, using patch elements that are more applicable to a higher frequency band plays a significant role in making it a bigger size to make component production simpler than before. Consequently, it is possible for the proposed SISL Butler matrix to use patch elements to achieve more novel high-frequency performance.

5.3.2 Architecture of the SISL Butler Matrix

5.3.2.1 Multicavity SISL Concept

The SISL circuit is made up of five double-sided PCB layers that are denoted by Substrate 1–Substrate 5 in Figure 5.22, and we use G_1–G_{10} to denote the 10 metal layers of the five PCB layers. The second layer and the fourth layer are hollowed out into the required shapes to form the air cavity in the suspended-line circuit when five layers are stacked up. The main circuit is designed on both sides of Substrate 3 on the metal layers G_5 and G_6. A thorough examination of performance and cost factors lead to the selection of Fr4, which has a thickness of 0.6 mm and a dielectric constant of 4.4 as the material of Substrates 1, 2, 4, and 5, and Rogers 5880 which have a thickness of 0.254 mm and a dielectric constant of 2.2 as Substrate 3.

When designing complex networks with multiple components, such as the Butler matrix, designing in a single cavity may cause mutual interference of electromagnetic fields between each component. This may cause problems with overall circuit design and system integration, particularly in the higher-frequency band.

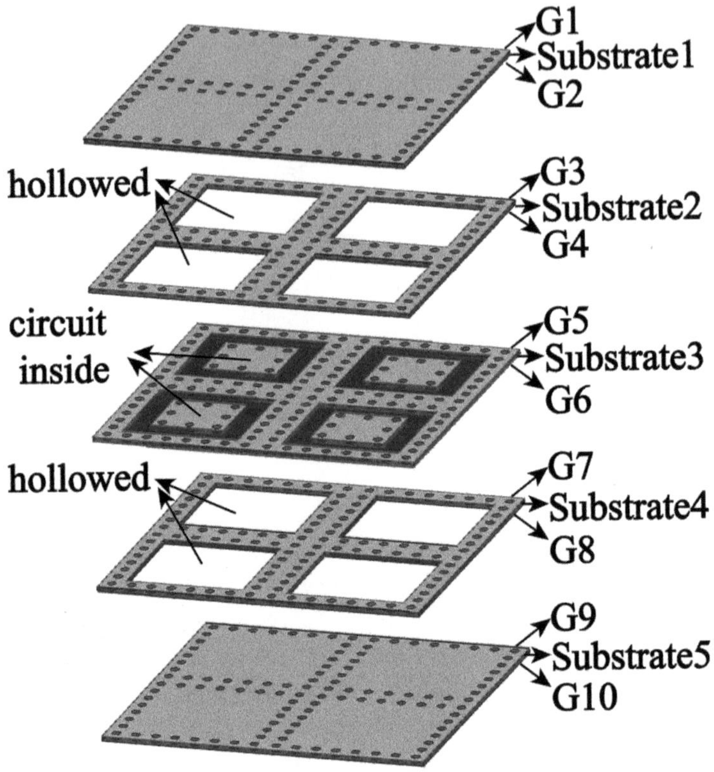

Figure 5.22 Multilayer structures of SISL in multicavity case [11].

We propose a honeycomb concept based on the multicavity structured SISL platform as shown in Figure 5.23(b). To isolate individual components, we make use of a large number of via holes in rows, as the metallic wall between the adjacent cavities prevents the electromagnetic energy from passing through. Each component has an individual cavity. Also, by cutting a small opening in the via fence and using a narrow stripline to meet the signal transmission needs of each component, mutual interference between each component can be minimized at the same time. Therefore, each component can be individually designed and adjusted making the overall circuit design more flexible.

The compact topology of the circuit is proposed on the basis of its practical use, as shown in Figure 5.23(a). This 4 × 4 Butler matrix consists of four 3-dB couplers with a 90° phase difference, two crossovers, two 45° phase shifters, and two 0° phase shifters. All these 10 components are based on the SISL platform and will be designed and adjusted separately. Since the crossovers and couplers of the Butler matrix are located in the middle of the entire circuit, their cavities are square-shaped, and the two ports of the phase shifters are bent at the right angle to facilitate connections between adjacent components and promote good integration.

5.3.2.2 SISL Patch-Based Crossover

The crossover located in the center of the whole circuit connects the couplers and the output ports, and its size, performance, and phase delay will have an impact on

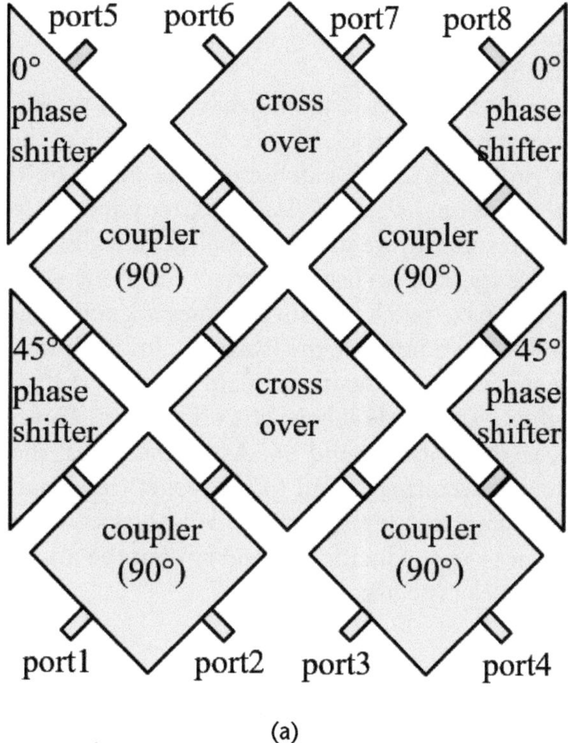

(a)

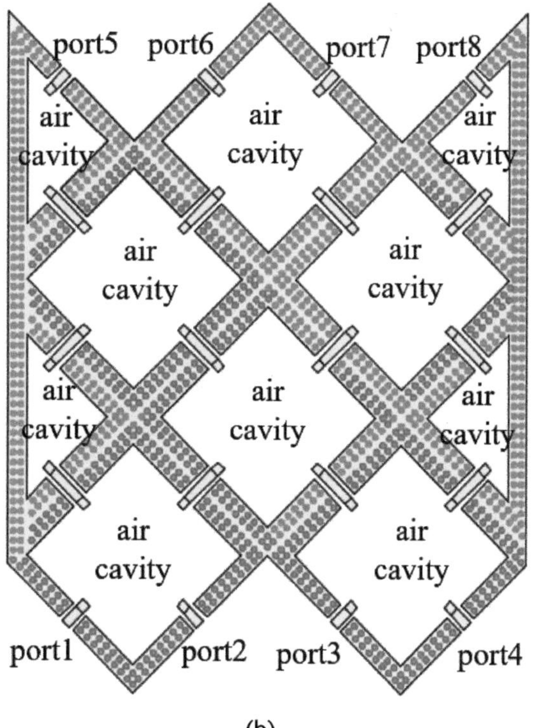

(b)

Figure 5.23 (a) Circuit topology of the Butler matrix, and (b) distribution of the isolated air cavities [11].

the design of other components. Based on the SISL platform, the electromagnetic shielding property can effectively alleviate the problem of strong radiation loss of patch components.

A crossover should satisfy the needs of both good transmission performance and high isolation properties. Figure 5.24(a) depicts the single metal layer structure of the proposed square patch crossover based on the SISL platform. During the normal situations, we use Wheeler's cavity model approximate analysis of the square patch resonator, we use perfect electric walls to equate the top and bottom sides of the cavity, and we use the perfect magnetic walls to equate the rest sides.

A simple square patch crossover achieves good return loss but relatively poor isolation. In order to improve the isolation, four symmetric slots are introduced in the square patch, the slot width w_s mainly affects the isolation of the components, and the radius r and angle θ have less effect on it. Figure 5.25 shows the relationship between the isolation and w_s. As w_s increases, the S_{21} and S_{41} curves move from the higher frequency band to the target frequency f_0. To further reduce the loss of the circuit, it is feasible to use via holes connecting the G_5 and G_6 metal layers to reduce the conductor loss and cut out the undesired parts of the substrate to reduce the dielectric loss.

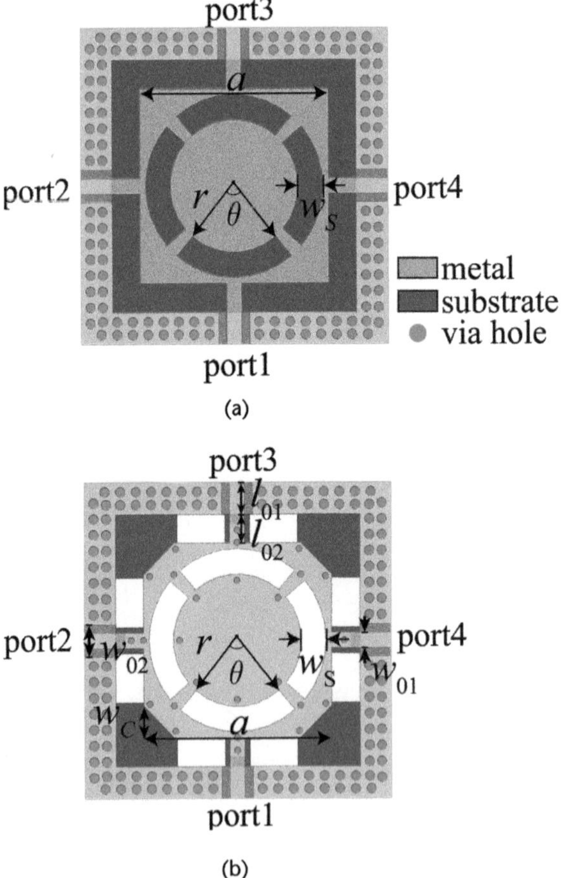

Figure 5.24 Main circuit of the SISL square patch crossover: (a) single-metal layer case, and (b) double-metal layers and patterned substrate case [11].

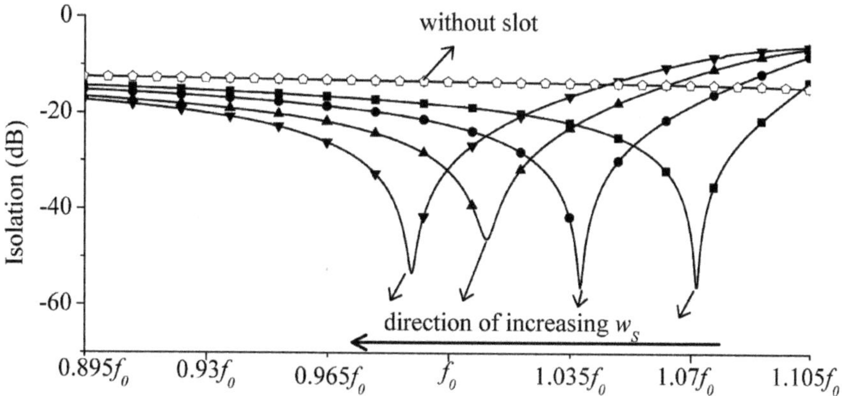

Figure 5.25 The relationship between the isolation and w_s of the SISL square patch crossover [11].

As shown in Figure 5.24(b), double-metal layers and a patterned substrate are used in the proposed square patch crossover based on the SISL platform. The final circuit is designed for the 24-GHz band and the parameters are as follows: $a = 6.14$ mm, $w_s = 0.8$ mm, $\theta = 81°$, $r = 2.07$ mm, $w_c = 1.05$ mm, $w_{01} = 0.48$ mm, $w_{02} = 1$ mm, $l_{01} = 1$ mm, and $l_{02} = 0.95$ mm. As shown in Figure 5.26, the simulated isolation is better than 28 dB and the return loss is over 15 dB from 24.7 to 26 GHz, the measured isolation is better than 25.5 dB and the return loss is over 10 dB.

5.3.2.3 SISL Patch-Based Coupler

A standard four-port Butler matrix requires four 3-dB couplers with a 90° phase difference. According to the above crossover analysis, using a simple square patch design may give rise to the signal flowing mainly into Port 3 and have a very weak coupling from Port 1 to Port 2 due to the symmetry structure. Consequently, we selected adding a rectangular slot to the square patch's center to control the coupling

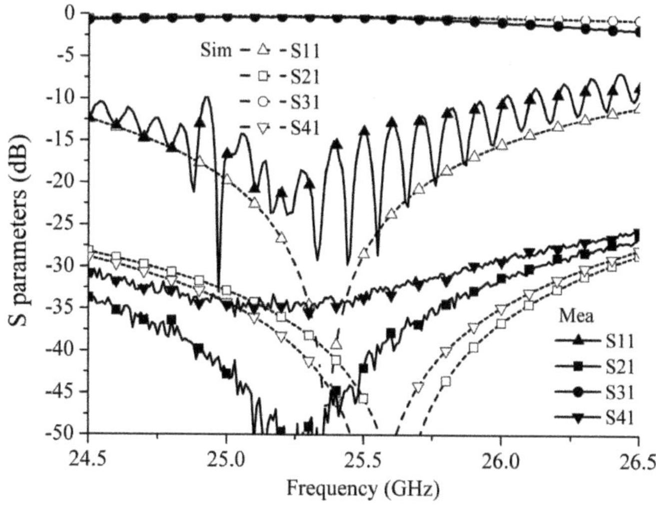

Figure 5.26 Simulated and measured S-parameters of the SISL patch-based crossover [11].

coefficient. The length of the rectangular slot is the main determining factor of the coupling coefficient, while the width of the slot has only a little influence. We can equate the circuit as shown in Figure 5.27(c), which consists of the series inductors and the shunt capacitors. By using the even-odd mode analysis, it is simple to

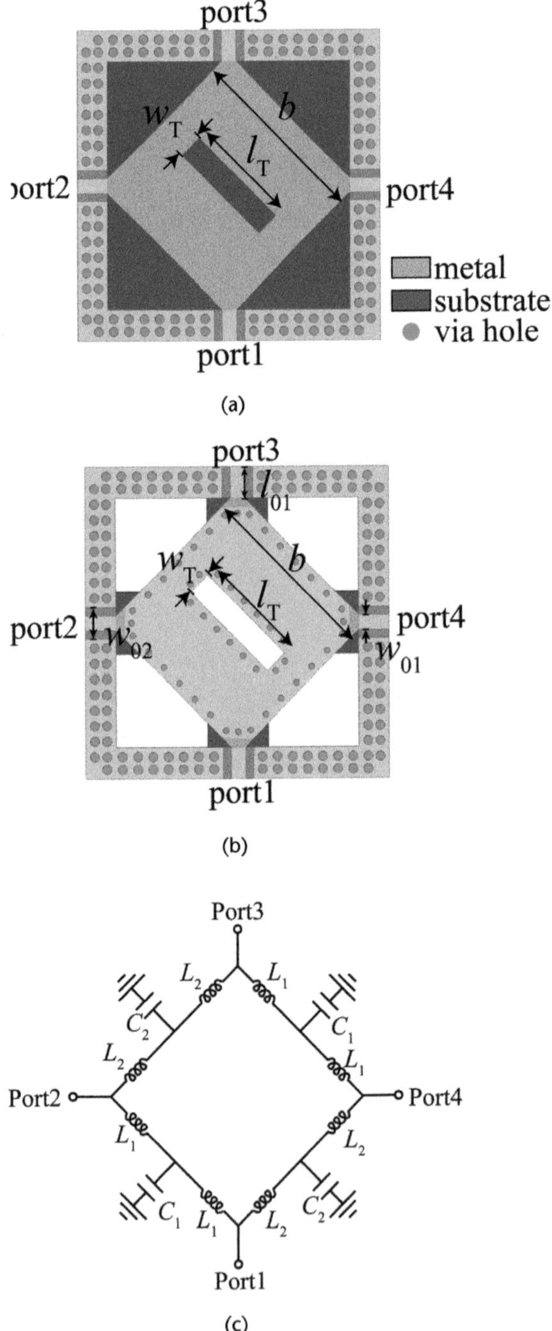

Figure 5.27 Main circuit of the SISL patch-based coupler: (a) single-metal layer case, (b) double-metal layers and patterned substrate case, and (c) equivalent circuit [11].

obtain the reflection coefficients and transmission coefficients of the even and odd modes, as well as the equivalent circuit's S-parameters, which may help to clarify the working principle of the SISL patch-based coupler.

Figure 5.27(b) is the patch-based coupler structure on SISL platform. Simultaneously, the structure makes use of double-metal layers to reduce the conductor loss and patterned substrate to reduce the dielectric loss, the final parameters are as follows: $b = 5.953$ mm, $w_T = 0.8$ mm, $l_T = 3.57$ mm, $w_{01} = 0.48$ mm, $w_{02} = 1$ mm, and $l_{01} = 1$ mm. As shown in Figure 5.28, the simulated isolation and return loss of the proposed coupler are better than 22 dB, and the simulated phase difference floats up and down at 90° from 24.5 to 26.5 GHz. The measured indicators of the return loss and isolation are both better than 17.8 dB, and the measured phase difference is 89° ± 1° from 24.5 GHz to 26.5 GHz.

5.3.2.4 The 0° and 45° Phase Shifters

Normally, a phase shifter is made up of the phase-shift circuit and the reference line. The phase difference between the two paths is the phase shift that we need, and the same slope between the phase-shift circuit and the reference crossover circuit that we designed above maintained the constant phase difference between them within the frequency band. This is the key to expanding the bandwidth of the phase shifter. We use meander lines to design these two phase shifters, and the final phase value depends on the length of the meander lines and the coupling between the adjacent lines. In order to make the circuit design more flexible, we add stubs at two ports of the phase shifters to tune the slope and value of the phase.

According to the reference crossover's phase property, the parameters of the 0° and 45° phase shifters are selected appropriately. The simulated return loss of the 0° phase shifter is better than 15 dB from 24.5 GHz to 26.3 GHz, and the simulated phase imbalance is within 1.2° in almost the whole frequency band. For the 45°

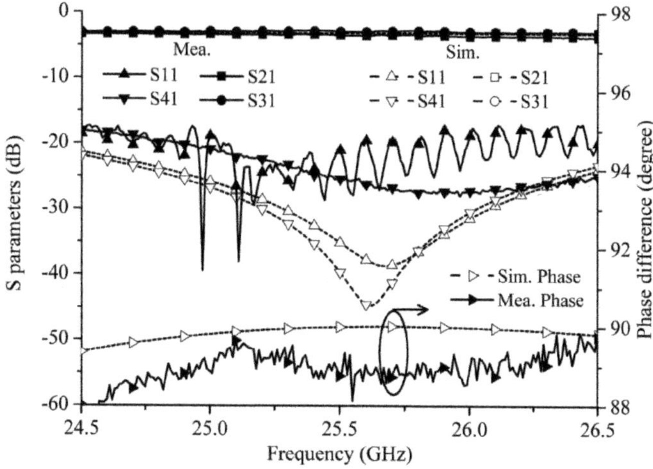

Figure 5.28 Simulated and measured S-parameters and phase difference of the SISL patch-based coupler [11].

phase shifter, the simulated return loss is over 22.5 dB, while the simulated phase imbalance is within 1° in almost the whole frequency band.

5.3.2.5 Integration of the Butler Matrix Using Previous Components

All the components of the Butler matrix based on the SISL platform, including the SISL patch-based crossovers, couplers, and the SISL meander line phase shifters, have been designed and adjusted in individual cavities. According to the compact topology of the whole circuit, the SISL Butler matrix can be obtained by combining all 10 components together. Figure 5.29 shows the planar view of the low-loss Butler matrix on the SISL platform.

On the basis of the multicavity SISL structure, the components inside are isolated from each other with a via fence between every adjacent cavity. Therefore, all of these 10 components will not influence each other when integrating them together, and the radiation loss, conductor loss, and dielectric loss of the whole circuit can be greatly reduced. Moreover, each substrate of the proposed Butler matrix on the SISL platform can be fabricated by the standard PCB process, and it is self-packaged with no need for any extra mechanical metal housing.

As shown in Figure 5.30(a), the simulated return loss of the SISL Butler matrix is better than 12 dB, and the isolation is better than 16 dB with a maximum of over 40 dB. From 24.5 to 26.5 GHz, the measured insertion loss is around 1.1 dB,

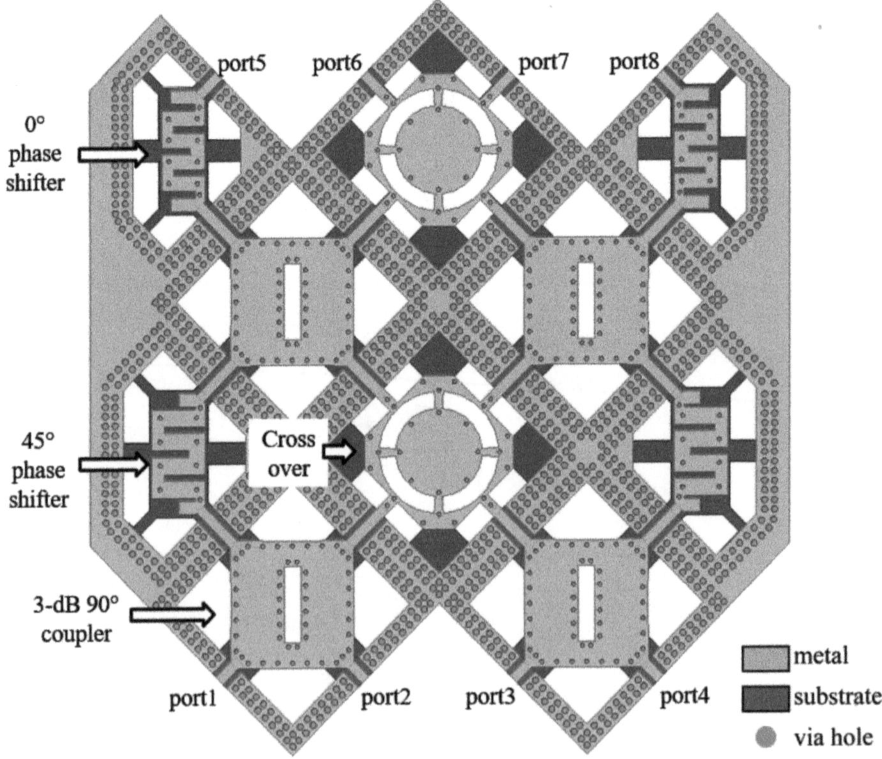

Figure 5.29 Planar view of SISL Butler matrix [11].

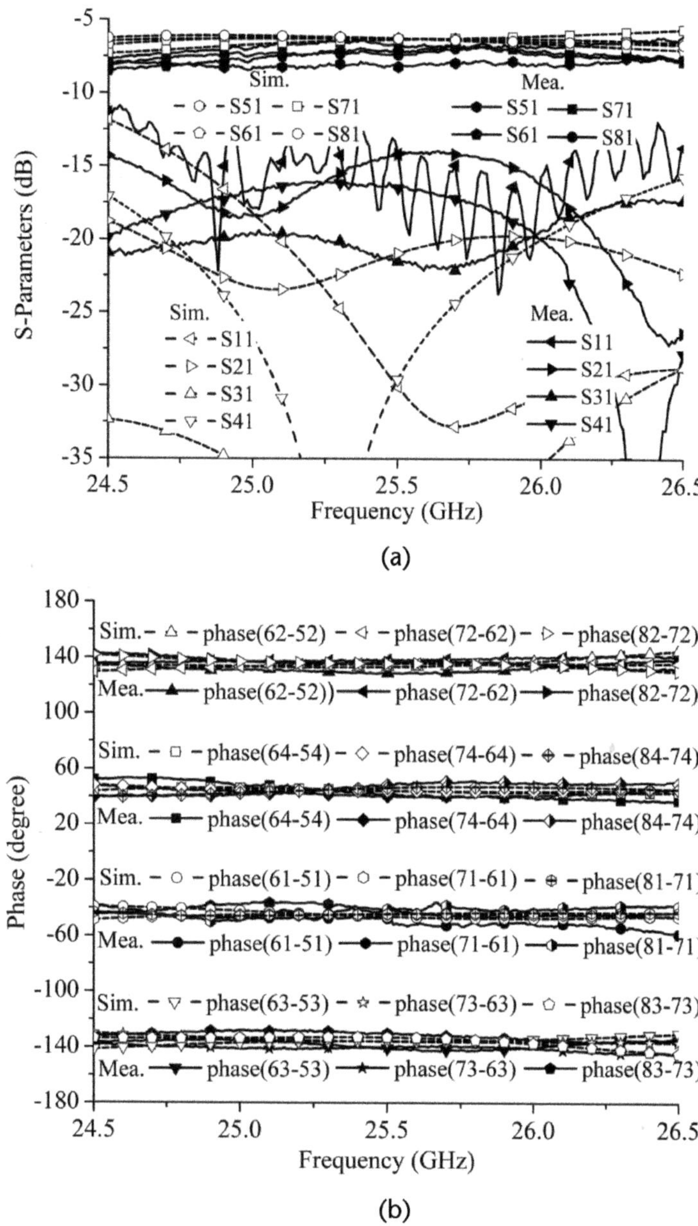

Figure 5.30 Simulated and measured results of the proposed SISL Butler matrix: (a) S-parameters, and (b) phase [11].

while the return loss and isolation are both better than 11 dB. The phase differences between adjacent output ports are given in Figure 5.30(b). The simulated phase imbalance is within 3°. The measured phase imbalance is within 8° in the whole frequency band. The photograph of the fabricated self-packaged SISL patched-based Butler matrix is shown in Figure 5.31.

In a nutshell, the SISL patch-based Butler matrix in the honeycomb concept has compact size, low loss, and high integration properties.

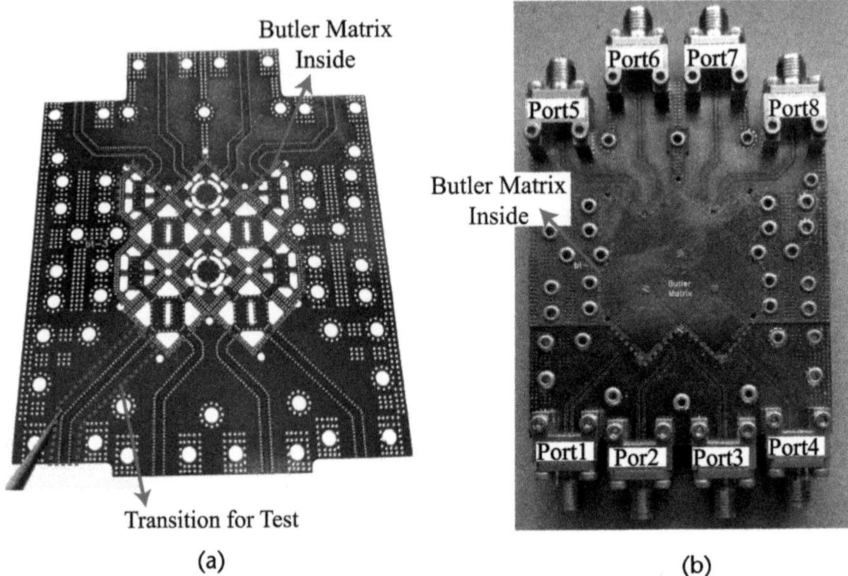

Figure 5.31 Photograph of the fabricated SISL patch-based Butler matrix with low loss and compact size: (a) Substrate 3, and (b) multilayer structure [11].

Acknowledgments

We would like to thank Hongen Cai, Faxian Zhang, Jixuan Ye, and Haozhen Huang for their valuable support to this chapter.

References

[1] Rotholz, E., "Transmission-Line Transformers," *IEEE Transactions on Microwave Theory and Techniques*, Vol. 29, No. 4, 1981, pp. 327–331.

[2] Kuylenstierna, D., and P. Linner, "Design of Broadband Lumped Element Baluns," *2004 IEEE MTT-S International Microwave Symposium Digest*, June 2004, pp. 899–902.

[3] Marchand, N., "Transmission Line Conversion Transformers," *Electron*, Vol. 17, No. 12, 1944, pp. 142–145.

[4] Wang, Y., K. Ma, and S. Mou, "A Compact SISL Balun Using Compensated Interdigital Capacitor," *IEEE Microwave and Wireless Components Letters*, Vol. 27, No. 9, 2017, pp. 797–799.

[5] Wang, Y., K. Ma, and M. Yu, "A Low-Cost Substrate Integrated Suspended Line Platform with Multiple Inner Boards and Its Applications in Coupled-Line Circuits," *IEEE Transactions on Components, Packaging and Manufacturing Technology*, Vol. 10, No. 12, 2020, pp. 2087–2098.

[6] Wang, Y., K. Ma, and N. Yan, "A Low Loss Patch-Based Phase Shifter Based on SISL Platform," *2018 IEEE/MTT-S International Microwave Symposium—IMS*, Philadelphia, PA, June 2018, pp. 281–284.

[7] Wang, Y., and K. Ma, "Low-Loss SISL Patch-Based Phase Shifters and the Mixer Application," *IEEE Transactions on Microwave Theory and Techniques*, Vol. 67, No. 6, 2019, pp. 2302–2312.

[8] Helszajn, J., and D. S. James, "Planar Triangular Resonators with Magnetic Walls," *IEEE Transactions on Microwave Theory and Techniques*, Vol. 26, No. 2, 1978, pp. 95–100.

[9] Bahl, I., and P. Bhartic, *Microwave Solid State Circuit Design*, New York: Wiley, 1988.

[10] Butler, J., and R. Lowe, "Beam-Forming Matrix Simplifies Design of Electronically Scanned Antennas," *Electron. Des.*, Vol. 9, No. 8, 1961, pp. 170–173.

[11] Wang, Y., K. Ma, and Z. Jian, "A Low-Loss Butler Matrix Using Patch Element and Honeycomb Concept on SISL Platform," *IEEE Transactions on Microwave Theory and Techniques*, Vol. 66, No. 8, 2018, pp. 3622–3631.

[12] Wang, Y., and K. Ma, "A Novel Self-Packaged SISL Butler Matrix for Automotive Radar Application," *2017 IEEE MTT-S International Microwave Workshop Series on Advanced Materials and Processes for RF and THz Applications (IMWS-AMP)*, Pavia, Italy, September 20–22, 2017, pp. 1–3.

CHAPTER 6

SISL Power Amplifiers and Low-Noise Amplifiers

This chapter provides an overview of power amplifiers (PAs) and low-noise amplifiers (LNAs) based on the SISL platform. PAs are primarily utilized in RF transmitters to amplify low-power RF signals generated by modulation oscillator circuits. LNAs are predominantly employed in RF receivers to amplify weak signals received by antennas. In wireless communication systems, the efficiency of PAs directly impacts the energy consumption of the entire communication system. Enhancing the efficiency of PAs can effectively reduce the overall energy consumption of system. Additionally, the noise figure of LNAs significantly influences the sensitivity of the communication system, while the degree of noise suppression in the subsequent circuit is determined by gain. Consequently, optimizing the performance of the LNA module is crucial for enhancing the overall system performance. This chapter focuses on the design of PAs and LNAs based on the SISL platform, further investigating the optimization effects that SISL offers for these amplifiers.

6.1 PA

The PA is a critical component in the transmission chain, responsible for amplifying low-power RF signals. The performance of PAs directly affects the quality of the entire communication system. Notably, PAs have a significant impact on the energy consumption of communication systems. Taking 5G base stations as an example, PAs account for only 25% to 30% of the base station cost but contribute to over 60% of the energy consumption. This highlights the importance of reducing power consumption and improving the efficiency of PAs.

6.1.1 Analysis Design of PA

6.1.1.1 Key Indicators

The following are key indicators:

- *Bandwidth:* Defined as the range within which the RF PA can meet or exceed various performance indicators:

$$BW = f_H - f_L \tag{6.1}$$

- *Power gain:* Defined as the ratio of the RF signal power P_{in} at the input to the signal power P_{out} at the output:

$$G\,(\text{dB}) = 10\lg\left(\frac{P_{out}\,(\text{W})}{P_{in}\,(\text{W})}\right) = P_{out}\,(\text{dBm}) - P_{in}\,(\text{dBm}) \tag{6.2}$$

When the PA operates within the linear working area, there exists a linear relationship between the output power of the amplifying transistor and the input power. The slope coefficient is defined as the gain, denoted as G. As the input power increases, the transistor's operating state enters a nonlinear region. In this region, the phenomenon of gain compression occurs, causing the value of G to gradually decrease. Once a certain critical point is reached, the output power no longer varies with the input power, and this critical point is known as the saturated power point. At the saturation power point, the drain efficiency of the power amplifier reaches its maximum, but the output voltage waveform is significantly distorted, resulting in a substantial compression of the gain. When the gain is compressed by 1 dB, the output power point is referred to as the 1-dB compression point. It can be observed that within the 1-dB compression point, there exists a linear relationship between the output power and input power of the PA, with a stable gain. For different types of transistors with the same saturated output power, the 1-dB compression point can be used to assess their linear performance. The larger the 1-dB compression point, the higher the degree of linearization of the transistor. The 3-dB compression point follows the same principle as the power 1-dB compression point.

- *Drain efficiency:* Defined as the ratio of the RF output power P_{out} to the dc consumed power P_{dc}:

$$\eta_{DE} = \frac{P_{out}}{P_{dc}} \times 100\% \tag{6.3}$$

- *Power added efficiency:* Combining the effect of input power on efficiency, defined as the ratio of the difference between RF output power P_{out} and RF input power P_{in} to the dc consumed power P_{dc}:

$$\eta_{DE} = \frac{P_{out} - P_{in}}{P_{dc}} \times 100\% \tag{6.4}$$

Under the condition of low gain or medium gain, the RF input power of the PA cannot be ignored. At this time, the efficiency of the PA can be better characterized by the power added efficiency.

- *Intermodulation distortion:* Under the condition of two-tone input signal, the combined signals of $m\omega_1 + n\omega_2$ $(m, n = 0, \pm1, \pm2, \pm3, \ldots)$ will be generated in the output spectrum, which is called intermodulation distortion. Among them, the three-ordered intermodulation component is near the frequency of the input signal ω_1 and ω_2, which is difficult to filter out, resulting in distortion of the output signal. This effect is called third-order intermodulation distortion.

- *Adjacent channel leakage ratio (ACLR):* When the PA works in the nonlinear region, the RF input signal will undergo spectrum diffusion when passing through the amplifying transistor, which will cause power leakage to adjacent and separated frequency bands. ACLR is defined as the logarithmic ratio result of the total power in adjacent and separated frequency bands to the total power in the central working frequency band f_c. If the PA acts on a broadband modulation signal, ACLR can be regarded as an important index parameter to measure the linearity of the current PA.

6.1.1.2 Advantages of the SISL Platform in PA Design

Heat dissipation is crucial for high-power PAs, and the proposal of the metal integrated and substrate integrated suspended line (MI-SISL) structure can greatly improve the heat dissipation capability of the PA. Figure 6.1 shows a comparison of the heat dissipation capability between the SISL and MI-SISL. It can be observed

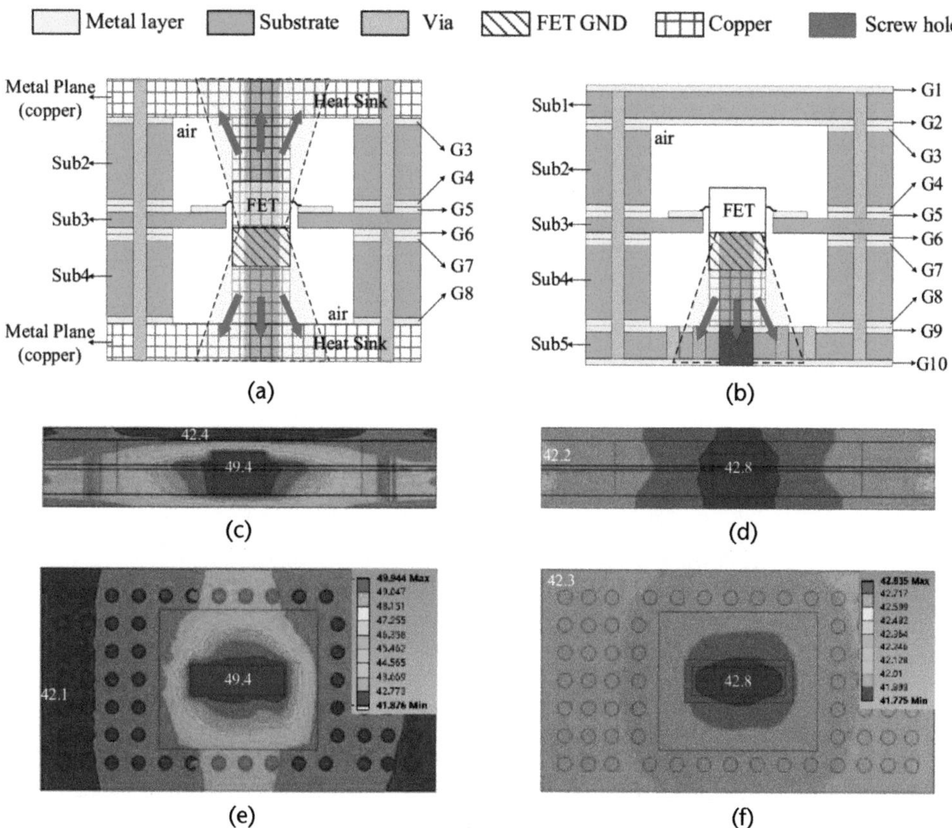

Figure 6.1 Schematic of (a) the SISL platform and (b) the MI-SISL platform and the simulated temperature distribution of (c) the side-cut view of SISL structure, (d) the side-cut view of MI-SISL structure, (e) the top view of the inside of the SISL structure, and (f) the top view of the inside of the MI-SISL structure [1].

that the maximum temperature of the MI-SISL structure is 6.6°C lower than that of the SISL structure. Additionally, the temperature difference of the MI-SISL structure between the inside and the outside is smaller compared to the SISL structure, indicating that the internal heat is more easily conducted to the outside, resulting in a better heat dissipation effect. The information regarding the material and height of each layer is listed in Table 6.1.

6.1.2 SISL PA Examples

The following are two examples of PA based on the SISL. The advantages of the SISL structure in an active circuit can be shown through the following examples.

6.1.2.1 Wideband Harmonic Suppression Bandpass-Filtering Power Based on the SISL Lumped Element

Figure 6.2(a) shows the schematic of the proposed BPF matching network, which comprises four lumped components used as the input impedance transformer of the bandpass-filtering PA. The scattering parameters of the BPF matching network can be expressed as follows [2–5]:

$$S_{11} = \frac{M_1M_3 - \omega^2 M_2M_4}{M_3^2 + \omega^2 M_4^2} + j\frac{\omega\left(M_1M_4 + M_2M_3\right)}{M_3^2 + \omega^2 M_4^2} \tag{6.5}$$

$$\left|S_{12}\right| = \sqrt{1 - \left|S_{11}\right|^2} \tag{6.6}$$

$$M_1 = \left(L_1\omega - X_g\right)F_1L_4Z_L\omega^3 + \left(F_2L_4R_g - F_3Z_L\right)\omega^2 + F_2X_gZ_L\omega + Z_L \tag{6.7}$$

$$M_2 = L_4\left(F_1R_gZ_L - L_1F_2\right)\omega^2 + L_4F_2X_g\omega - F_2R_gZ_L + L_4 \tag{6.8}$$

$$M_3 = \left(L_1\omega + X_g\right)F_1L_4Z_L\omega^3 - \left(L_4R_gF_2 + F_3Z_L\right)\omega^2 - F_2X_gZ_L\omega + Z_L \tag{6.9}$$

$$M_4 = L_4\left(F_1R_gZ_L + L_1F_2\right)\omega^2 + L_4F_2X_g\omega - F_2R_gZ_L - L_4 \tag{6.10}$$

Table 6.1 Information of Materials and Height of Each Layer

Layer	*Height*	*Material*	*Characteristic*
G3-G8	0.035 mm	Copper	Cond = 58000000 Siemens/m
Metal planes	2 mm	Copper	Cond = 58000000 Siemens/m
Sub2, Sub4	1.93 mm	FR4	$\varepsilon_r = 4.4$, $\tan\delta = 0.02$
Sub3	0.254 mm	Rogers_RT_Duroid5880	$\varepsilon_r = 2.2$, $\tan\delta = 0.0009$

Source: [1].

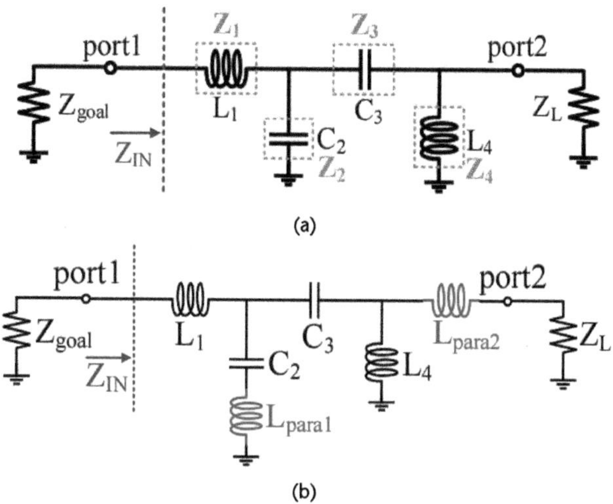

Figure 6.2 (a) Schematic of the BPF matching network, and (b) modified schematic of the BPF matching network [2].

$$F_1 = C_2 C_3 \tag{6.11}$$

$$F_2 = C_2 + C_3 \tag{6.12}$$

$$F_3 = L_1 C_2 + L_1 C_3 + C_3 L_4 \tag{6.13}$$

To realize the expected frequency response, the S-parameters should satisfy the following equations:

$$\mathrm{Re}(S_{11}) = \mathrm{Re}(S_{11}) = 0 \tag{6.14}$$

$$|S_{12}| = 1 \tag{6.15}$$

Then the following relationships can be obtained:

$$L_1 = \frac{C_3 L_4 X_g Z_L \omega^2 + \sqrt{R_g Z_L \Delta_1}}{C_3 L_4 Z_L \omega^3} \tag{6.16}$$

$$C_2 = \frac{C_3 R_g \left[Z_L^2 \left(C_3 L_4 \omega^2 - 1 \right) - L_4^2 \omega^2 \right] + C_3 L_4 \omega \sqrt{R_g Z_L \Delta_3}}{R_g \left(C_3^2 L_4^2 Z_L^2 \omega^4 + \Delta_2 + Z_L^2 \right)} \tag{6.17}$$

$$\Delta_1 = \left(C_3 L_4 \omega^2 - 1 \right)^2 Z_L^2 - C_3^2 L_4^2 R_g Z_L \omega^4 + L_4^2 \omega^2 \tag{6.18}$$

$$\Delta_2 = L_4\omega^2\left(L_4 - 2C_3Z_L^2\right) \tag{6.19}$$

$$\Delta_3 = C_3^2L_4^2Z_L\omega^4\left(Z_L - R_g\right) + \Delta_2 + Z_L^2 \tag{6.20}$$

Taking into account the actual layout, the soldering pad of the SMT capacitor incorporates numerous via holes, resulting in the introduction of a parasitic inductance referred to as L_{para1}. Furthermore, the connection line between the output matching networks and the transition introduces another parasitic inductance known as L_{para2}. Consequently, it is necessary to include these parasitic parameters in the schematic of the matching BPF, which has been modified accordingly, as depicted in Figure 6.2(b).

Figure 6.3(a) depicts the physical structure of the proposed MI-SISL power synthesis amplifier, which consists of two MI-SISL BPF PAs and two SISL LPF power dividers. The power synthesis amplifier utilizes two MI-SISL BPF PAs as power units, while an SISL LPF power divider is connected to the input terminal as a power divider, dividing the input signal into two channels. These two channels of signals then enter their respective MI-SISL bandpass filters for amplification. The amplified signals from both channels are combined using another SISL LPF power splitter connected to the output terminal, resulting in a greater output power compared to a single PA unit. Within the white dotted circle, four SISL-to-stripline transitions are used to connect the LPF power divider and BPF PA unit. Two transitions, from the SISL to CBCPW, are connected to the RF input and output of the power synthesis amplifier, enabling the connection of the circuit inside the MI-SISL air cavity with the external SMA connector for easy testing. The top and bottom metal plates, along with the dense metal vias surrounding the air cavity, form six independent small units. These units enclose the power splitter, combiner, input matching circuit, and output matching circuit of the unit PA in the two branches, effectively preventing mutual interference among the various components. The core circuit size of the MI-SISL power synthesis amplifier is 79.02 mm × 47.2 mm. Considering heat dissipation concerns and ease of assembly, both the LPF power divider and BPF PA in the synthesis PA branch, proposed in this section, are implemented using the MI-SISL structure. Figure 6.3(b) displays split photos of the MI-SISL power synthesis amplifier, showcasing the top metal plate, Sub2, Sub3, and Sub4, down to the bottom metal plate.

Figure 6.4(a) presents the small-signal S-parameter simulation and test curves of the proposed MI-SISL power synthesis amplifier. It can be observed that the PA exhibits a good BPF response, primarily due to the utilization of the LPF function. The splitter demonstrates excellent harmonic suppression. From dc to 0.18 GHz and from 1.32 GHz to 8 GHz, the MI-SISL power combining amplifier achieves a high out-of-band rejection level with an attenuation of 53.7 dBc, effectively suppressing the ninth harmonic. In the frequency range from 0.8 GHz to 1 GHz, the measured $|S_{11}|$ of the MI-SISL power synthesis amplifier is less than −10 dB, and the measured $|S_{21}|$ fluctuates between 11.4 dB and 13.35 dB. Figure 6.4(b) illustrates the simulation and test curves of the saturation point gain, efficiency, and output power of the MI-SISL synthetic amplifier in the frequency band from 0.8 GHz to 1 GHz.

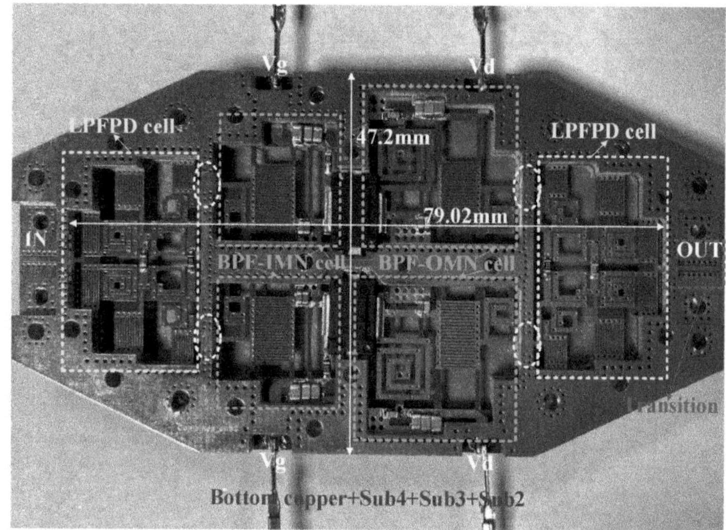

(a)

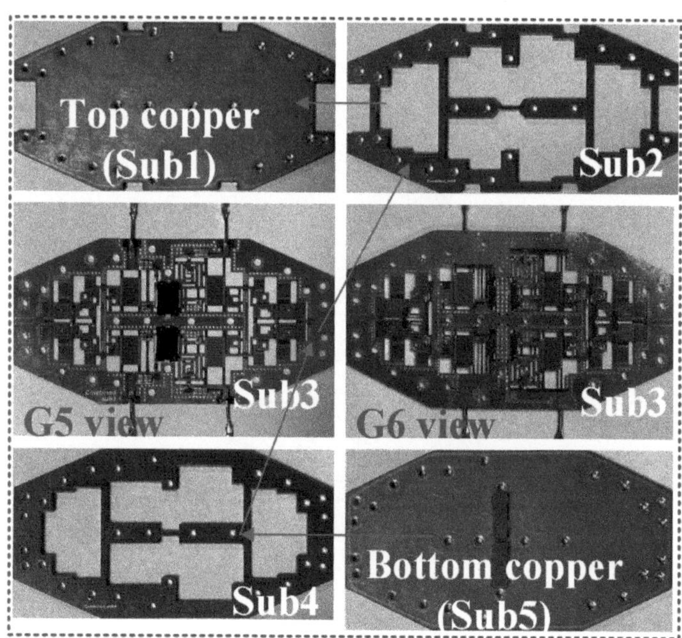

(b)

Figure 6.3 Photograph of: (a) center circuit and (b) split layers of MI-SISL power combined amplifier [2].

As depicted in the figure, the saturation point gain of the MI-SISL power synthesis amplifier fluctuates between 10.3 dB and 13.3 dB; the measured saturated output power ranges from higher than 42.24 dBm to lower than 43.81 dBm; the measured drain efficiency ranges from 36% to 47.2%. Figure 6.4(c) displays the simulation and test curves of the output power P_{out}, gain, power added efficiency PAE, and dc

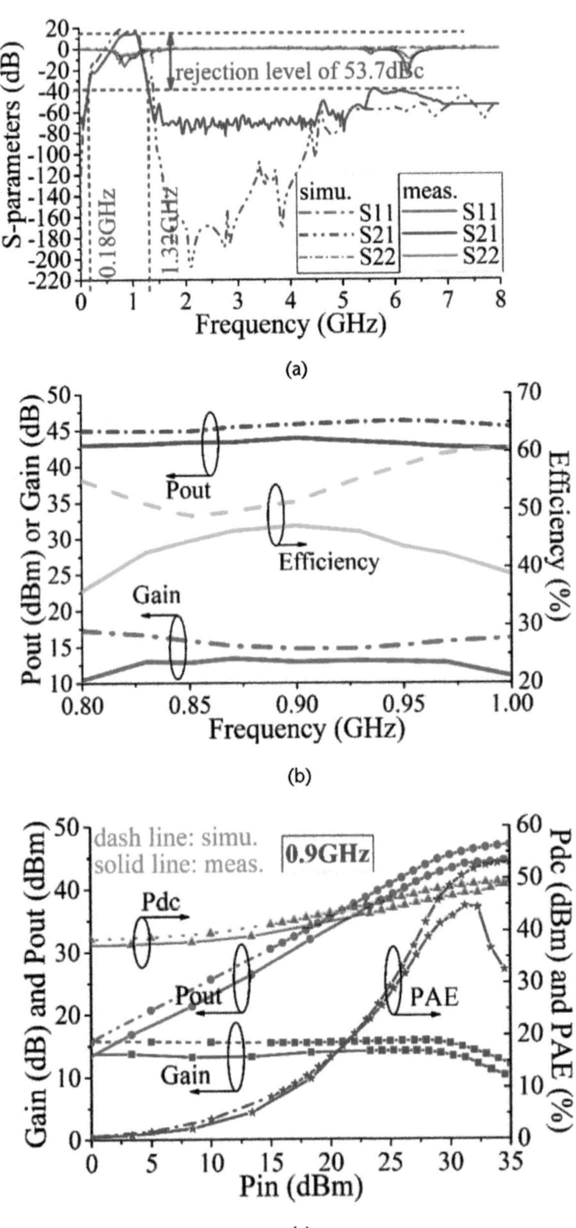

Figure 6.4 Simulated and measured: (a) S-parameters, (b) output power, gain, and efficiency versus frequencies, and (c) output power, gain, PAE, and P_{dc} versus input power of MI-SISL power combined amplifier [2].

power P_{dc} of the MI-SISL power synthesis amplifier as the input power P_{in} varies at the center frequency of 0.9 GHz. It can be observed that the peak PAE of the MI-SISL power combining amplifier is 44.6% at 0.9 GHz.

To assess the nonlinearity of the MI-SISL power synthesis amplifiers, two-tone signals are utilized, with a frequency interval of 10 MHz. Figures 6.5(a, b) display the IMD3 and IMD5 tests of the MI-SISL power synthesis amplifier, respectively. As depicted in Figure 6.5(a), the measured IMD3 when the output power of the

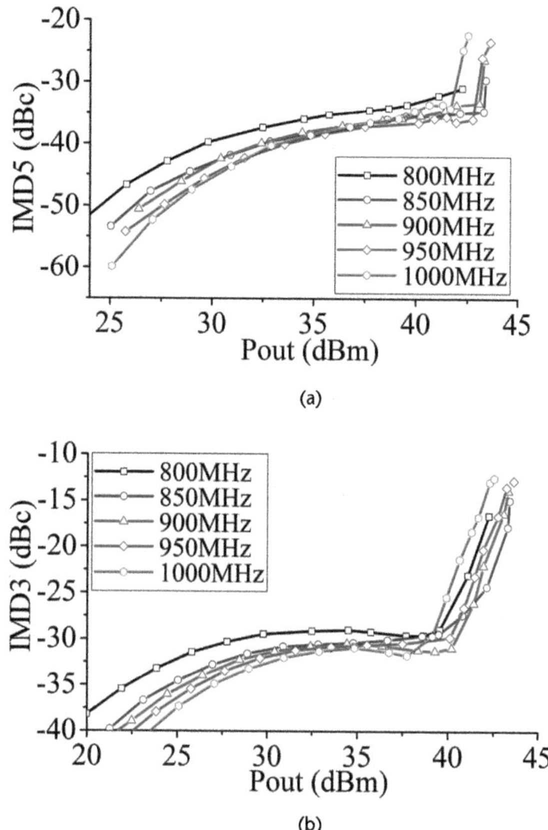

Figure 6.5 Measured (a) IMD3 and (b) IMD5 of the proposed MI-SISL power combined amplifier [2].

MI-SISL power combining amplifier is backed off by 3 dB is −26 dBc. Similarly, Figure 6.5(b) shows that the measured IMD5 of the MI-SISL power combining amplifier is −34.7 dBc when the output power of the amplifier is backed off by 3 dB. These results indicate that the proposed MI-SISL BPF power synthesis amplifier exhibits good linearity.

Table 6.2 compares the performance of the proposed MI-SISL BPF power synthesis amplifier with other filter PAs published in recent years. The performance comparison and discussion are as follows:

1. *Efficiency:* Table 6.2 demonstrates that the efficiency of the filter PA proposed in this chapter is lower compared to the PAs in [6, 7]. The filter PAs in the aforementioned references employ high-performance GaN power transistors CGH40010, which typically achieve an efficiency as high as 65%. However, this chapter utilizes low-cost MOSFET power transistors, such as the MW6S010N, which have a typical efficiency of only 32%. Despite the lower efficiency, the price of the MW6S010N power transistor is less than half of the CGH40010. Although the efficiency of the filter PA designed in this chapter does not match that of the PAs based on GaN power transistors, it offers a more affordable cost [6, 7]. Moreover, the proposed filter PAs in this chapter achieve efficiencies of 57.5% and 47.2%, respectively.

Table 6.2 Performance Comparison of Recently Published Filtering PAs

Reference		*Bandwidth (GHz)/(%)*	*Filter Type*	*Class of Operation*	*Small Signal Gain (dB)*	*Transistor Type*	*Selected Transistor/ Typical Saturation Efficiency*	*Efficiency (%)*	P_{out} *(dBm)*	*Bandpass Response*	*Harmonic Suppression*	*40-dB Rectangle Coefficient**	*Size*	*Self-Packaging*
[7]	2020	1.6–2.8/54.5	Coupled microstrip BPF	NA	NA	GaN	CGH40010F/65%	49–65 (D)**	38–41	NA	NA	NA	0.79 × 0.62	No
[8]	2020	2–2.4/18.2	Microstrip BPF	NA	13.4–15.1	GaN	CGH40010/65%	69–78.2 (D)	39–40.4	Yes	31-dBc rejection up to 2.9 f_0	NA	0.98 × 0.74	No
[12]	2019	5.4–5.6/3.6	SIW filtering balun	Push-pull	12.5	GaAs	FLM53594F/37%	48.7 (P)***	38.5	Yes	40-dBc rejection up to 5f_0	14.3	188 mm × 157 mm	No
[9]	2017	3-dB FBW 2.3	Dielectric resonator BPF	Class-F	18.3	GaN	CGH40010F/65%	70.7 (P)	40–41.06	Yes	NA	NA	NA	No
[10]	2016	1.72–1.85/7.2	Bandpass quadrature coupler	Doherty	14.3	MOSFET	MW6S004NT1/33%	41.8 (P)	38.1–40	Yes	40-dBc rejection up to 3f_0	NA	1.37 × 0.67	No
[6]	2014	2.1–2.7/25	Microstrip BPF	NA	16.4	GaN	CGH40010F/65%	69.8 (P)	40.1	Yes	Poor	NA	0.65 × 0.48	No
This work	Power combined amplifier	0.8–1/22.5	SISL BPF + SISL LPFPD	Class AB	11.4–13.35	MOSFET	MW6S010N/32%	36–47.2 (D)	42.24–43.81	Yes	53.7-dBc rejection up to 9 f_0	3.12	0.25 × 0.15	Yes

Source: [5].
*40-dB rectangle coefficient: the ratio of 40-dB bandwidth over 3-dB bandwidth of small signal $|S_{21}|$.
**(D): drain efficiency.
***(P): power added efficiency (PAE).

This represents a significant improvement when compared to the typical efficiency of 32% for the power transistor. In a similar vein, literature [7] also has employed low-cost metal-oxide-semiconductor field-effect transistor (MOSFET) power transistors with a typical efficiency of 33%, yet the filter PA in literature [7] has achieved an efficiency of only 41.8%.

2. *Bandwidth, power, gain, and size:* Table 6.2 reveals that the designed filter PA in this chapter possesses a narrower bandwidth compared to literature [6, 7], while being wider than the other PAs listed in Table 6.2. Despite this, the filter PA presented in this chapter achieves the maximum output power. Although the small-signal gain of the two proposed PAs is not the highest, the utilization of high-capacitance density SISL double-layer interdigitated capacitors and high-inductance density, along with high Q-value SISL double-layer spiral inductors, allows for a significantly smaller size of the filter PA. Even when including two power splitters and two filter PA units in the power-combining amplifier, the overall size remains much smaller than other filter PAs in Table 6.2. To our knowledge, the MI-SISL BPF PA proposed in this chapter is the most compact in size among board-level filter PAs.
3. *Filter performance and harmonic suppression:* Due to the effects of BPF-impedance transformer networks and BPF-output matching networks, Table 6.2 demonstrates that the proposed MI-SISL filter PA in this chapter achieves an out-of-band rejection level with an attenuation of up to 40 dBc in the fifth-harmonic frequency range. This performance exceeds that of literature [8–10]. Furthermore, thanks to the excellent harmonic suppression effect of the LPF power splitter, the MI-SISL filter power synthesis amplifier in this chapter achieves a remarkable suppression level of 53.7 dBc in the frequency range of the ninth harmonic, as indicated in Table 6.2. This represents the highest level of harmonic suppression. The square coefficient is a measure of the BPF's filtering ability. A square coefficient close to 1 indicates a steep sideband and a high filtering ability. The squareness coefficients of the MI-SISL filter PA and MI-SISL filter power synthesis amplifier proposed in this chapter are 3.45 and 3.12, respectively, while [7] has reported 5.6, [11] has reported 3.4, and [12] has reported 14.3. Only the two BPF PAs in this chapter and the filter PA in [11] exhibit squareness coefficients closest to 1, reflecting their high filtering abilities.
4. *Self-packaging:* Table 6.2 reveals that only the MI-SISL filter PA and the MI-SISL filter power synthesis amplifier proposed in this chapter possess the characteristic of self-packaging. In terms of application, these two PAs, based on MI-SISL, can be directly connected to other devices after processing and assembly. However, other filter PAs listed in Table 6.2 require additional processing, such as the installation of metal shielding shells, to achieve electromagnetic shielding. Literature [13] has highlighted the presence of a metal cover plate effect after installing a metal shielding shell in the microstrip PA. If the height of the metal shielding shell is set unreasonably, it can result in performance degradation of the PA. However, the MI-SISL filter PA and MI-SISL filter power synthesis amplifier with self-packaging characteristics are designed collaboratively, considering both the circuit and package aspects. Hence, these issues can be addressed during the design process.

In summary, the proposed MI-SISL filter PA and MI-SISL filter power synthesis amplifier offer several advantages, including BPF response, high-frequency selectivity, miniaturization, high output power, high efficiency, high-order harmonic suppression, and self-packaging.

6.1.2.2 Dual-Band and Dual-State Doherty PA Using Metal-Integrated and Substrate-Integrated Suspended-Line Technology

In a practical dual-band design, the optimal impedance of the transistor at the dual band differs from that of a single-band power amplifier due to the presence of non-negligible nonlinear parasitic components associated with the operating frequency in high-power packaged devices. Figure 6.6(a) illustrates the package model of the power transistor at saturation and back-off. The dual-path amplifier (DPA) combines a carrier amplifier and a peaking amplifier to enhance back-off efficiency through load modulation. By replacing the transistors with current generators (CGs), the schematic of the DPA is depicted in Figure 6.6(b).

The matching network typically incorporates output matching networks and impedance transformer networks, as illustrated in Figure 6.7(a).

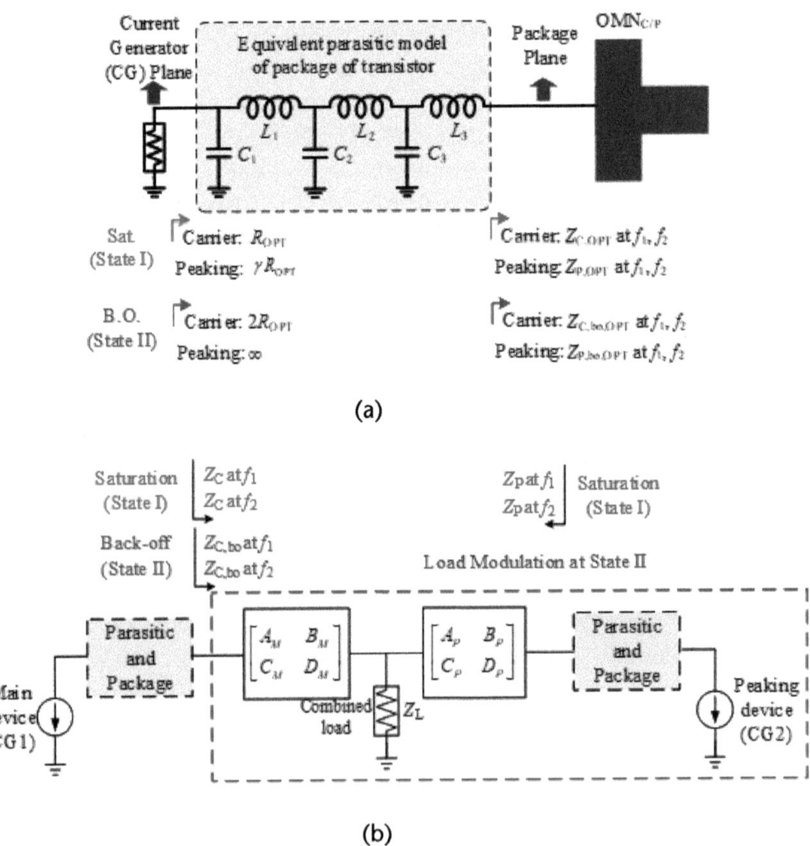

Figure 6.6 (a) Package model of the power transistor at saturation and back-off, and (b) architecture of proposed dual-band and dual-state Doherty PA [4].

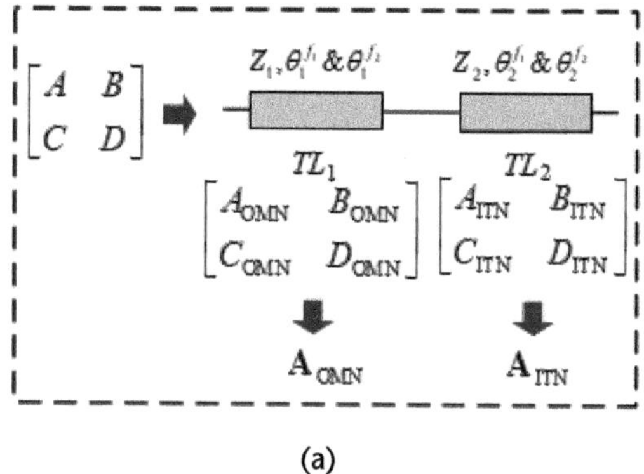

(a)

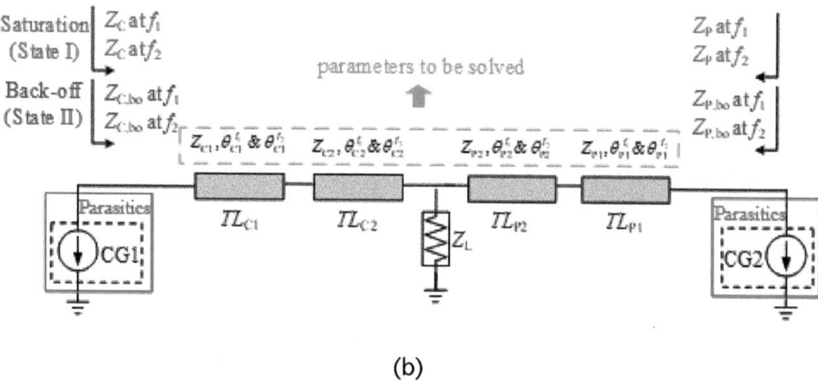

(b)

Figure 6.7 (a) Matching network composed of output matching networks and impedance transformer networks, and (b) equivalent model of proposed dual-band DPA [4].

The ABCD matrix of the output matching network and impedance transformer network has a similar expression to that of a transmission line, as follows:

$$\begin{bmatrix} A & B \\ C & D \end{bmatrix} = \begin{bmatrix} \cos\theta^{f_1(f_2)} & jZ\sin\theta^{f_1(f_2)} \\ j\dfrac{\sin\theta^{f_1(f_2)}}{Z} & \cos\theta^{f_1(f_2)} \end{bmatrix} \tag{6.21}$$

Then output matching networks and impedance transformer networks can be analyzed using the ideal transmission-line model, as shown in Figure 6.7(b). To more accurately represent the actual network with transmission lines, we define phase dispersion μ as the representation of the phase slope of the equivalent transmission line, where μ is greater than or equal to 1 ($\mu = 1$ for an ideal transmission line (TL)). Thus, the phase slopes of TL_{C1}, TL_{C2}, TL_{P1}, and TL_{P2} are represented by μ_{C1}, μ_{C2}, μ_{P1}, and μ_{P2}, respectively. The frequency-dependent phase of transmission lines defined as:

$$\theta^{f_2}_{C1/C2/P1/P2} = \theta^{f_1}_{C1/C2/P1/P2} - \mu_{C1/C2/P1/P2}\theta^{f_1}_{C1/C2/P1/P2}(1-k) \tag{6.22}$$

$$k = \frac{f_2}{f_1} \tag{6.23}$$

The matching network composed of transmission lines is frequency-dependent. In order to achieve a dual-band DPA, the input impedance of the matching network in Figure 6.7(b) needs to match the optimal impedance in two states at the dual band. The optimal impedance Z_{OPT} can be obtained by combining the parasitic parameters of the transistor and package, as shown in Figure 6.6, defined as follows:

$$Z_{OPT} = F_1\left(L_1, L_2, L_3, C_1, C_2, C_3, \gamma R_{OPT}, f\right) \tag{6.24}$$

$$F_1 = \frac{\left(\begin{array}{l}\frac{1}{Z_A} + j\left(L_1, L_2, L_3\right)\omega - \frac{\left(C_2L_2 + C_2L_3 + C_3L_3\right)}{Z_A}\omega^2 \\ -j\left(C_2L_1L_2 + C_2L_1L_3 + C_3L_1L_3 + C_3L_2L_3\right)\omega^3 + \frac{C_2C_3L_2L_3}{Z_A}\omega^4 + jC_2C_3L_1L_2L_3\omega^5\end{array}\right)}{\left(1 + j\frac{\left(C_2 + C_3\right)}{Z_A}\omega - \left(C_2L_1 + C_3L_1 + C_3L_2\right)\omega^2 - j\frac{C_2C_3L_2}{Z_A}\omega^3 + C_2C_3L_1L_2\omega^4\right)} \tag{6.25}$$

$$Z_A = \frac{1}{\gamma R_{OPT}} + j\omega C_1 \tag{6.26}$$

In Figure 6.7(b), $Z_{T,C}$ and $Z_{T,P}$ can be calculated as:

$$Z_{T,C/T,P} = Z_{C2/P2}\frac{2Z_L + jZ_{C2/P2}\tan\theta^f_{C2/P2}}{Z_{C2/P2} + 2jZ_L\tan\theta^f_{C2/P2}} \tag{6.27}$$

At saturation state, define the combining load of the two branches as Z_L, and the impedance of two branches in Figure 6.7(b) has the following relationship:

$$Z_{C2} = Z_{P2} = 2Z_L \tag{6.28}$$

At this state, to satisfy the Doherty operation conditions, Z_C and Z_P should match $Z_{C,OPT}$ and $Z_{P,OPT}$, and the unknown parameter Z_{C1} and Z_{P1} can be calculated as follows:

$$Z_{C1/P1} = j\frac{2Z_L - Z^{f_1}_{C/P,OPT} \pm \left(-\sqrt{Z^{f_1}_R}\right)}{2\tan\theta^{f_1}_{C1/P1}} \text{ at } f_1 \tag{6.29}$$

$$Z_R^{f_1} = \left(2Z_L - Z_{C/P,\mathrm{OPT}}^{f_1}\right)^2 - 8Z_L Z_{C/P,\mathrm{OPT}}^{f_1}\left(\tan\theta_{C1/P1}^{f_1}\right)^2 \tag{6.30}$$

$$Z_{C1/P1} = j\frac{2Z_L - Z_{C/P,\mathrm{OPT}}^{f_2} \pm \left(-\sqrt{Z_R^{f_2}}\right)}{2\tan\theta_{C1/P1}^{f_2}} \text{ at } f_2 \tag{6.31}$$

$$Z_R^{f_2} = \left(2Z_L - Z_{C/P,\mathrm{OPT}}^{f_2}\right)^2 - 8Z_L Z_{C/P,\mathrm{OPT}}^{f_2}\left(\tan\theta_{C1/P1}^{f_2}\right)^2 \tag{6.32}$$

Since the impedance conditions need to satisfy both (6.29) and (6.31), the following relationship can be obtained:

$$j\frac{2Z_L - Z_{C/P,\mathrm{OPT}}^{f_1} \pm \left(-\sqrt{Z_R^{f_1}}\right)}{2\tan\theta_{C1/P1}^{f_1}} = j\frac{2Z_L - Z_{C/P,\mathrm{OPT}}^{f_2} \pm \left(-\sqrt{Z_R^{f_2}}\right)}{2\tan\theta_{C1/P1}^{f_2}} \tag{6.33}$$

$$\begin{aligned} &2Z_L\left(\tan\theta_{C1/P1}^{f_2} - \tan\theta_{C1/P1}^{f_1}\right) - \tan\theta_{C1/P1}^{f_2}\left(Z_{C/P,\mathrm{OPT}}^{f_1} \pm \sqrt{Z_R^{f_1}}\right) \\ &+ \tan\theta_{C1/P1}^{f_1}\left(Z_{C/P,\mathrm{OPT}}^{f_2} \pm \sqrt{Z_R^{f_2}}\right) = 0 \end{aligned} \tag{6.34}$$

$$\tan\theta_{C1/P1}^{f_2} = -j\frac{\left(\cos\theta_{C1/P1}^{f_1} + j\sin\theta_{C1/P1}^{f_1}\right)^{2\left[1-\mu_{C1/P1}(1-k)\right]} - 1}{\left(\cos\theta_{C1/P1}^{f_1} + j\sin\theta_{C1/P1}^{f_1}\right)^{2\left[1-\mu_{C1/P1}(1-k)\right]} + 1} \tag{6.35}$$

Then with the help of calculation software, the value of $\theta_{C1/P1}^{f_1}$ and $\theta_{C1/P1}^{f_2}$ can be finally obtained.

At 6-dB back-off state, only the carrier PA is active, while the peak PA is closed. Therefore, $Z_{C,\mathrm{bo}}$ can be calculated as follows:

$$Z_{C,\mathrm{bo}} = F_2\left(Z_{P,\mathrm{bo,OPT}}, Z_{C1}, Z_{P1}, \theta_{C1}^f, \theta_{P1}^f, \theta_{C2}^f, \theta_{P2}^f\right) \text{ at } f_1 \;\&\; f_2 \tag{6.36}$$

$$F_2 = \frac{Z_{C1}\left(Z_{P1}Z_L C_{O1} + jZ_{C1}Z_L\tan\theta_{C1}^f C_{O2} + jZ_{P1}Z_{C1}\tan\theta_{C1}^f C_{O3} + Z_L^2 C_{O4}\right)}{\left(jZ_{P1}Z_L\tan\theta_{C1}^f C_{O1} + Z_{C1}Z_L C_{O2} + Z_{P1}Z_{C1}C_{O3} + jZ_L^2\tan\theta_{C1}^f C_{O4}\right)} \tag{6.37}$$

$$C_{O1} = \left(Z_{P,\mathrm{bo,OPT}} + jZ_{P1}Z_L\tan\theta_{P1}^f\right)\cdot\left(2\tan\theta_{P2}^f\tan\theta_{C2}^f - 4j\tan\theta_{C2}^f - 2\right) \tag{6.38}$$

$$C_{O2} = \left(Z_{P1} + jZ_{P,\mathrm{bo,OPT}}\tan\theta_{P1}^f\right)\cdot\left(2\tan\theta_{P2}^f\tan\theta_{C2}^f - 4j\tan\theta_{P2}^f - 2\right) \tag{6.39}$$

$$C_{O3} = \left(Z_{P1}\tan\theta_{P1}^{f} - jZ_{P,bo,OPT}\right)\cdot\left(\tan\theta_{P2}^{f} + \tan\theta_{C2}^{f} - 2j\right) \tag{6.40}$$

$$C_{O4} = \left(Z_{P1} - jZ_{P,bo,OPT}\tan\theta_{P1}^{f}\right)\cdot\left(8\tan\theta_{P2}^{f}\tan\theta_{C2}^{f} - 4j\tan\theta_{P2}^{f} - 4j\tan\theta_{C2}^{f}\right) \tag{6.41}$$

When $Z_{C,bo}$ is matched to $Z_{C,bo,OPT}$ at both f_1 and f_2, $\theta_{C2}^{f_1}$ and $\theta_{P2}^{f_1}$ can be calculated.

The second harmonic of the amplifier is modulated using the harmonic control network, as shown in Figure 6.8, which consists of two parts. Two open stubs at the first part are used to control the dual-band second-harmonic impedance. Here, the impedance and phase of the two open stubs are represented by $\{Z_{H1}, Z_{H2}\}$ and $\{\theta^{f}_{H1}, \theta^{f}_{H2}\}$, respectively. The second part collaborates with the first part to ensure impedance matching at the two desired fundamental frequencies, which is composed of four transmission lines and a capacitor. The characteristic impedance and phase of the four matching transmission lines are marked by $\{Z_1, Z_2, Z_3, Z_4\}$ and $\{\theta_1^f, \theta_2^f, \theta_3^f, \theta_4^f\}$, respectively. The ABCD matrix of the proposed dual-band output matching networks (OMNs) can be expressed as follows:

$$A_{OMN} = \begin{bmatrix} A_{OMN} & B_{OMN} \\ C_{OMN} & D_{OMN} \end{bmatrix} = \begin{bmatrix} A_{OMN_{part1}} & B_{OMN_{part1}} \\ C_{OMN_{part1}} & D_{OMN_{part1}} \end{bmatrix} \cdot \begin{bmatrix} A_{OMN_{part2}} & B_{OMN_{part2}} \\ C_{OMN_{part2}} & D_{OMN_{part2}} \end{bmatrix} \tag{6.42}$$

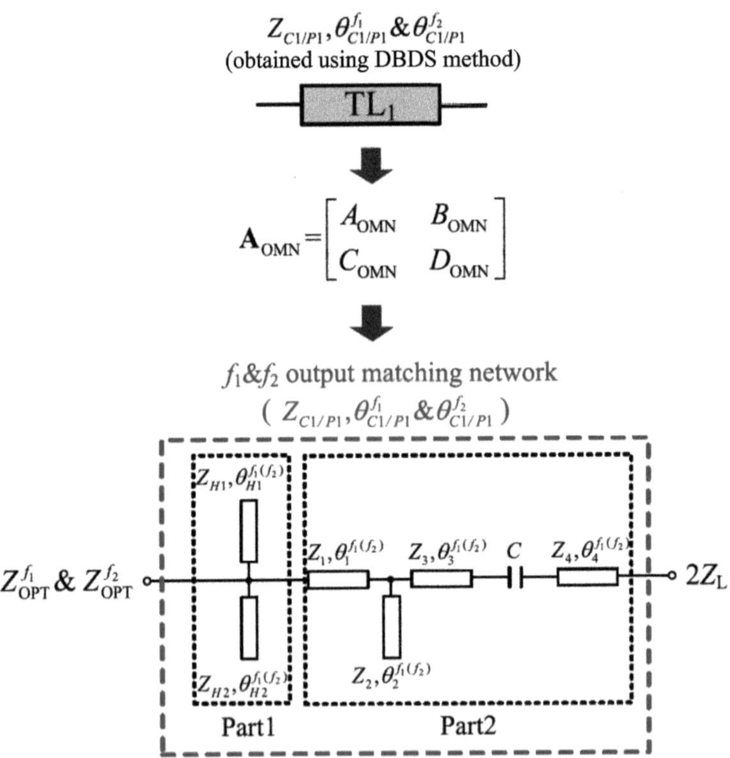

Figure 6.8 Proposed dual-band output matching networks circuit model with second-harmonic-tuned [4].

$$\begin{bmatrix} A_{\text{OMN}_{\text{part1}}} & B_{\text{OMN}_{\text{part1}}} \\ C_{\text{OMN}_{\text{part1}}} & D_{\text{OMN}_{\text{part1}}} \end{bmatrix} = \begin{bmatrix} 1 & 0 \\ j\dfrac{Z_{H2}\tan\theta_{H1}^{f} + Z_{H1}\tan\theta_{H2}^{f}}{Z_{H1}Z_{H2}} & 1 \end{bmatrix} \tag{6.43}$$

$$\begin{aligned} \begin{bmatrix} A_{\text{OMN}_{\text{part2}}} & B_{\text{OMN}_{\text{part2}}} \\ C_{\text{OMN}_{\text{part2}}} & D_{\text{OMN}_{\text{part2}}} \end{bmatrix} &= \begin{bmatrix} \cos\theta_1^f & jZ_1\sin\theta_1^f \\ j\dfrac{\sin\theta_1^f}{Z_1} & \cos\theta_1^f \end{bmatrix} \cdot \begin{bmatrix} 1 & 0 \\ j\dfrac{\tan\theta_2^f}{Z_2} & 1 \end{bmatrix} \\ &\cdot \begin{bmatrix} \cos\theta_3^f & jZ_3\sin\theta_3^f \\ j\dfrac{\sin\theta_3^f}{Z_3} & \cos\theta_3^f \end{bmatrix} \cdot \begin{bmatrix} 1 & \dfrac{1}{j\omega C} \\ 0 & 1 \end{bmatrix} \cdot \begin{bmatrix} \cos\theta_4^f & jZ_4\sin\theta_4^f \\ j\dfrac{\sin\theta_4^f}{Z_4} & \cos\theta_4^f \end{bmatrix} \end{aligned} \tag{6.44}$$

According to the conversion between ABCD and S-parameters for complex impedances, the forward transmission coefficient S_{21} of output matching networks can be calculated using the following equation:

$$\theta_{21}^{f_1} = F_3\left(Z_{H1}, Z_{H2}, Z_1, Z_2, Z_3, Z_4, Z_L, Z_{\text{OPT}}^{f_1}, \theta_{H1}^{f_1}, \theta_{H2}^{f_1}, \theta_1^{f_1}, \theta_2^{f_1}, \theta_3^{f_1}, \theta_4^{f_1}\right) \tag{6.45}$$

$$\theta_{21}^{f_2} = F_3\left(Z_{H1}, Z_{H2}, Z_1, Z_2, Z_3, Z_4, Z_L, Z_{\text{OPT}}^{f_2}, \theta_{H1}^{f_2}, \theta_{H2}^{f_2}, \theta_1^{f_2}, \theta_2^{f_2}, \theta_3^{f_2}, \theta_4^{f_2}\right) \tag{6.46}$$

$$F_3 = \frac{2\left(A_{\text{OMN}}D_{\text{OMN}} - B_{\text{OMN}}C_{\text{OMN}}\right)\left(R_{01}R_{02}\right)^{1/2}}{2A_{\text{OMN}}Z_L + B_{\text{OMN}} + 2C_{\text{OMN}}Z_{\text{OPT}}^{f}Z_L + D_{\text{OMN}}Z_{\text{OPT}}^{f}} \tag{6.47}$$

Among them, R_{01} and R_{02} are the real part of Z_{OPT}^{f} and Z_L. To obtain the remaining unknown values, here the parameter optimization method is used, and four optimal goals are given as:

$$\left|S_{21}^{f_1}\right| \to 1 \tag{6.48}$$

$$\left|S_{21}^{f_2}\right| \to 1 \tag{6.49}$$

$$\angle S_{21}^{f_1} \to \theta_{C1/P1}^{f_1} \tag{6.50}$$

$$\frac{\angle S_{21}^{f_2} - \angle S_{21}^{f_1}}{(k-1)\cdot\angle S_{21}^{f_1}} \to \mu_{\text{C1/P1}} \tag{6.51}$$

The schematic circuit diagram of the proposed dual-band Doherty PA based on the MI-SISL are illustrated as Figure 6.9(a). The relevant design parameters

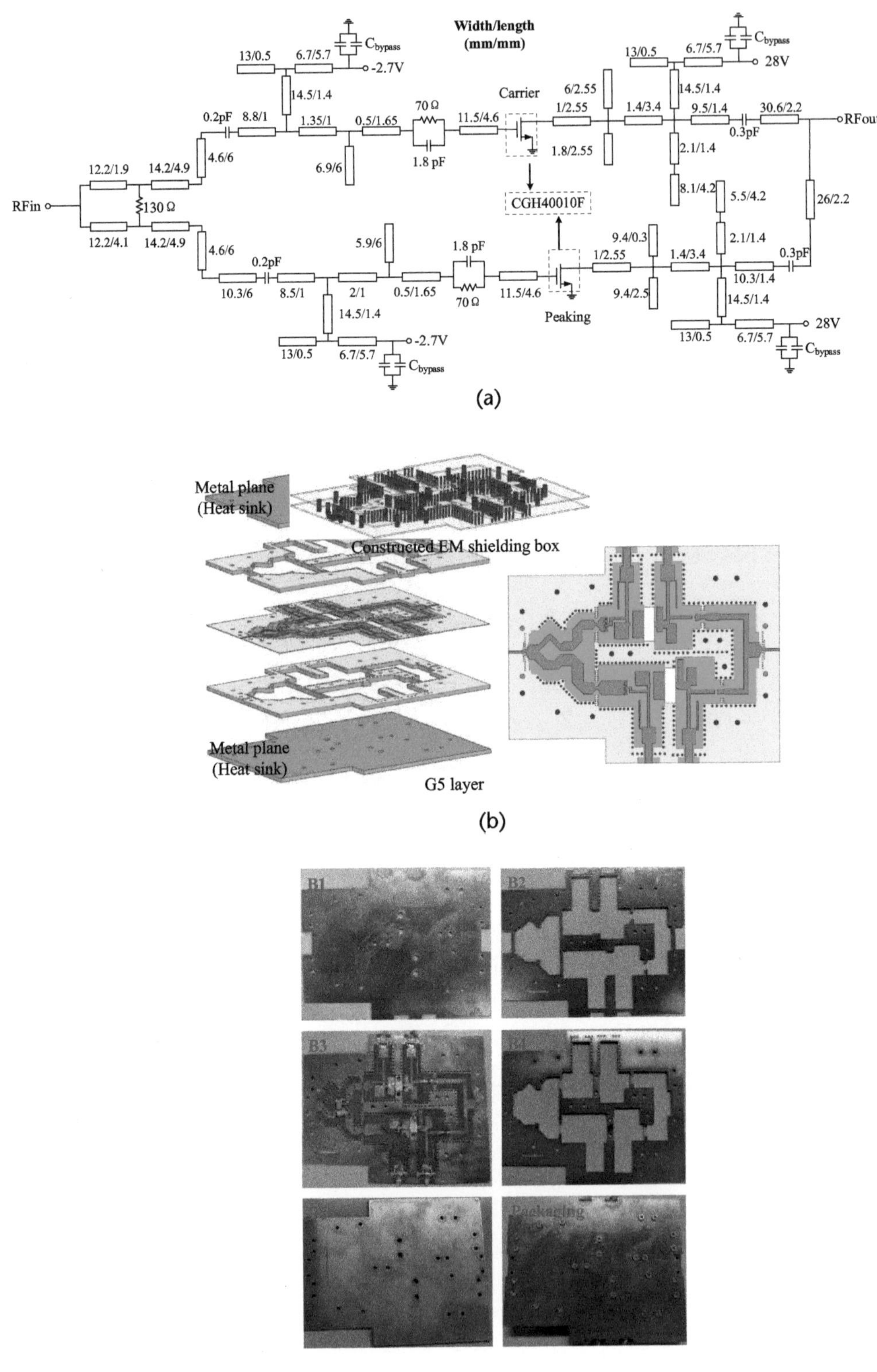

Figure 6.9 (a) Schematic, (b) 3D electromagnetic model, and (c) photograph of the proposed MI-SISL dual-band DPA [4].

have been marked in the figure. Commercial GaN HEMTs CGH40010F from Wolfspeed were selected for the design of dual-band DPA with the center frequencies of 3.4G/4.9G. The 3D electromagnetic model and photograph of the designed MI-SISL PA simulation model are shown in Figures 6.9(b, c). This model applies the codesign of dual-band DPA and package based on the MI-SISL platform. This design has excellent electromagnetic compatibility defense thanks to its five levels, which together create electromagnetic shielding efficacy. Four transitions are used to prevent cavity resonance inside the circuit, and two transitions are used to link the signal trace to SMA connections. The complete transition consists of conductor-backed coplanar waveguide, stripline, and suspended line. Figures 6.10(a, b) illustrate the simulated gain and drain efficiency (DE) of the proposed DPA at the two operating bands, respectively, as a function of output power.

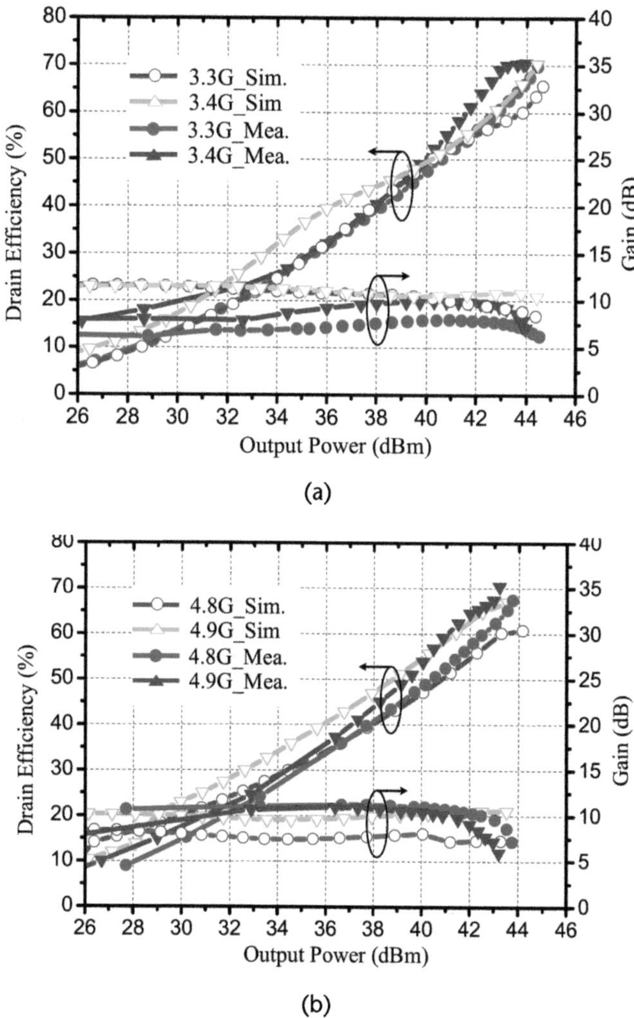

Figure 6.10 Simulated and measured gains and DE versus output power: (a) 3.4 GHz, and (b) 4.9 GHz [4].

Dimensions Via CW Signal

To evaluate the power and efficiency performance of the manufactured DPA, CW measurement under large-signal operation was conducted based on the aforementioned study. Performance at saturation and back-off was tested from 3.3 to 3.4 GHz and 4.8 to 4.9 GHz, respectively, with a frequency step of 0.05 GHz to accurately depict the DPA's performance across various frequency bands. At 3.3–3.4 GHz and 4.8–4.9 GHz, the suggested DPA achieved saturation powers ranging from 44 to 44.5 dBm and 43.2 to 43.75 dBm, respectively. Figure 6.10 illustrates that at maximum output power and a 6-dB back-off area, the DPA achieved DE of 70.7%–73.2%/67.5%–70.4% and 39%–42.4%/40%–42.2%, respectively. It is observed that the Doherty operation of the designed DPA is not straightforward based on the simulation and measurement results mentioned above. Several factors, including basic impedance mismatch and the impact of harmonic control in different states, may contribute to this. Specifically, the nonlinear region causes the harmonic components to increase when the DPA operates at saturation, and the second-harmonic tuning significantly enhances peak efficiency. However, the linear region causes the harmonic components to occasionally arise when the DPA operates at back-off, resulting in back-off efficiency comparable to DPAs in a similar frequency range in Table 6.3. Consequently, the peaking efficiency is significantly higher than the back-off efficiency, indicating the nontrivial nature of the Doherty operation in the designed DPA. Second-harmonic tuning has a more pronounced effect at higher operating frequencies. The comparison of simulation and measurement results at various frequencies is presented in Figure 6.11. It can be observed that the output power decreases at high frequencies compared to the simulated results. This deviation can be attributed to inaccuracies in the large-signal model and the degraded performance of the transistor at these frequencies. The performance comparison of dual-band DPA designs is summarized in Table 6.3. This design optimizes the circuit as a whole by employing direct matching and, when used in conjunction with second-harmonic impedance tuning, it achieves good DE.

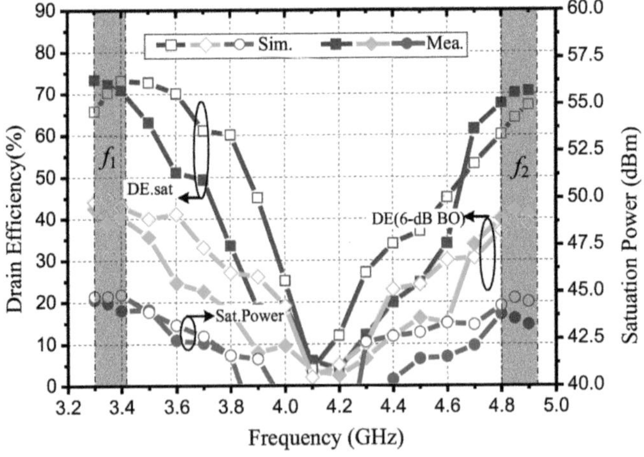

Figure 6.11 Simulated and measured performance versus frequency [4].

Table 6.3 Performance of the Proposed DPA Under the Stimulation of a 20-/100-MHz Modulated Signal with a 6-dB PAPR

	20-MHz Modulated Signal		*100-MHz Modulated Signal*	
Frequency (GHz)	*Average DE (%)*	*ACPR (dBc)*	*Average DE (%)*	*ACPR (dBc)*
3.3	43.7	−27.4/−28.9	43.9	−29.3/−29.7
3.35	39.8	−28.9/−27.8	39.9	−29.7/−28.4
3.4	37.9	−29.1/−31.4	38.3	−29.4/−29.7
4.8	41.8	−30.5/−29.3	41.1	−31.3/−33.6
4.85	43.2	−30.9/−28.3	42.8	−28.1/−28.9
4.9	40.9	−29.3/−31.9	40.5	−31.7/−29.6

Source: [4].

Detection of Modulated Signals

The DPA prototype was evaluated using modulated signal stimulation at various signal bandwidths and operating frequencies for further investigation. First, a 20-MHz modulated signal with 6-dB peak-to-average power ratio (PAPR) was used to assess the performance of the manufactured DPA at the working frequencies. The corresponding output spectrum is shown in Figures 6.12(a, b). Additionally, the capability of the proposed DPA was evaluated using a 100-MHz, 6-dB modulated

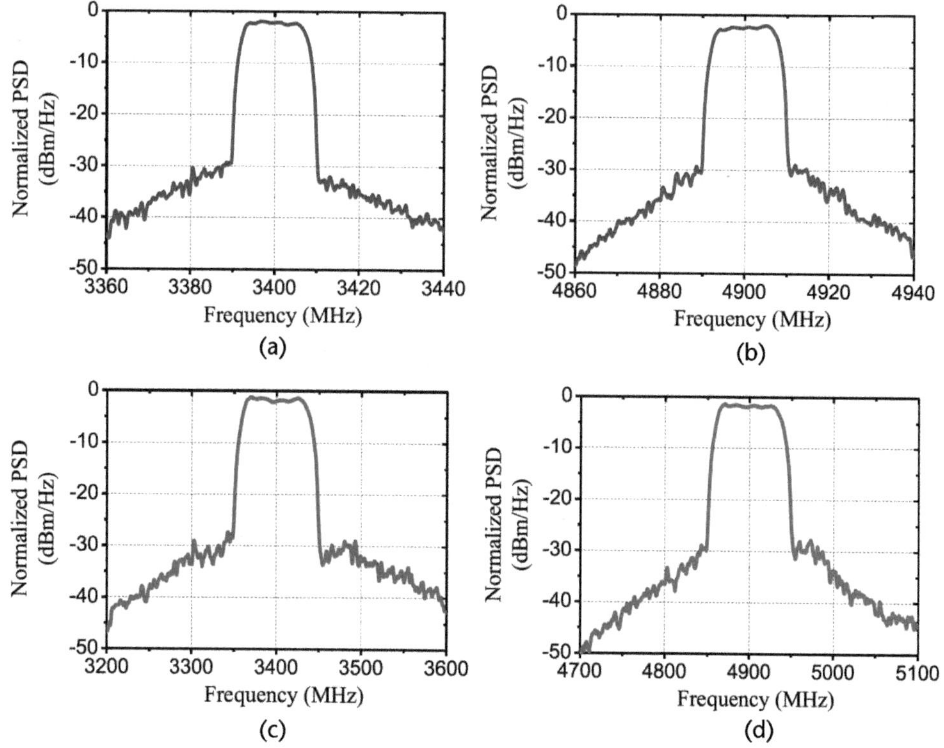

Figure 6.12 Normalized output spectrum of the proposed design under different bandwidth 6-dB PAPR modulated signal stimulation at the different frequency under: (a) 20 MHz at 3.4 GHz, (b) 20 MHz at 4.9 GHz, (c) 100 MHz at 3.4 GHz, and (d) 100 MHz at 4.9 GHz [4].

signal stimulation, and the corresponding output spectra at the dual frequencies are depicted in Figures 6.12(c, d). An overview of the results under modulated signal stimulation is provided in Table 6.4. It is evident that the suggested DPA exhibits excellent linearity and moderate DE performance. At the dual-band operating frequencies, the suggested DPA achieves an average DE of 37.9%–43.9%/40.5%–43.2% with an average output power of about 38/37 dBm. The matching ACPR is maintained below 27 dBc. Measurements were also conducted during the concurrent signal stimulus in the dual band with a frequency interval of 1,500 MHz, as shown in Figure 6.13. The DPA achieves an average DE of 34.4%/32.9% when simultaneously stimulated with a 20-MHz modulated signal, and an average DE of 35.9%/34.5% when simultaneously stimulated with a 100-MHz modulated signal. The matching ACPR values are 28/23.5 dBc and 26/21 dBc, respectively.

6.2 LNA

LNA, being the first active component of the RF receiver end, plays a decisive role in the performance of the whole wireless communication receiver system [25]. In the RF front-end receiving system, to make the weak received signal suitable for channel processing, gain amplification and noise reduction are needed before the signal enters the processing unit. This function is realized by the LNA. Not only can the LNA only improve the sensitivity of the system by reducing the interference of clutter, but it can also amplify the weak signals received by the system through the antenna from the wireless space. It has been widely used in radar, electronic countermeasures, satellite correspondence, and other fields.

6.2.1 Analysis Design of LNA

In the design of an LNA, some strict requirements must be met by the amplifier itself. This includes the reduction of return loss through the use of broadband input matching, suppression of mixer noise by increasing gain and enhancing receiver sensitivity through low noise figure (LNF), as well as increasing battery duration through lower energy consumption [26]. Next, we will briefly introduce the noise, gain, isolation, and other key indicators of LNA.

6.2.1.1 Key Indicators

For electronic circuits, noise is the result of the random movement of charges at the atomic level. Due to the presence of noise, the sensitivity of the receiver will be poor, so it will not be able to detect signals that are too low in energy. At this point, the LNA is particularly important, because it can reduce the noise factor of the whole system and thus improve the sensitivity of the receiver. The common noise in the circuit mainly includes thermal noise, shot noise, and flicker noise [27]. Thermal noise is the most basic kind of noise, which is produced by the thermal motion of electrons. Above absolute zero, random motion of electrons occurs due to their thermal energy. Therefore, almost all devices produce thermal noise. Shot noise is generated by the random movement of charge carriers in active devices. For

Table 6.4 Performance Comparison of Dual-Band DPA Designs and Some DPA Designs Operating at Similar Frequency Bands

Reference		Frequency (GHz)		Platform Type	Method/Combiner Structure	P_{SAT} (dBm)	Second-Harmonic Tuned	η_{sat} (%)	η_{6dB} (%)	Self-Packaging
[14]	2012	3–3.6		Microstrip	Output compensator	43–44	Yes	55–66	38	No
[15]	2018	1.5–3.8		Microstrip	Linear simulations	42.3–43.4	No	42–62	33–55	No
[16]	2019	2.2–4.8		Microstrip	Reciprocal Gate bias	39.2–41	No	50–72	35–49.7	No
[17]	2019	2.7–4.3		Microstrip	Class-J	38.5–39.2	Yes	48–61	40–43	No
[18]	2019	4.7–2.3		Microstrip	Synthesis methodology	39–39.5	No	52–57	29.7–33.1*	No
[19]	2012	1.96/3.5		Microstrip	Dual-band 90° (Pi-network)	42.5/41.6	No	59.5/49.6	39.2/29.1**	No
[20]	2014	0.85/2.33		Microstrip	Dual-band 90° (T-network)	44/42.5	No	60/54	38/36.5	No
[21]	2016	1.8/2.6		Microstrip	Direct matching	43.7/43.9	No	72/60	63/51	No
[22]	2017	1.5/2.14 or 1.85/2.35		Microstrip	Reconfigurable	42.4–43.2	No	67/70 or 62/65	52/56 or 43/48	No
[23]	2020	1.4/3.5		Microstrip	Dual-band GCN†	42.5/41.5	Yes	70/68	55/55.1*	No
[24]	2020	Mode	1.8–2.2/3.9–4.3	Microstrip	Dual-band 90°	39.6–41.5	Yes	61.2–67.3	49.2–54.4	No
		Mode	1.52–1.72/ 2.38–2.53/ 3.67–3.82/ 4.53–4.68			39.6–41.3		56/60	50/55, 43/40	
2020		1.8–2.7/2.7–3.4		Microstrip	Reconfigurable	42.3–43.9/ 42.2–43.7	No	45–62.3/ 51.8–62.1	37.8–48.5/ 36.7–43.2	No
This work		3.4/4.9		MI-SISL	Direct matching	44/43.2	Yes	70.7/70.4	38/42	Yes

Source: [4].
*9-dB back-off efficiency.
**6.6-dB back-off efficiency.
†GCN: generalized combining network.

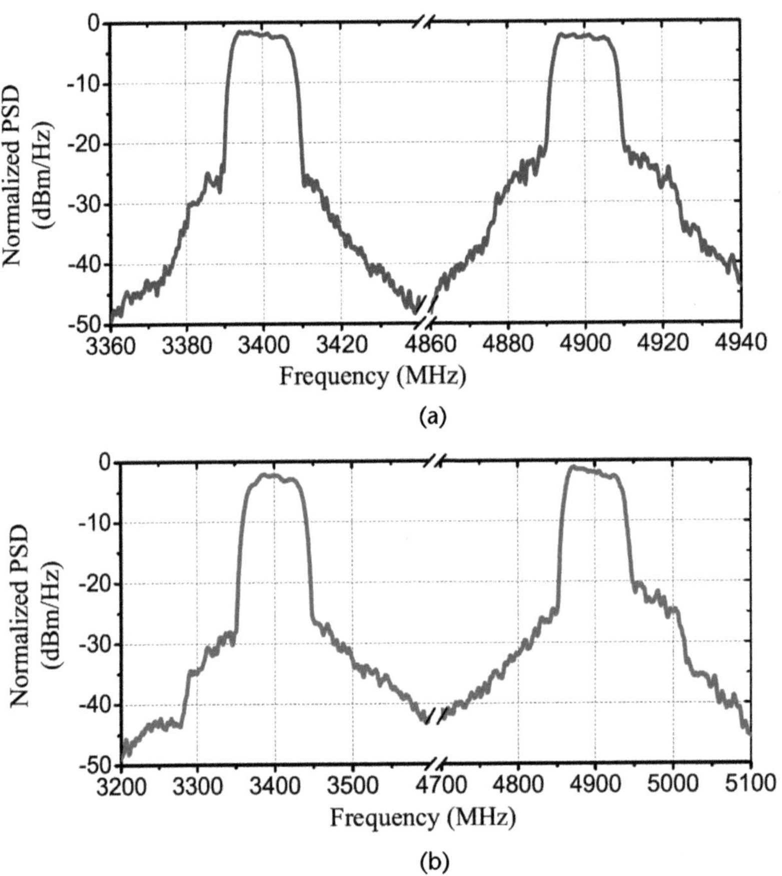

Figure 6.13 Normalized output spectrum of the proposed design under the dual-band concurrent: (a) 20-MHz and (b) 100-MHz 6-dB PAPR modulated signal stimulation [4].

example, when the transistor works in the saturated region, there will be movement of electrons and holes. This process will produce loose particle noise. The above two types of noise are white noise because their power spectral density does not change with frequency. Flicker noise generally exists in semiconductors with surface defects. It is produced by slow random fluctuations of electrons. Flicker noise is mainly concentrated in a low-frequency band. Its power spectral density decreases with the increase of frequency, so it is also called 1/*f* noise. The noise figure is the most important indicator of LNA. The parameter used to describe the noise performance of the system is defined as

$$F = \frac{SNR_{\mathrm{i}}}{SNR_{\mathrm{o}}} = \frac{S_{\mathrm{i}}/N_{\mathrm{i}}}{S_{\mathrm{o}}/N_{\mathrm{o}}} = 1 + \frac{N_x}{N_{\mathrm{i}}G} \tag{6.52}$$

where S_i and S_o are input and output signal power. G is the gain of the two-port device. N_i and N_o are input and output noise power. N_x is the noise power of the two-port device itself. The ratio of the voltage to the noise voltage of the input or output signal at the same port is called the input or output SNR. The ratio of SNR of the input to the output is the noise factor. Specifically, when the echo signal enters the input end of the LNA together with the noise, it is amplified and finally output from the output end. In addition, there is also noise inside the LNA, so the output noise becomes larger, and the output SNR becomes smaller. The formula for cascade circuit noise is given by

$$F_{total} = F_1 + \frac{F_2 - 1}{G_1} + \frac{F_3 - 1}{G_1 G_2} + \ldots + \frac{F_n - 1}{G_1 G_2 \ldots G_{n-1}} \quad (6.53)$$

where F_{total} is the total noise factor after the cascade of N-class circuits. As can be seen from the cascade formula, the closer the circuit is to the previous stage, the greater the impact of its noise factor on the overall noise. At the same time, to further reduce the overall noise coefficient of the system, LNA needs to achieve a certain gain.

As shown in Figure 6.14, the two-port network of a single-stage amplifier is given. S_{11} is equal to the voltage reflection coefficient of Port 1 when Port 2 is connected to a matching load. The better Port 1 matches, the closer Γ_{11} is to 0.

$$S_{11} = \left.\frac{b_1}{a_1}\right|_{a_2=0} = \Gamma_{11}\big|_{a_2=0} \quad (6.54)$$

S_{11} is generally required to be less than −15 dB, to ensure that the weak signals received by the antenna can enter the LNA to complete the amplification to the greatest extent. The definition of S_{11} can be deduced to Port 2 by analogy to get the definition of S_{22}. Port 2 needs to be well-matched so that the amplified signal can smoothly enter the downconverter at the rear stage. The gain of an LNA is introduced as

$$S_{21} = \left.\frac{b_2}{a_1}\right|_{a_2=0} = T_{21}\big|_{a_2=0} \quad (6.55)$$

The ratio of the outgoing wave voltage of Port 2 to the incoming wave voltage of Port 1 when Port 2 is connected to a matching load is S_{21}. In the LNA, T_{21} is usually a number greater than 1, indicating that the input signal from Port 1 is amplified through the LNA. The amplification process occurs in the active component of the LNA. Based on the input signal, the energy provided by the dc bias

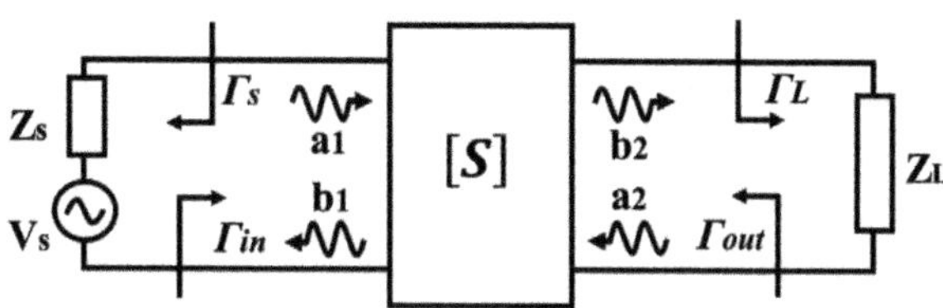

Figure 6.14 Two-port networks for single-stage amplifiers.

source is converted into the energy of the output signal. So we can measure the size of LNA gain with S_{21}. The reverse isolation of LNA is designed to suppress the local vibration leakage of the rear-stage downconverter. Poor isolation often makes the amplifier circuit unstable and may cause shock. This will cause the amplifier to lose amplification function and may even damage the circuit structure.

6.2.1.2 Advantages of the SISL Platform in an LNA Design

The new SISL proposed by Professor Kaixue Ma inherits the excellent characteristics of the traditional suspension wire and overcomes the shortcomings of the traditional waveguide suspension wire circuit, such as high processing cost, complex assembly, and poor compatibility. This new type of transmission line combines the characteristics of the PCB process with the traditional suspension-line structure and creatively proposes a planarization processing method for the suspension line. As a new board-level circuit processing technology, the SISL has many advantages, such as high Q value, low loss, no dispersion, self-packaging, compact structure, small size, low cost, and high integration. Using the SISL platform to design LNAs can effectively reduce circuit loss, to achieve the superior performance of ultralow noise. Next, we will analyze and discuss the advantages of the SISL platform in an LNA circuit design from two aspects: structural and performance properties.

SISL circuit consists of several PCB circuit boards by solder paste bonding, riveting, bolts, and nuts tightening [28]. The typical structure of the SISL platform is composed of five PCB layers with double-sided metal layer. Each PCB is a dielectric substrate board coated with copper (or other metals) above and below. Generally, the main circuit is designed on Sub3. Active and passive devices can be integrated on the upper and lower metal layers G_5 and G_6, to achieve high integration, high design freedom, and miniaturization. By hollowing out Sub2 and Sub4, the upper and lower cavities of the suspension line are realized, which can make the electromagnetic field of the SISL mainly distributed in the air, thus reducing the requirements of the circuit on the dielectric substrate. At the same time, it also creates enough space for the implantation of various discrete devices, monolithic microwave integrated circuit (MMIC) chips, and bias control circuits. The PCBs (except the core circuit layer Sub3) provide mechanical support, they can choose cheap materials like low-cost FR4. Compared with the use of expensive IC processing processes such as 90-nm GaAs p-HEMT and 0.13-μm SiGe BiCMOS, the SISL platform enables the manufacturing cost of LNA to be significantly reduced [29, 30].

In addition, unlike other microstrips, LNAs require additional mechanical packaging [31]. Using the SISL platform to design LNA can realize self-packaging. The metalized through-hole and the PCB surface is coated with copper to realize electromagnetic shielding, so the SISL circuit requires no metal box and assembly and is self-packaged. This structural feature provides favorable conditions for the design and implementation of miniaturized and low-cost LNAs.

The noise factor F of a common-source input topology LNA can be expressed as

$$F = 1 + \gamma \frac{1}{R_s} \frac{1}{g_m} \frac{1}{4Q_{\text{in}}^2} \tag{6.56}$$

where the parameter definition can refer to [32]. The noise factor can be mitigated when the Q factor of the input stage increases. According to Figure 6.15, the simulation result of the Q value of the SISL can reach 900. The SISL platform improves Q value by reducing loss, thus achieving the purpose of reducing circuit noise. Circuit loss generally contains radiation loss, conductor loss, and substrate loss.

The via fence and the covers constitute an electromagnetic shielding environment to reduce the radiation loss of the circuit inside the multilayer structure [34]. The signal line of the SISL is wider than the microstrip signal line with the same impedance, so the conductor loss is smaller. Most of the electromagnetic field of the SISL is distributed in the air cavity rather than the medium, so the substrate loss is small. Moreover, conductor and substrate losses can be further reduced through the use of double metal layers wiring and substrate-cutting design methods. The SISL performance advantages mentioned above make it possible to implement LNAs with ultralow noise.

6.2.2 SISL LNA Examples

6.2.2.1 An Ultralow Noise Dual-Band LNA for 5-GHz WLAN Application

In this section, we will present an ultralow noise concurrent 2.45-/5.25-GHz dual-band LNA based on the SISL platform for 5-GHz WLAN application. The whole circuit can be divided into three parts: RF input and output port, SISL to CBCPW transition, and core circuit. The core circuit is composed of input and output matching, dc bias, and transistor. According to the basic circuit structure analysis of LNA mentioned in Section 6.2.1, we decided to adopt the structure of cascade common source (CS) and cascade for the dual-band LNA design.

The transition was designed to guide the electromagnetic field from encapsulated SISL to conductor-backed CPW, which will provide convenience for measurement. The input transition has been modified to provide a specific real part approaching optimal noise impedance [35].

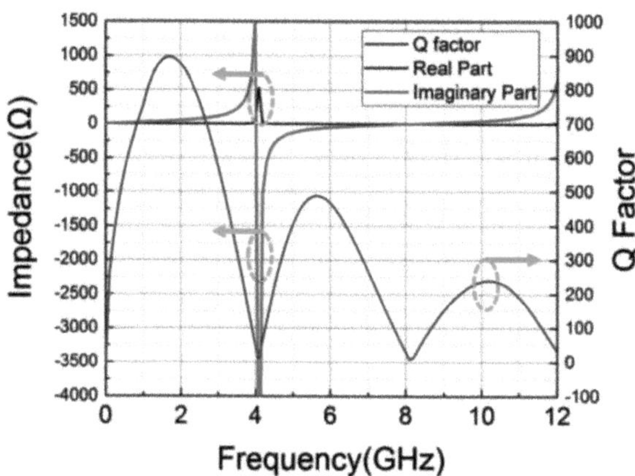

Figure 6.15 The Q-factor and its corresponding impedance of an SISL-based transmission line [33].

We define the impedance seen from the inside core circuit to the transition as Z_{trans}. The two key parameters angle α and the width W affect Z_{trans} by altering equivalent capacitors and inductors in Figure 6.16(a). W changes the plate capacitor C_6 by changing the overlap area between the signal path and the background. α alters the distance between the signal path and coplanar ground to change the parasitic capacitor C_5. The objective of the optimization of W and α is to make the impedance Z_{trans} close to the optimal noise impedance in the two communication frequency bands to achieve the lowest noise. The SISL structure itself contains many design variables. Without introducing complex additional circuits, we can adjust the impedance Z_{trans} at the transition through these two parameters, as shown in Figures 6.16(b, c). This can effectively reduce the area of the circuit.

Based on the ADS model of transistor ATF36163, its optimum impedance for minimum NF Z_{opt} can be obtained as (6.57) and (6.58). To achieve the goal of noise matching, we must make (6.59) work at both the passbands.

$$Z_{\text{opt}} = (38.0 + j159.5)\Omega \quad @\ 2.45\ \text{GHz} \tag{6.57}$$

$$Z_{\text{opt}} = (21.9 + j55.6)\Omega \quad @\ 5.25\ \text{GHz} \tag{6.58}$$

$$Z_{\text{back}} = Z_{\text{opt}} \quad @\ 2.45\ \text{GHz}\ \&\ 5.25\ \text{GHz} \tag{6.59}$$

The initial value of Z_{back} comes from Z_{trans}. In the process of optimizing the transition structure, let the real part of the impendence Z_{trans} approach Z_{opt}. The imaginary part of the impendence Z_{back} can be adjusted by connecting the LC branch in the transitional foundation to complete the noise matching.

Figure 6.17 shows the basic topology of the input-matching network. An LC tank is introduced to suppress the spurious signal between two passbands. The LC tank functions equivalently to an inductor L_E and a capacitor C_E. The input impedance can be written as [36]:

$$Z_{\text{in}} = \left(\frac{1}{j\omega C_1} \middle\| j\omega L_1 \right) + \frac{1}{j\omega C_2} + j\omega L_g + \frac{1}{j\omega C_{gs}} + \frac{g_m L_s}{C_{gs}} + j\omega L_s \tag{6.60}$$

At 2.45 GHz, the expression of input impedance Z_{in} can be simplified to:

$$Z_{\text{in}} = \frac{1}{jw}\left(\frac{1}{C_{gs}} + \frac{1}{C_2} \right) + j\omega\left(L_g + L_E + L_s \right) + \frac{g_m L_s}{C_{gs}} \quad @\ 2.45\ \text{GHz} \tag{6.61}$$

At 5.25 GHz, the expression of input impedance Z_{in} can be simplified to:

$$Z_{\text{in}} = \frac{1}{jw}\left(\frac{1}{C_{gs}} + \frac{1}{C_E} + \frac{1}{C_2} \right) + j\omega\left(L_g + L_s \right) + \frac{g_m L_s}{C_{gs}} \quad @\ 5.25\ \text{GHz} \tag{6.62}$$

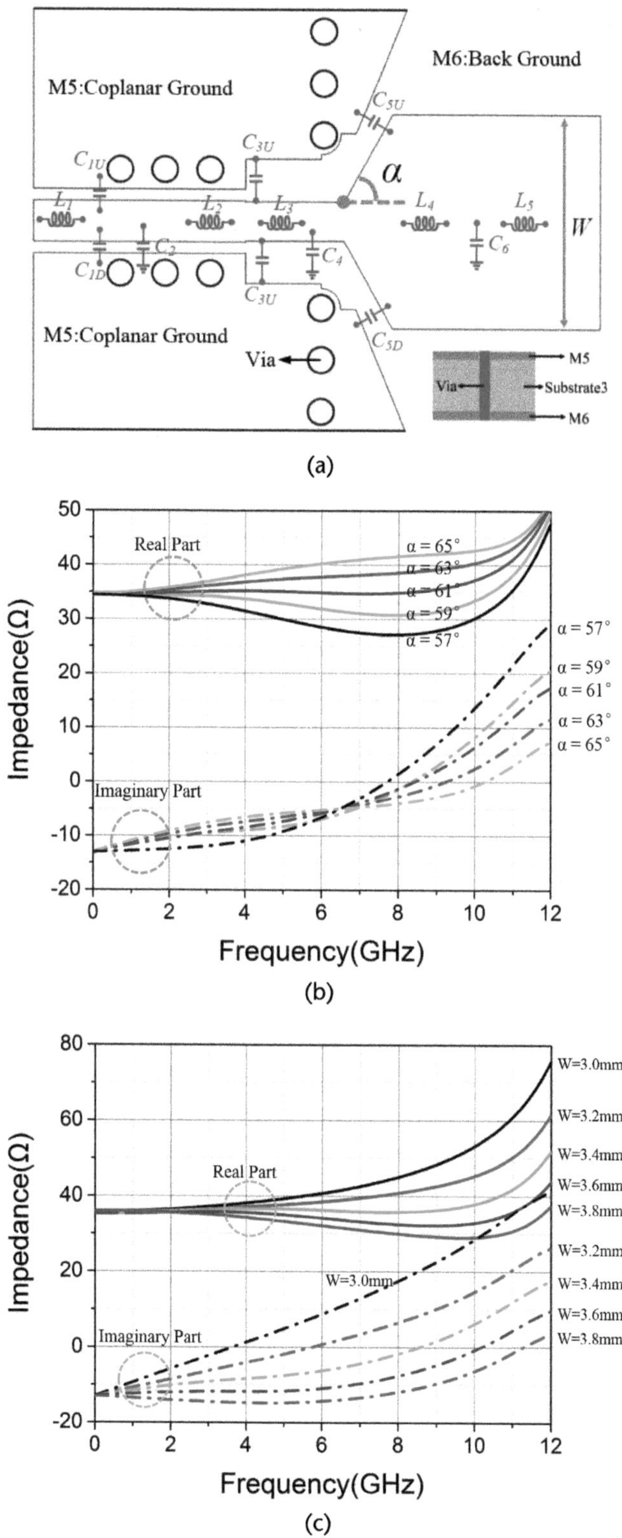

Figure 6.16 (a) Location of each component in the transition's equivalent circuit, (b) the affection of the angle α on the impedance Z_{trans}, and (c) the affection of the width W on the impedance Z_{trans} [35].

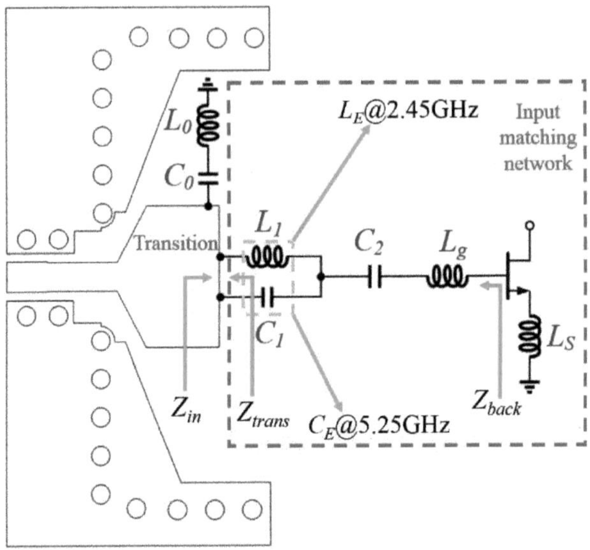

Figure 6.17 Transition and basic topology of the input matching network [35].

To get a good match at the transition, Z_{in} should be conjugated with Z_{trans}. Their real parts should be approximately equal, and their imaginary parts should be negative as:

$$\text{Re}(Z_{\text{in}}) = \frac{g_m L_s}{C_{gs}} \approx \text{Re}(Z_{\text{trans}}) \tag{6.63}$$

$$\begin{aligned} \text{Im}(Z_{\text{in}}) &= \frac{1}{jw}\left(\frac{1}{C_{gs}} + \frac{1}{C_2}\right) + j\omega\left(L_g + L_E + L_s\right) = j8.4\Omega \\ &= -\text{Im}(Z_{\text{trans}}) \text{ @ 2.45 GHz} \end{aligned} \tag{6.64}$$

$$\begin{aligned} \text{Im}(Z_{\text{in}}) &= \frac{1}{jw}\left(\frac{1}{C_{gs}} + \frac{1}{C_E} + \frac{1}{C_2}\right) + j\omega\left(L_g + L_s\right) = -j17.0\Omega \\ &= -\text{Im}(Z_{\text{trans}}) \text{ @ 5.25 GHz} \end{aligned} \tag{6.65}$$

The initial value of Z_{back} comes from Z_{trans}, and the relationship between the two is:

$$Z_{\text{back}} = \left(\frac{1}{j\omega C_1} \middle\| j\omega L_1\right) + \frac{1}{j\omega C_2} + j\omega L_g + Z_{\text{trans}} \tag{6.66}$$

The real part of the impedance Z_{back} from the transistor to the input matching network equal to the optimal noise impedance Z_{opt} by optimizing the size of the transition and introducing an LC branch. To achieve the lowest noise, the imaginary parts of Z_{back} and Z_{opt} should also be equal.

$$\begin{aligned}\operatorname{Im}\left(Z_{\text{back}}\right) &= \frac{1}{jwC_2} + j\omega\left(L_g + L_E\right) - j8.4 = j159.5\Omega \\ &= \operatorname{Im}\left(Z_{\text{opt}}\right) \text{ @ 2.45 GHz}\end{aligned} \tag{6.67}$$

$$\begin{aligned}\operatorname{Im}\left(Z_{\text{back}}\right) &= \frac{1}{jw}\left(\frac{1}{C_E} + \frac{1}{C_2}\right) + j\omega L_g + j17.0 = j55.6\Omega \\ &= \operatorname{Im}\left(Z_{\text{opt}}\right) \text{ @ 5.25 GHz}\end{aligned} \tag{6.68}$$

Solving the equations from (6.63), (6.64), (6.65), (6.67), and (6.68) to get the values of L_E, C_E, C_2, L_g, and L_s. Then solve the equations in (6.69) and (6.70) and convert L_E and C_E into L_1 and C_1, which constitutes an LC-tank, and the values of each element in the basic topology can be obtained.

$$jwC_E = jwC_1 + \frac{1}{jwL_1} \text{ @ 2.45 GHz} \tag{6.69}$$

$$\frac{1}{jwL_E} = jwC_1 + \frac{1}{jwL_1} \text{ @ 5.25 GHz} \tag{6.70}$$

As shown in Figure 6.18(a), all parasitic capacitors near node X are equivalent to C_X. The parasitic capacitance reduces the impedance of the node, thereby reducing the circuit gain. The noise of the common gate stage can also be coupled to the front stage of the circuit through C_X, thus worsening the overall noise figure. The expression for the noise factor caused by C_X is as [37]:

$$F_{C_X} = 4R_S\gamma_3 g_{d02}\left(\frac{w_0^2}{w_T^2}\right)\left(\frac{C_x^2}{g_{m2}^2}w_0^2\right) \tag{6.71}$$

The noise factor increases sharply with the increase of frequency, which worsens the system noise in the 5.25-GHz band in (6.71).

To counteract this noise, we introduce an interstage inductance L at the X node. Assuming that the drain-source impedance of transistor M_1 is infinite, and the impedance Z_x looking from transistor M_2 toward node X is [37]:

$$Z_x = \frac{1}{sC_{p1}} \left\| \left(sL + \frac{1}{sC_{p0}}\right) = \frac{s^2LC_{p0} + 1}{s\left(C_{p0} + C_{p1} - w^2LC_{p0}C_{p1}\right)}\right. \tag{6.72}$$

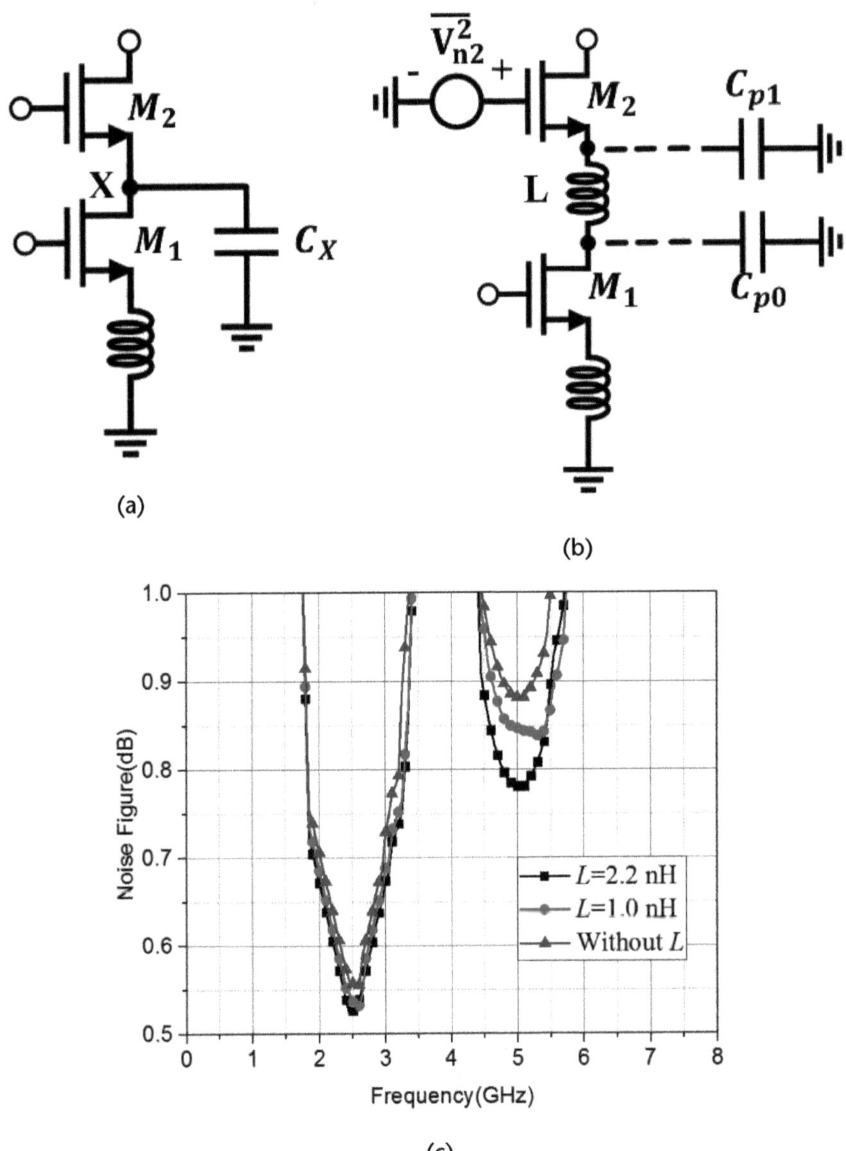

Figure 6.18 (a) Parasitic capacitance of *X* node, (b) schematic diagram of the introduction of L_X, and (c) the effect of interstage inductance *L* on noise figure.

$$C_x = C_{p0} + C_{p1} \tag{6.73}$$

When the denominator in (6.72) is 0, Z_x will take the maximum value; therefore, the inductance L between the stages is

$$L = \frac{C_{p0} + C_{p1}}{w^2 C_{p0} C_{p1}} \tag{6.74}$$

Figure 6.18(c) compares the overall noise factor of the circuit without introducing the interstage inductance L, L = 1.0 nH, and L = 2.2 nH. The appropriate selection of interstage inductance L can improve the circuit noise factor above 0.1 dB. The interstage inductor L can also be resonated with C_X to block the coupling path of the output stage reflected wave voltage leakage to the input stage, improving the isolation and stability of the circuit.

A quarter-wavelength open stub is connected between the two stages to further restrain the spurious between two passbands. In the output stage of the circuit, a step impedance LPF is designed using the transmission line to suppress the high-frequency stray signal of the circuit and carry out impedance transformation to achieve good output matching. The schematic diagram of the proposed dual-band SISL LNA is shown in Figure 6.19(a). The fabricated multilayer SISL LNA is shown in Figure 6.19(b).

The circuit is implemented with pseudomorphic high-electron-mobility transistors (pHEMT) ATF36163, GJM series capacitors, and LQW series inductors from

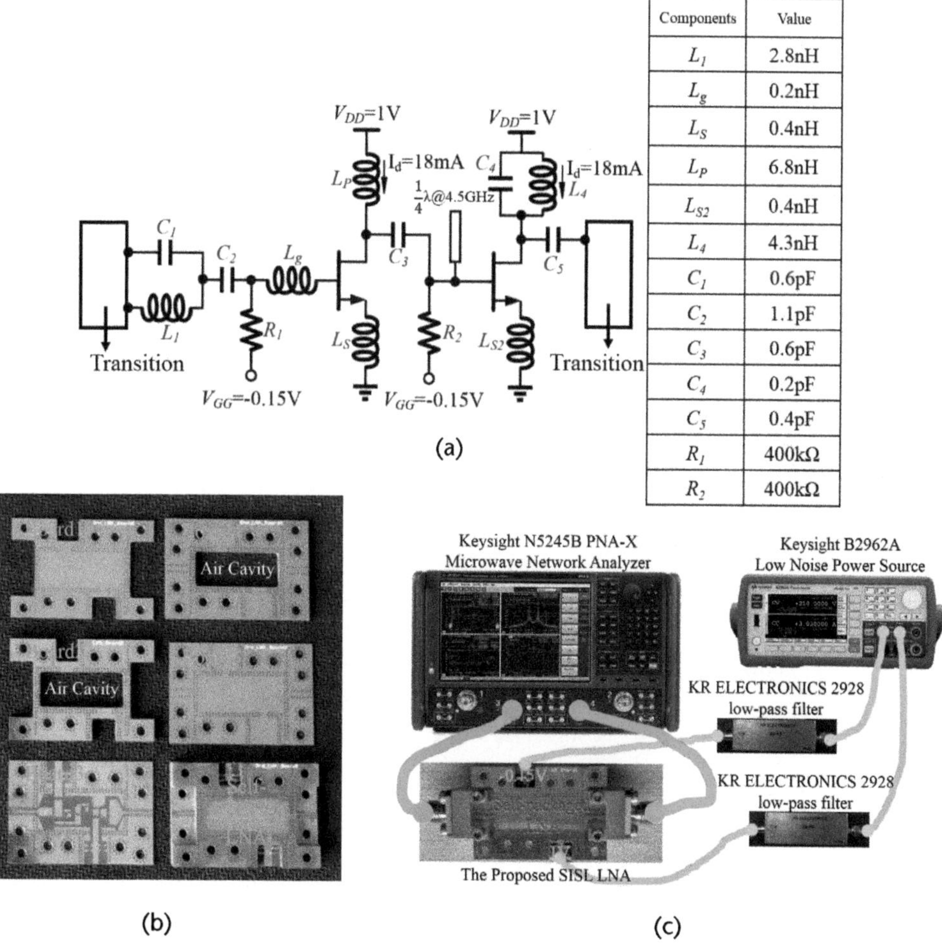

Components	Value
L_1	2.8nH
L_g	0.2nH
L_S	0.4nH
L_P	6.8nH
L_{S2}	0.4nH
L_4	4.3nH
C_1	0.6pF
C_2	1.1pF
C_3	0.6pF
C_4	0.2pF
C_5	0.4pF
R_1	400kΩ
R_2	400kΩ

Figure 6.19 (a) Schematic of the dual-band SISL LNA, (b) photograph of the fabricated LNA, and (c) measurement setup for NF [35].

Murata. The NF was measured with the Keysight N5245B PNA-X Microwave Network Analyzer. Figure 6.20 shows the measurement results of the S-parameter and the measurement results of the noise figure. The drain voltage is 1V, the drain current is 18 mA, and the power consumption is 36 mW. The concurrent dual-band LNA delivers 28.4 and 28.8-dB gain with NF of 0.7 and 1.1 dB at 2.45 and 5.25 GHz, respectively. In terms of matching, S_{11} in the two passbands is less than −13 dB and −20 dB, and S_{22} is less than −13 dB and −12 dB. Compared with the simulation results, the overall test curve is offset to high frequency.

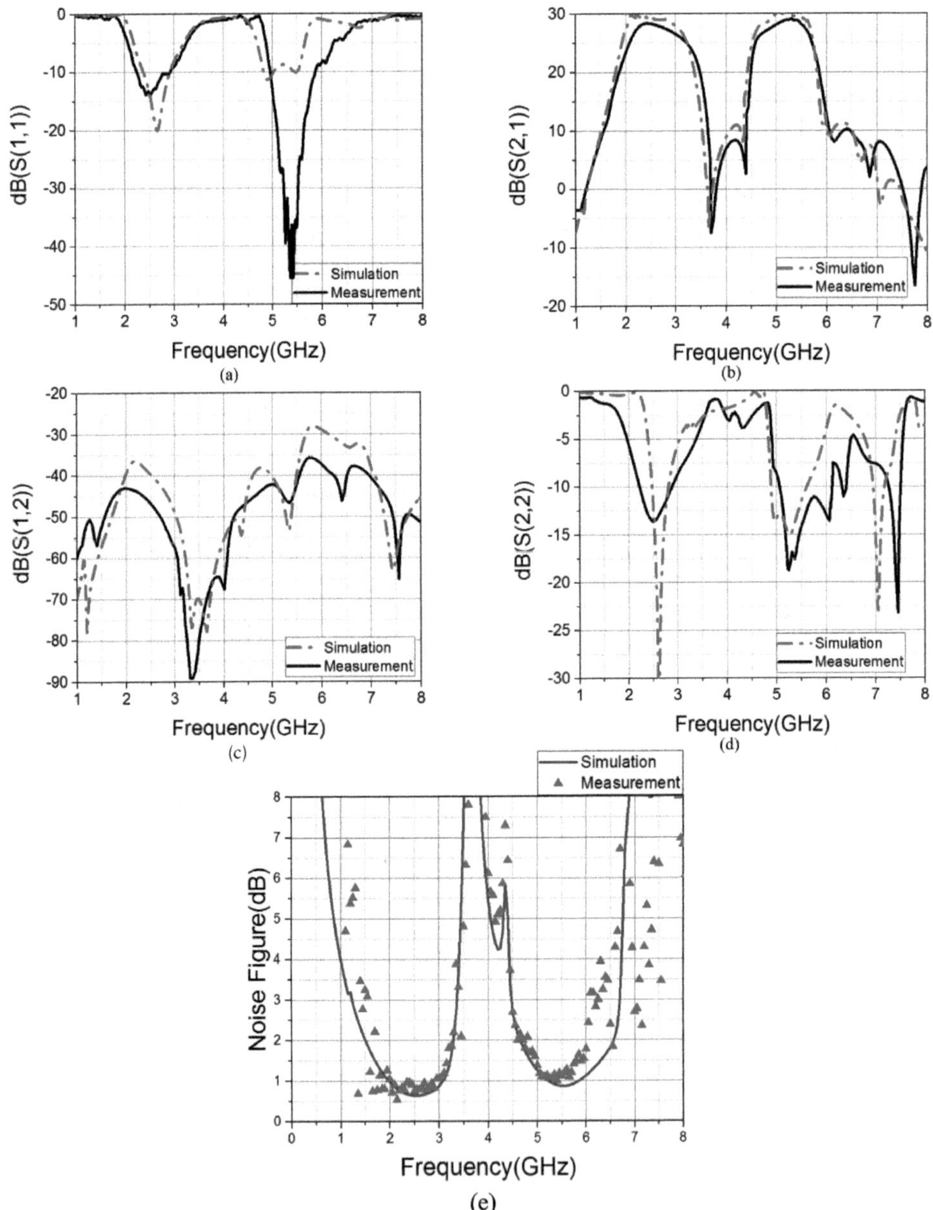

Figure 6.20 Simulated and measured S-parameters and NF of the dual-band SISL LNA: (a) S11, (b) S21, (c) S12, (d) S22, and (e) noise figure [35].

This experimental error is caused by the discrete devices of low self-resonant frequency (SRF). In the case of the Springfield GJM1555C1H2R0WB01 capacitor, the SRF of this capacitor with a nominal value of 2 pF is only 6.2 GHz. As the frequency gradually approaches the SRF, the actual capacitance value decreases and is less than the nominal value, resulting in the overall deviation of the test curve to high frequency. However, the protection bandwidth has been reserved during the circuit design, so the noise figure and power gain curves can still cover the two WLAN bands after the offset.

Comparing the performance of WLAN dual-band LNA based on the SISL process with similar designs based on microstrip line and IC process, SISL LNA has outstanding performance in noise figures. Of all the designs listed in Table 6.5, the LNA of the SISL process achieves the lowest noise figure at 5.25 GHz. Meanwhile, compared with the microstrip that needs a bulky mechanical house, the dimension of self-packaged SISL LNA is just 25 mm × 34 mm × 9 mm. SISL circuit is easy to manufacture with the PCB process, so its size and fabrication cost significantly decrease.

6.2.2.2 An FR4-Based K-Band 1.0-dB Noise Figure LNA

As the core sensor module of the Advanced Driving Assistance System (ADAS), automobile anticollision radar is widely manufactured and applied. In this section,

Table 6.5 Summary of the SISL LNA and Some Published Designs

Platform	*Frequency (GHz)*	*NF (dB)*	*Size*	S_{11} *(dB)*	S_{21} *(dB)*	S_{22} *(dB)*	*IIP3 (dBm)*	*Power (mW)*
0.35-μm CMOS	2.45	2.3	0.6 mm²*	−25	14	NA	0	10
	5.25	4.5		−15	15.5		5.6	
0.13-μm CMOS	2.45	2.8	0.6 mm²*	−12.62	9.4	NA	−4.3	2.79
	6	3.8		−21	18.9		−5.6	
0.15-μm pHEMT	2.4	2.2	1.5 mm²*	−18	20	−8	−8.5	37.8
	5.0	2.0		−7	15	−7	−4	
Microstrip Infineon BFP640	2.45	1.5	30 mm × 30 mm**	−20	22	NA	−10	7.5
	5.2	1.6		−21	12		−4	
Microstrip Avago ATF36163	2.4	0.5	55 mm × 60 mm**	−8.5	12.2	NA	NA	41.25
	4.4	2.5		−15	12.9			
Microstrip Avago ATF36163	2.44	4.34	60 mm × 80 mm**	−10.54	7.15	NA	NA	35.1
	5.25	4.69		−15.982	7.80			
Microstrip Avago ATF36163	2.4	3.96	34 mm × 120 mm**	−25	11.6	NA	NA	56
	5.7	2.85		−12	8.92			
SISL Avago ATF36163	2.45	0.7	25 mm × 34 mm†	−13	28.4	−13	−6.6	56
	5.25	1.1		−20	28.8	−12	−5.1	

Source: [35].
*LNAs in IC process.
**Without mechanical house.
†With package.

we will present an FR4-based 1.0-dB noise figure LNA based on the SISL platform for a 24-GHz vehicle-mounted radar application. To better integrate with the radar system, the input and output terminal of this design will use a common coaxial interface. The design is mainly divided into RF_{in} and RF_{out} port, transition structure, matching network, core transistor, and dc bias circuit. The overall design idea is the same as the previous LNA example. The low-cost K-band LNA is realized by selecting appropriate discrete transistors, plates, and topological structures.

To pursue low noise in the K-band, this LNA is designed with a two-stage common source topology. GaAs pHEMT has a higher cutoff frequency, better noise performance, and better thermal stability. After a comprehensive investigation, we chose California Eastern Laboratories (CEL's) CE3520K3 GaAs field effect transistor (FET) as the core transistor to carry out the design. It possesses a minimum NF of 0.8 dB and an associated gain of 13.9 dB at 24 GHz from the data sheet.

The main function of the dc bias circuit is to minimize the impact on the RF main circuit while providing the appropriate bias voltage for the active device. The dc bias circuit is mainly composed of two parts: the dc blocking capacitor and the radio frequency choke (RFC). Due to the parasitic effect of lumped capacitance at high frequencies, it is difficult to use it to realize the dc blocking function at the K-band. This design uses coupled lines to achieve a dc locking capacitor, as shown in Figure 6.21.

It consists of two sections of 50Ω lines and a pair of coupled lines with a quarter-wavelength. As seen from the simulation results of the S-parameter, this structure plays a role in blocking dc and forms a BPF response from 21 to 32 GHz, and the insertion loss is less than 0.3 dB [38].

For the design of RFC, there are two methods commonly used. One is to integrate RFC into matching networks, such as the π-type network formed by matching two parallel inductors with a series capacitor between them. The two inductors can be used as the gate and drain RFCs, respectively. This method is relatively economical

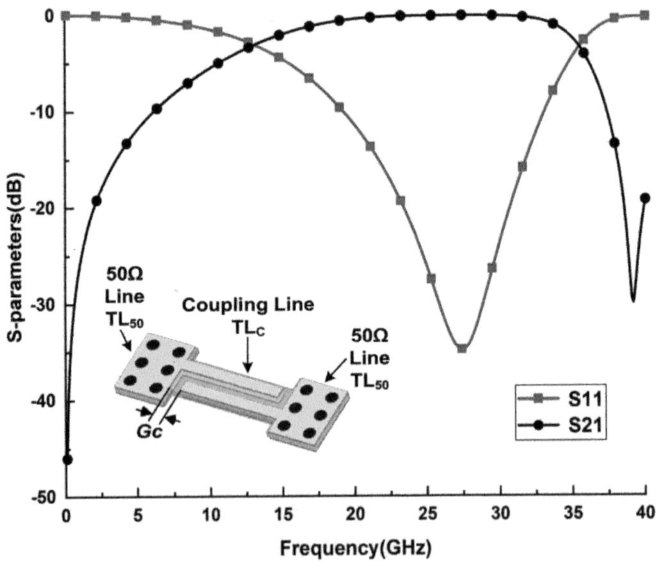

Figure 6.21 Simulated results of the proposed dc blocking capacitor [38].

in the layout area. Another method is to use discrete inductors with large inductance to achieve the choke effect, but the self-resonant frequency of discrete inductor devices generally does not support the application of the K-band.

$$Z_{\text{in}} = Z_0 \frac{0 + Z_0 j \tan(\pi/2)}{Z_0 + 0j\tan(\pi/2)} = \infty \tag{6.75}$$

$$Z_{\text{in}} = Z_0 \frac{\infty + Z_0 j \tan(\beta l)}{Z_0 + \infty j\tan(\beta l)} = -jZ_0 \frac{1}{\tan(\beta l)} \tag{6.76}$$

Equation (6.75) shows a quarter-wavelength transmission line with a short-circuiting terminal that has the same characteristics as an open circuit [38]. As seen from (6.76), an open terminal transmission line is equivalent to a capacitor and exhibits a short circuit when $l = \lambda/4$. The RFC can be realized by using the high-resistance line of $\lambda/4$ and the open line of $\lambda/4$. To widen the bandwidth of the RFC, fan lines are generally used to replace the open route. As the angle of the fan line increases, its equivalent characteristic impedance decreases and the bandwidth of the bias circuit becomes wider. As shown in Figure 6.22(a), TL_F and RS together

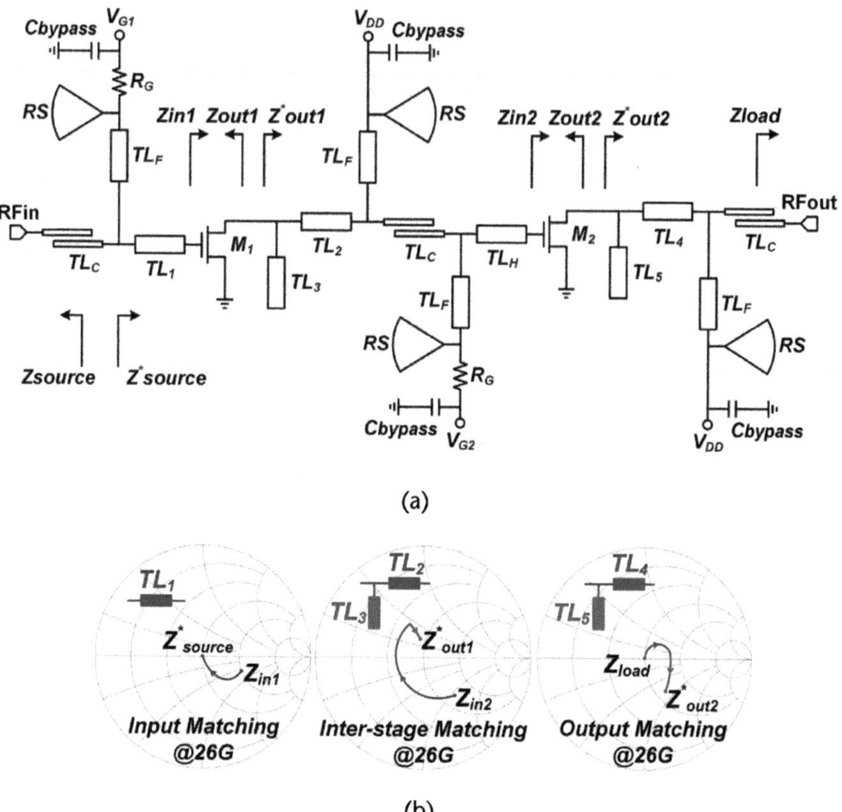

Figure 6.22 (a) Schematic of the proposed LNA, and (b) illustration of the matching network [38].

form the RFC. To suppress potential low-frequency instability, we add an additional resistor *Rg* in each *Vg* path.

To minimize NF, the input matching network should be simplified as much as possible. The impedance transformation from 50Ω to the optimum noise impedance Z_{opt} is realized with only a single line *TL1* in Figure 6.22(b). The load impedance of the first stage will generally affect the input impedance because of the poor isolation of the CS structure. To achieve the best possible power match, we optimize the interstage matching to provide an optimal load impedance for the first stage.

Figure 6.23 shows the core circuit layer of the K-band SISL LNA. The design is based on the SISL multicavity honeycomb concept, which places the ground hole

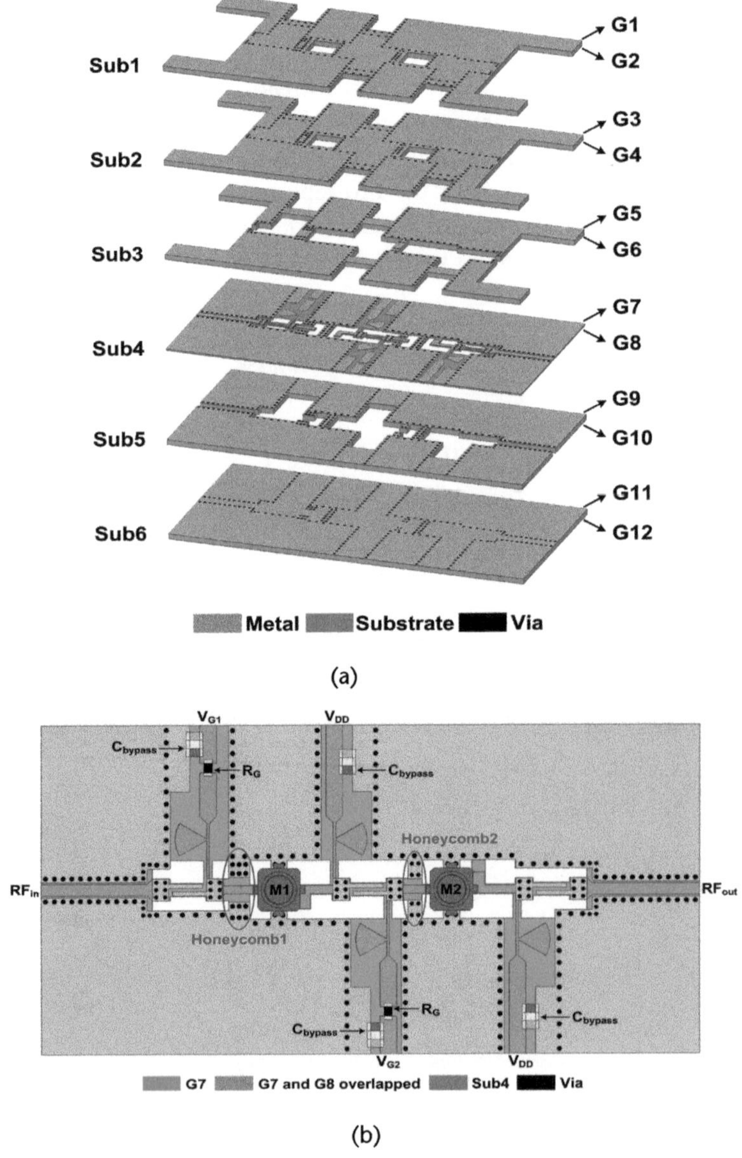

Figure 6.23 (a) The 3D view and the core circuit layer of K-band SISL LNA and (b) the core circuit layer of SISL LNA [38].

close to the input matching branch *TL1* at *Honeycomb1* and adjusts the structure to form a suspended coplanar waveguide (CPW) [40]. In this way, effective isolation between adjacent cavities is realized, which is very beneficial to the realization of high-frequency circuits on SISL. Similar to *Honeycomb1*, an additional 50Ω transmission line for cavity isolation is added at *Honeycomb2*, providing a pad for the transistor while not affecting the matching [38].

The circuit is composed of 6 layers of PCB, and the medium materials are low-cost FR4. *Sub1* and *Sub2* play the role of shielding boards and are partially windowed to be compatible with the height of the transistor package. *Sub3* and *Sub5* are designed with specific hollow shapes to achieve the cavities of the SISL circuit. The core circuit is implemented on *Sub4*. Finally, the 6 layers of PCB board are riveted together by rivets to form a self-encapsulated K-band LNA based on the SISL platform, with an overall size of 40.2 mm × 20 mm. Figure 6.24 shows the fabrication photograph of the proposed SISL LNA.

After testing, the gain bandwidth of this LNA is 25–27.4 GHz, and the highest small signal gain is 27.7 dB at 26.6 GHz. The minimum noise value is 1 dB at 26.5 GHz. In the passband, the return loss of both input and output is better than 5 dB. There is a certain degree of inconsistency between the simulation results and the test results in Figure 6.25. After analysis, these inconsistencies are due to not taking into account the effect of series inductance and parasitic capacitance between the FET source pin and the ideal ground. After these factors are substituted into the

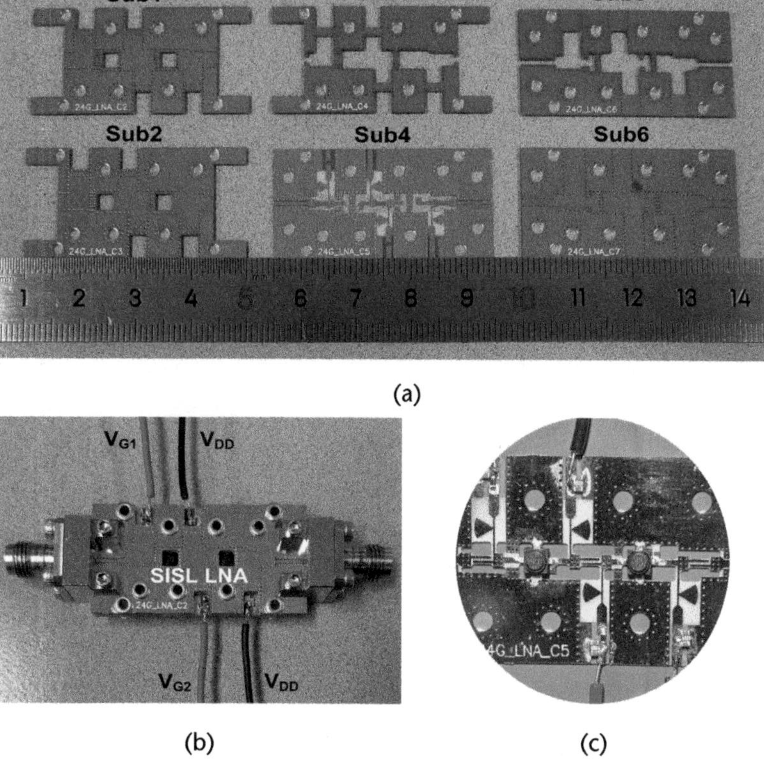

Figure 6.24 (a) Photograph of the fabricated LNA, (b) self-packaged K-band LNA, and (c) core circuit layer with soldered components [38].

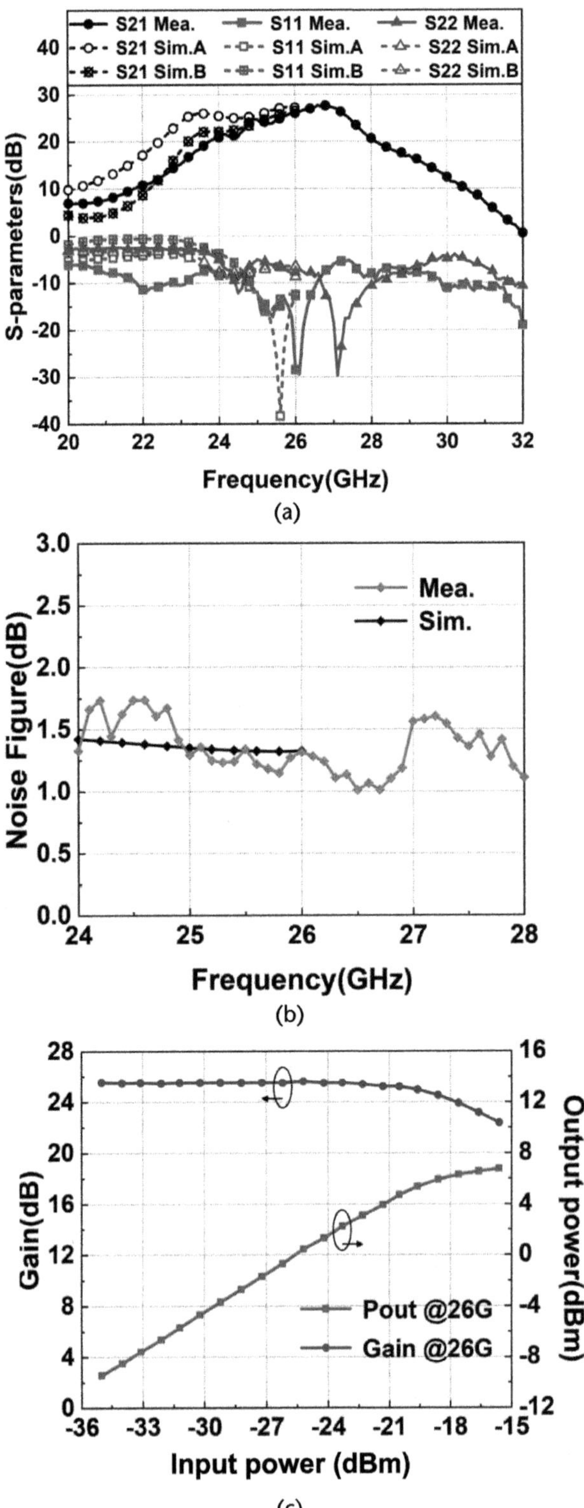

Figure 6.25 Measured and simulated results. (a) S-parameters, (b) noise figure, and (c) gain and output power [38].

simulation, the Sim. B curve in the figure is obtained, which is in good agreement with the measured results.

K-band SISL LNA employs the simplified purely distributed parameter matching network to realize the simultaneous matching of noise and impedance with the minimum loss. Due to the high Q value of the SISL platform, this LNA has the advantages of high power gain, ultralow noise, low cost, and self-packaging. It demonstrates the potential of the SISL platform to design active circuits in the millimeter-wave frequency band.

Acknowledgments

We would like to thank Zhan Yu, Xiaoyu Hong, JiXuan Ye, and Haozhen Huang for their valuable support in this chapter.

References

[1] "Quasi-Planar Circuits with Air Cavities (2006) | Kaixue Ma | 89 Citations," https://typeset.io/papers/quasi-planar-circuits-with-air-cavities-4jqlfo07zf.

[2] Ke, Z., et al., "A 0.7/1.1-dB Ultra-Low Noise Dual-Band LNA Based on SISL Platform," *IEEE Transactions on Microwave Theory and Techniques*, Vol. 66, No. 10, October 2018, pp. 4576–4584.

[3] Ma, K., et al., "A Novel Compact Self-Packaged SPDT Switchable BPFs Based on SISL Platform," *IEEE Transactions on Industrial Electronics*, Vol. 66, No. 9, September 2019, pp. 7239–7249.

[4] Zhang, L., et al., "A Dual-Band and Dual-State Doherty Power Amplifier Using Metal-Integrated and Substrate-Integrated Suspended Line Technology," *IEEE Transactions on Microwave Theory and Techniques*, Vol. 70, No. 1, January 2022, pp. 402–415.

[5] Feng, T., et al., "Bandpass-Filtering Power Amplifier with Compact Size and Wideband Harmonic Suppression," *IEEE Transactions on Microwave Theory and Techniques*, Vol. 70, No. 2, February 2022, pp. 1254–1268.

[6] Gao, L., et al., "Compact Power Amplifier with Bandpass Response and High Efficiency," *IEEE Microwave Wireless Component Letters*, Vol. 24, No. 10, October 2014, pp. 707–709.

[7] Haider, M. F., et al., "Co-Design of Second Harmonic-Tuned Power Amplifier and a Parallel-Coupled Stub Loaded Resonator," *IEEE Transactions on Circuits and Systems II: Express Briefs*, Vol. 67, No. 12, December 2020, pp. 3013–3017.

[8] Su, Z., et al., "Bandpass Filtering Power Amplifier with Extended Band and High Efficiency," *IEEE Microwave and Wireless Components Letters*, Vol. 30, No. 2, February 2020, pp. 181–184.

[9] Xu, J. -X., X. Y. Zhang, and X. -Q. Song, "High-Efficiency Filter-Integrated Class-F Power Amplifier Based on Dielectric Resonator," *IEEE Microwave and Wireless Component Letters*, Vol. 27, No. 9, September 2017, pp. 827–829.

[10] Zheng, S. Y., et al., "Bandpass Filtering Doherty Power Amplifier with Enhanced Efficiency and Wideband Harmonic Suppression," *IEEE Transactions on Circuits and Systems I: Regular Papers*, Vol. 63, No. 3, March 2016, pp. 337–346.

[11] Guo, Q. -Y., et al., "Bandpass Class-F Power Amplifier Based on Multifunction Hybrid Cavity–Microstrip Filter," *IEEE Transactions on Circuits and Systems II*, Vol. 64, No. 7, July 2017, pp. 742–746.

[12] Feng, W., et al., "A Bandpass Push–Pull High Power Amplifier Based on SIW Filtering Balun Power Divider," *IEEE Transactions on Plasma Science*, Vol. 47, No. 9, September 2019, pp. 4281–4286.

[13] Lin, L., et al., "Improvement in Cavity and Model Designs of LDMOS Power Amplifier for Suppressing Metallic Shielding Cover Effects," *IEEE Transactions on Electromagnetic Compatibility*, Vol. 58, No. 5, October 2016, pp. 1617–1628.

[14] Rubio, J. M., et al., "3–3.6-GHz Wideband GaN Doherty Power Amplifier Exploiting Output Compensation Stages," *IEEE Transactions on Microwave Theory and Techniques*, Vol. 60, No. 8, August 2012, pp. 2543–2548.

[15] Moreno Rubio, J. J., et al., "Design of an 87% Fractional Bandwidth Doherty Power Amplifier Supported by a Simplified Bandwidth Estimation Method," *IEEE Transactions on Microwave Theory and Techniques*, Vol. 66, No. 3, March 2018, pp. 1319–1327.

[16] Li, M., et al., "Ultra-Wideband Dual-Mode Doherty Power Amplifier Using Reciprocal Gate Bias for 5G Applications," *IEEE Transactions on Microwave Theory and Techniques*, Vol. 67, No. 10, October 2019, pp. 4246–4259.

[17] Li, Y., et al., "Two-Port Network Theory-Based Design Method for Broadband Class J Doherty Amplifiers," *IEEE Access*, Vol. 7, 2019, pp. 51028–51038.

[18] Fang, X., A. Chung, and S. Boumaiza, "Linearity-Enhanced Doherty Power Amplifier Using Output Combining Network with Predefined AM–PM Characteristics," *IEEE Transactions on Microwave Theory and Techniques*, Vol. 67, No. 1, January 2019, pp. 195–204.

[19] Rawat, K., and F. M. Ghannouchi, "Design Methodology for Dual-Band Doherty Power Amplifier with Performance Enhancement Using Dual-Band Offset Lines," *IEEE Transactions on Industrial Electronics*, Vol. 59, No. 12, December 2012, pp. 4831–4842.

[20] Chen, W., et al., "A Concurrent Dual-Band Uneven Doherty Power Amplifier with Frequency-Dependent Input Power Division," *IEEE Transactions on Circuits and Systems I: Regular Papers*, Vol. 61, No. 2, February 2014, pp. 552–561.

[21] Pang, J., et al., "Novel Design of Highly-Efficient Concurrent Dual-Band GaN Doherty Power Amplifier Using Direct-Matching Impedance Transformers," *2016 IEEE MTT-S International Microwave Symposium (IMS)*, May 2016, pp. 1–4.

[22] Kalyan, R., K. Rawat, and S. K. Koul, "Reconfigurable and Concurrent Dual-Band Doherty Power Amplifier for Multiband and Multistandard Applications," *IEEE Transactions on Microwave Theory and Techniques*, Vol. 65, No. 1, January 2017, pp. 198–208.

[23] Liu, H. -Y., C. Zhai, and K. -K. M. Cheng, "Novel Dual-Band Equal-Cell Doherty Amplifier Design with Extended Power Back-Off Range," *IEEE Transactions on Microwave Theory and Techniques*, Vol. 68, No. 3, March 2020, pp. 1012–1021.

[24] Pang, J., et al., "Multiband Dual-Mode Doherty Power Amplifier Employing Phase Periodic Matching Network and Reciprocal Gate Bias for 5G Applications," *IEEE Transactions on Microwave Theory and Techniques*, Vol. 68, No. 6, June 2020, pp. 2382–2397.

[25] Jyothi Banu, A., G. Kavya, and D. Jahnavi, "Performance Analysis of CMOS Low Noise Amplifier Using ADS and Cadence," *Materials Today: Proceedings*, Vol. 24, 2020, pp. 1981–1986.

[26] Liu, J., "Analysis of Low-Noise Amplifier," *2021 6th IEEE International Conference on Intelligent Computing and Signal Processing (ICSP)*, Xi'an, China, April 2021, pp. 1427–1432.

[27] Howard, R. M., *Principles of Random Signal Analysis and Low Noise Design*, New York: John Wiley & Sons, 2002.

[28] Li, L., et al., "A Novel Transition from Substrate Integrated Suspended Line to Conductor Backed CPW," *IEEE Microwave Wireless Component Letters*, Vol. 26, No. 6, June 2016, pp. 389–391.

[29] Nikandish, G., A. Yousefi, and M. Kalantari, "A Broadband Multistage LNA with Bandwidth and Linearity Enhancement," *IEEE Microwave Wireless Component Letters*, Vol. 26, No. 10, October 2016, pp. 834–836.

[30] Kumar, A., and N. P. Pathak, "Coupled Stepped-Impedance Resonator (CSIR) Based Concurrent Dual Band Filtering LNA for Wireless Applications," *2015 IEEE MTT-S International Microwave and RF Conference (IMaRC)*, Hyderabad, India, December 2015, pp. 262–265.

[31] Menzel, W., and A. Balalem, "Quasi-Lumped Suspended Stripline Filters and Diplexers," *IEEE Transactions on Microwave Theory and Techniques*, Vol. 53, No. 10, October 2005, pp. 3230–3237.

[32] Rudell, J. C., et al., "A 1.9-GHz Wide-Band IF Double Conversion CMOS Receiver for Cordless Telephone Applications," *IEEE J. Solid-State Circuits*, Vol. 32, No. 12, December 1997, pp. 2071–2088.

[33] Ke, Z., et al., "A Compact 0.8 dB Low Noise and Self-Packaged LNA Using SISL Technology for 5 GHz WLAN Application," *2018 IEEE/MTT-S International Microwave Symposium—IMS*, Philadelphia, PA, June 2018, pp. 1507–1510.

[34] Wang, Y., and K. Ma, "Loss Mechanism of the SISL and the Experimental Verifications," *IET Microwaves, Antennas & Propagation*, Vol. 13, No. 11, September 2019, pp. 1768–1772.

[35] Ke, Z., et al., "A 0.7/1.1-dB Ultra-Low Noise Dual-Band LNA Based on SISL Platform," *IEEE Transactions on Microwave Theory and Techniques*, Vol. 66, No. 10, October 2018, pp. 4576–4584.

[36] Shaeffer, D. K., and T. H. Lee, "A 1.5-V, 1.5-GHz CMOS Low Noise Amplifier," *IEEE Journal of Solid-State Circuits*, Vol. 32, No. 5, 1997, pp. 745–759.

[37] Samavati, H., H. R. Rategh, and T. H. Lee, "A 5-GHz CMOS Wireless LAN Receiver Front End," *IEEE Journal of Solid-State Circuits*, Vol. 35, No. 5, 2000, pp. 765–772.

[38] Zhang, J., et al., "An FR4-Based *K*-Band 1.0-dB Noise Figure LNA Using SISL Technology," *IEEE Microwave Wireless Component Letters*, Vol. 32, No. 2, February 2022, pp. 129–132.

[39] Edwards, T. C., and E. Steer, "Transmission Line Theory," in *Foundations for Microstrip Circuit Design*, Fourth Edition, New York: John Wiley & Sons, 2016, pp. 76–88.

[40] Wang, Y., K. Ma, and Z. Jian, "A Low-Loss Butler Matrix Using Patch Element and Honeycomb Concept on SISL Platform," *IEEE Transactions on Microwave Theory and Techniques*, Vol. 66, No. 8, August 2018, pp. 3622–3631.

CHAPTER 7

Oscillators, VCOs, Mixers, Multipliers, and the Front-End System

This chapter covers oscillators, voltage-controlled oscillators (VCOs), mixers, multipliers, and a 24-GHz front-end system. Oscillators are fundamental in RF systems, converting DC to RF power and providing stable sinusoidal signals. Both receivers and transmitters use oscillators, with small signal ones in receiver local oscillators (LO) and large signal ones in transmitter LO. Mixers facilitate frequency conversion (upconversion or downconversion) by blending input signals with local oscillation signals, preserving original characteristics. The chapter also explores a 24-GHz radar front-end system, demonstrating its ability to determine movement, direction, speed, angle, and position of targets, offering advantages over traditional sensing technologies.

7.1 Oscillator

The 5G applications and other new-generation wireless communication solutions have put forward higher requirements for system indicators. Among them, as an important part of wireless communication systems, the RF front-end circuit is one of the key problems in developing the performance and integration of wireless communication systems. The oscillator, as an LO source in the RF front-end circuit, its phase noise, output power, and other indicators have a decisive impact on the performance of the whole system. So, the analysis and design of oscillators have been a hot topic in industry and academia.

The necessity of oscillator research is mainly reflected in the following aspects:

1. Noise has always been an important indicator of communication systems and radar systems; for a communicating system, the noise will affect the speed of the data transmission system. For radar systems, the noise will affect measurement accuracy. In many cases, the system has to increase the extra power dissipation in order to meet the noise target, which leads to the system efficiency reduction. As a device that provides the signal source in the process of upconversion and downconversion in communication and radar systems, the phase noise of the oscillator in the frequency domain is shown as a gradually decreasing noise sideband on both sides of the carrier signal, which will lead to the noise and useful signals appearing at the demodulation

terminal at the same time, thus reducing the baseband SNR and interfering with the performance of the system. Therefore, reducing the phase noise of the oscillator is one of the key steps to realizing high-performance wireless communication systems.

2. Miniaturization and high integration are the inevitable development trend of communication systems at present. Meanwhile, in order to adapt to industrial mass production, RF front-end circuit modules are required to have the characteristics of low cost, easy processing, compact structure, and easy integration. The traditional metal waveguide resonator and dielectric resonator used to be the mainstream choice of oscillator design to some extent, but they are large, have high processing costs, and are difficult to integrate with other circuits, so they cannot be applied to the current integrated high degree of an RF front-end system. How to achieve high integration while inheriting a high Q factor is a key problem.
3. The essence of the working process of the oscillator is nonlinear and time-varying, and the analysis of the oscillator is mostly based on the linear or time-invariant principle, so there are certain application limitations. Therefore, it is necessary to summarize and further discuss the theoretical research results of existing oscillators. In conclusion, the research and design of high-performance, high-integration, and low-cost oscillators have important engineering applications and theoretical value.

As a new type of nonplanar transmission line, SISL has the advantages of low loss, high Q factor, and high integration, which is very suitable for the design of high-performance oscillators.

7.1.1 High Q Resonators Based on SISL Technology

According to the classic Leeson formula [1],

$$L(\Delta\omega) = \frac{2FkT}{P_{\text{sig}}}\left(1+\left(\frac{\omega_0}{2Q\Delta\omega}\right)^2\right) \tag{7.1}$$

The Q factor of the frequency selection network in the oscillator circuit has an important contribution to phase noise, so the design of a resonator with a high Q factor is the main research direction of realizing a low-phase noise oscillator.

The air-filled cavity resonator with a high Q factor and easy integration with the planar circuit is realized based on SISL technology [2], where two rectangular slots of the same size were removed for the medium layers S_2 and S_4, and S_1 was also partially removed to install for the purpose of installing the test adapter.

When the five layers of dielectric substrate are pressed together by rivets, two air cavities can be formed between S_2 and S_3 and S_3 and S_4. The cavity is lined with dense metal through-holes, which form an almost ideal electromagnetic wall. In the previous SISL circuit design, the above two air cavities are mainly used as cavity structures other than the core circuit, but in the design of this chapter, the cavity will be directly excited to achieve a high Q factor resonator. In this design, in order to balance the requirements of high performance and low cost, the low-loss Rogers 5880 (dielectric constant is 2.2; loss tangent is 0.0009) medium is selected as the substrate of the core circuit layer S_3, while the dielectric layers S_1, S_2, S_4, and S_5 mainly act as the cover plate and have little impact on the circuit performance. Therefore, the relatively cheap FR4 (dielectric constant is 4.4; loss tangent is 0.02) medium is selected as the substrate material. In this way, the high-performance circuit can be realized and processing costs can be controlled. The thickness of the five layers of the dielectric substrate from top to bottom is 0.6 mm, 0.6 mm, 0.254 mm, 0.6 mm, and 0.6 mm, and each layer of the dielectric plate is double-sided.

Considering that the energy of the resonator needs to be excited before it can be transmitted to the load, the electric field excitation is chosen to stimulate the SISL resonator in the design of this chapter. The implementation method is to add a rectangular probe to the metal layer M_5, as shown in Figure 7.1(a). Because the

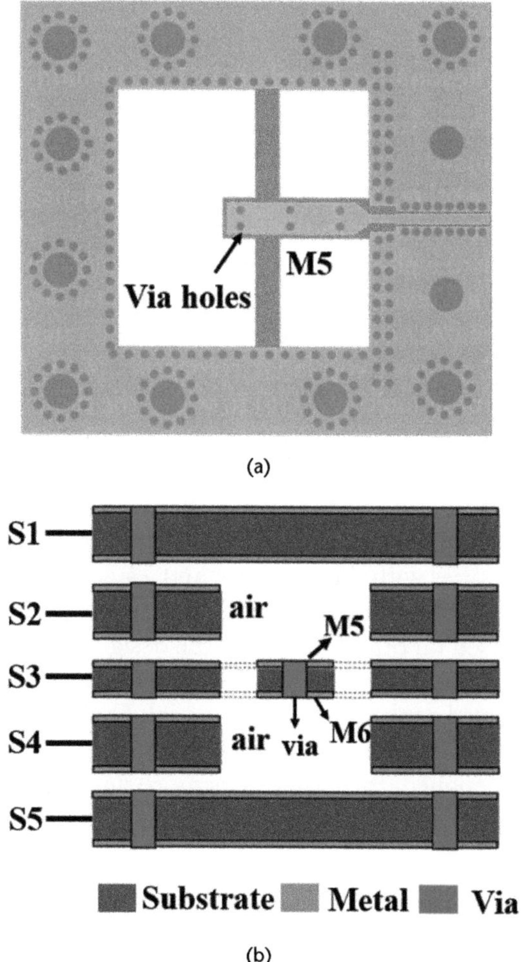

Figure 7.1 (a) Layer 3 of the SISL SMCR with substrate excavation, and (b) the side view of the DMCR with substrate excavation [2].

electric and magnetic fields of the rectangular or cylindrical resonator TE101 mode are strongest in the center of the cavity, the position of the probe end is set close to the exact center of the SISL resonator. In order to more accurately analyze the performance of the SISL, the whole SISL single-metal cavity resonator (SMCR) is modeled in electromagnetic simulation software HFSS, and the cavity is resonant by optimizing the length of the rectangular probe. The final optimized probe size was $L_P \times W_P = 8.3$ mm × 2 mm. In addition, for the convenience of testing, the SISL-to-CPW transition structure is connected to the output end of the probe. The whole transition structure is equivalent to a 50Ω line, so there is almost no impact on the core circuit. The transition structure was introduced in detail in the literature [3], which can realize low-loss energy transmission from SISL to CPW [3].

Q can be further improved by partial dielectric excision (SISL SMCR with substrate excavation) and double-metal cavity resonator (DMCR). The side view of the DSISL and the top view of the resonator dielectric layer S_3 is shown in Figure 7.1(b). The dielectric layer S_3 in the middle of the resonator is partially cut out, leaving only the supporting part of the probe.

7.1.2 Planar Resonators Based on SISL Technology

Combined with the structural characteristics of the SISL resonator, the varactor can be directly coupled inside the SISL SMCR in this chapter, and the feeding probe is added from the other side of the resonator. This method can be equivalent to coupling a variable capacitor in the resonator.

In addition to high-Q-factor resonators implemented using cavity structures, the SISL can also implement some traditional planar structures, such as transformers or planar resonators, to achieve a wider tuning range.

Figure 7.2 shows two planar resonators [4, 5] based on SISL. Figure 7.2(a) uses the concept of multitanks [6] to widen the tunable bandwidth and raise the Q factor. Figure 7.2(c) uses the TRS resonator and a $\lambda/4$ open stub as the frequency-selective element.

The basic idea is to use the high Q characteristic of the SISL and design the resonator using a cavity or planar structure. If a tuning range is required, the varactor is connected in series with the resonator and a suitable bias should be designed for the varactor. In this way, the SISL-based tunable resonator can be obtained. Similar to resonators based on other processes, the designer also needs to make the trade-off between a high Q factor and a wide tuning range. Generally, in the SISL process, if the cavity is used as the resonator, the high Q characteristics are more concerned, and they are usually used in the design of single frequency oscillator. If the planar resonator is designed, the structure is more flexible and the tuning range is wider, which is mostly used to make the VCOs.

7.1.3 Oscillators Using Cavity Resonators

The oscillator using a cavity resonator is mainly based on a negative resistance structure, which requires a one-port resonator, or a parallel feedback topology, which requires a two-port bandpass response resonator.

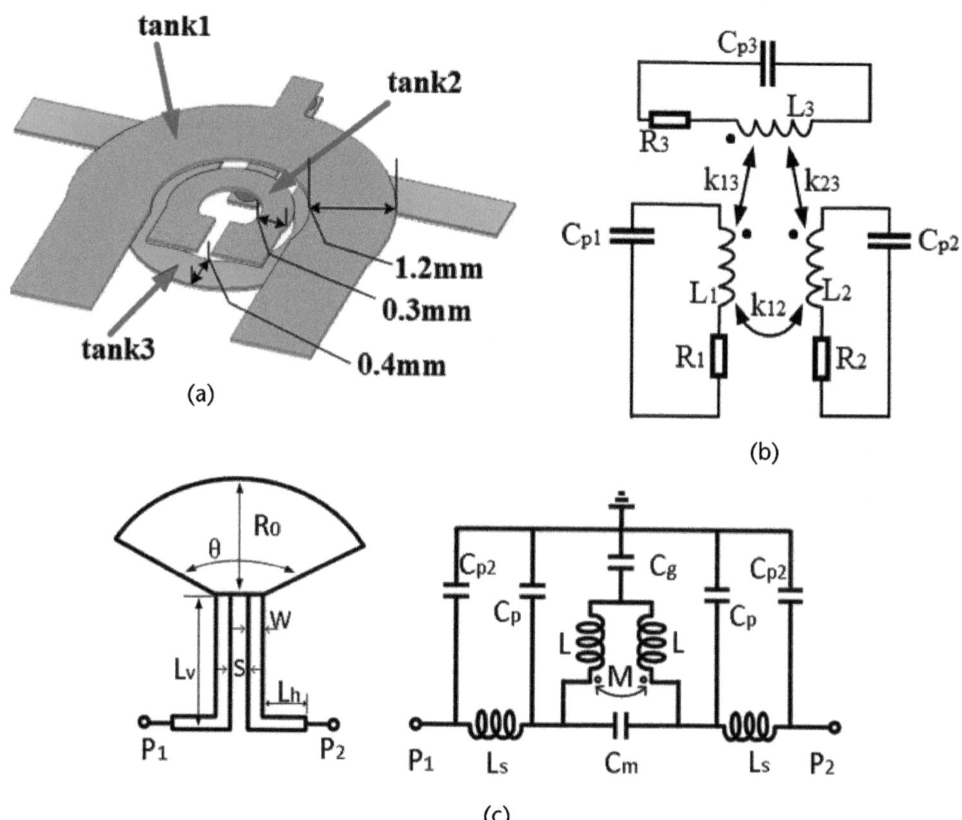

Figure 7.2 (a) SISL planar resonator using triple tanks, (b) the equivalent circuit of the SISL planar resonator using triple tanks [4], and (c) SISL TRS resonator and its equivalent circuit [5].

Figure 7.3 shows the circuit configuration of the negative-resistance oscillator using SISL DMCR [2]. For the emitter feedback network, an open stub and a grounded stub are used to generate the negative resistance, which can guarantee the oscillation state. The start-up and stable oscillation need to satisfy the following conditions:

$$\begin{aligned} |\Gamma_r| \times |\Gamma_{\text{in}}| &\geq 1 \\ Arg(\Gamma_r) &= Arg(\Gamma_{\text{in}}) \end{aligned} \tag{7.2}$$

The carrier frequency is 11.86 GHz, and the output power is 3.63 dBm. The measured phase noise is up to −133.91 dBc/Hz at 1-MHz offset, and the measured figure-of-merit (FOM) is −203.35 dBc/Hz at 1 MHz.

Figure 7.4 shows the circuit configuration of the parallel feedback oscillator using SISL weakly coupled cavities [7]. The weakly coupled cavities are used to raise the Q factor of the single cavity. The Q factor is increased by 80%, and the measured phase noise is up to −137.36 dBc/Hz at 1-MHz offset. The operation carrier frequency is 12.35 GHz, and the output power is 0.72 dBm. The measured FOM is −206.93 dBc/Hz at a 1-MHz offset.

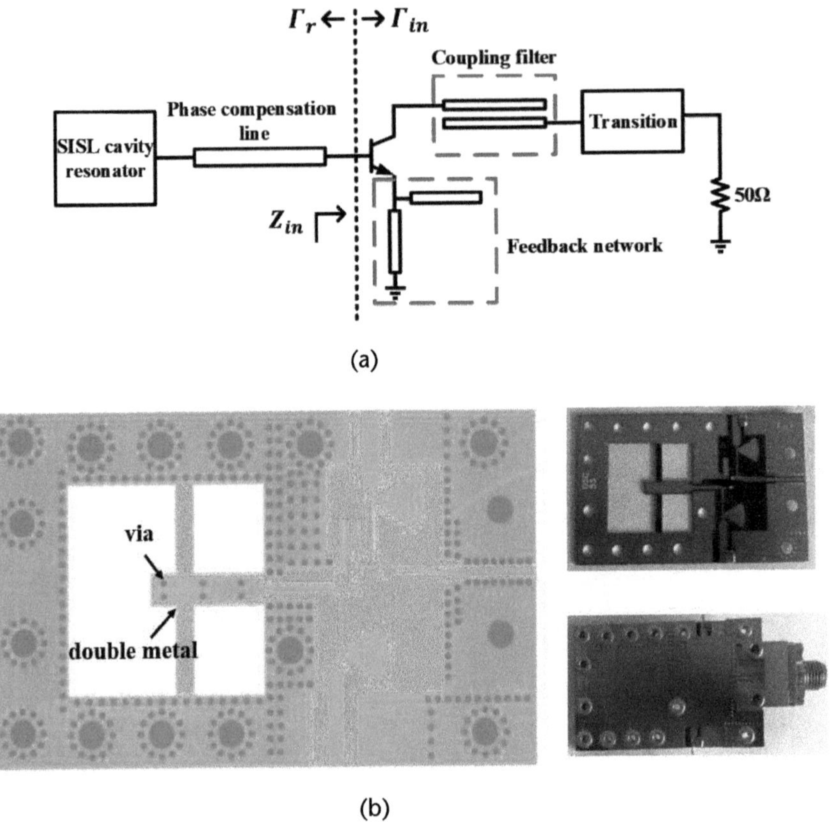

Figure 7.3 (a) Basic configuration of series feedback oscillator, and (b) oscillator using double-metal with substrate excavation cavity resonator [2].

The oscillators using an SISL cavity resonator not only make the 3D cavity resonators easy to integrate with the planar circuit, but also greatly improves the Q factor of the resonator and effectively reduces the phase noise. It also has the characteristics of compact size and self-packaged properties.

7.1.4 VCOs Based on SISL Technology

The schematic diagram of cross-coupled VCO based on a three-coupled resonator [4] is shown in Figure 7.5. Two of the transistors, Q_1 and Q_2, form a cross-coupled pair to provide energy. The emitter resistance R_e and base voltage V_b together provide dc bias for the transistor, and the collector capacitor C acts as a dc block. Three coupled resonators based on transformers are used for coupling the resonant loop and outputting the RF signal, and two sets of varactor diodes C_{v1} and C_{v2} are used for frequency tuning. The coupling of Tank 1 in the transformer-based three-coupling cavity as the primary resonant loop and Tank 2 in the secondary resonant loop can enhance the loop Q factor and improve the phase noise. At the same time, since Tank 2 is not directly connected to the cross-coupled pair, the influence of the transistor parasitic capacitance on the tuning ability of C_{v2} is avoided,

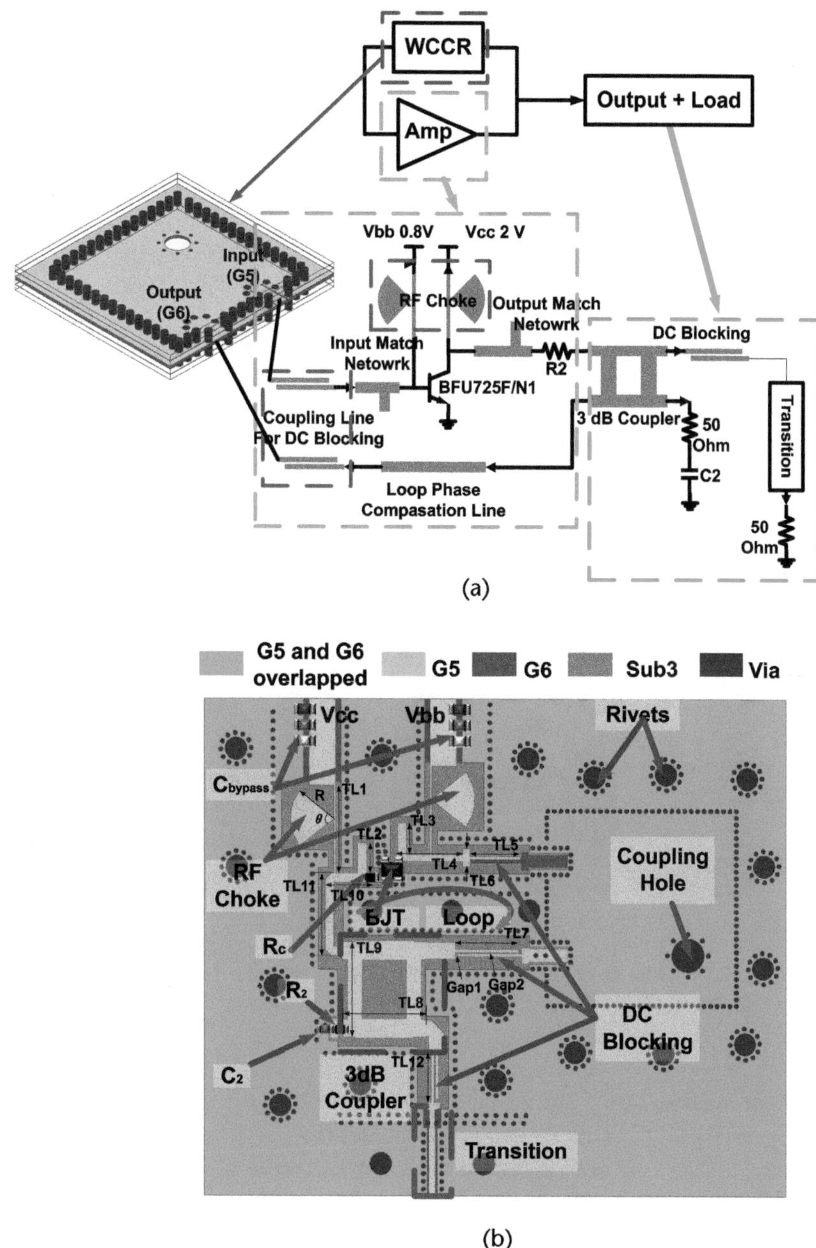

Figure 7.4 (a) Basic configuration of parallel feedback oscillator using SISL weakly coupled cavities, and (b) oscillator using SISL weakly coupled cavities [7].

thus broadening the tuning range. Tank 3, as a Barron coupling with Tank 1, can convert differential signals into single-ended signal output.

The circuit can be simplified as Figure 7.5(b) for ease of analysis. Among them, resistors R_1, R_2, and R_3, respectively, represent the loss of each coil; C_{p1}, C_{p2}, and C_{p3}, respectively, represent the parasitic capacitance of each resonant loop; R_{c1} and R_{c2}, respectively, represent the loss of varactor C_{v1} and C_{v2}; $-Gm$ represents the

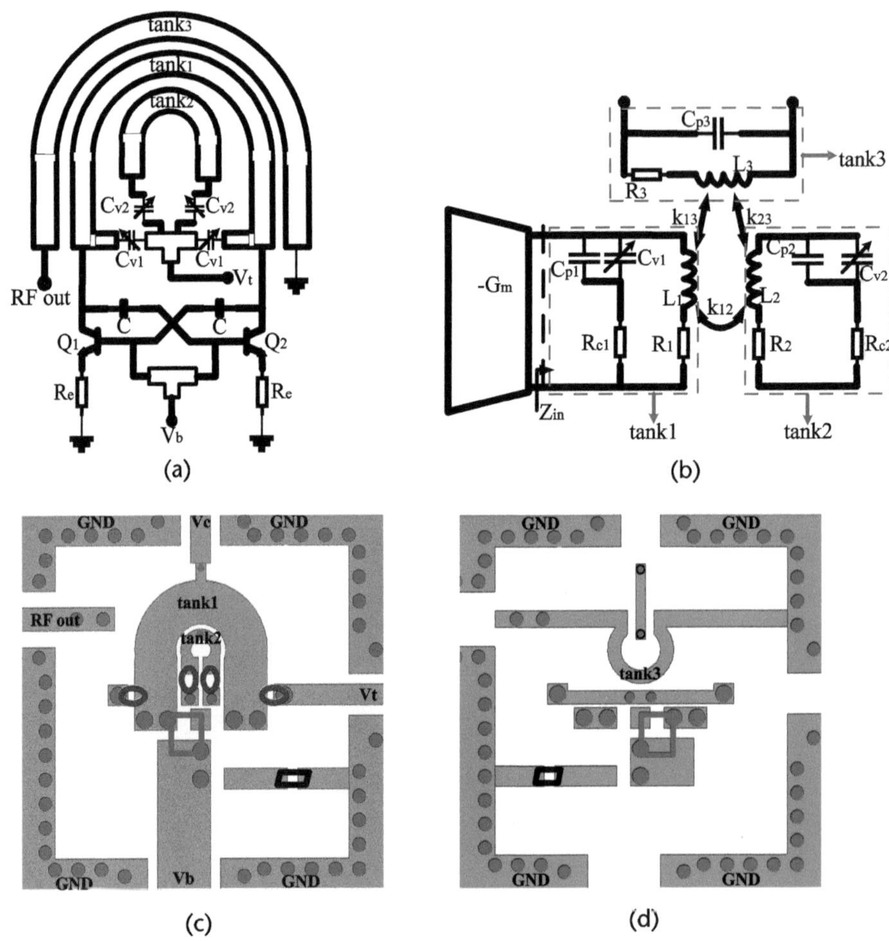

Figure 7.5 Cross-coupled VCO: (a) topology, (b) simplified equivalent circuit, (c) layout of M_5, and (d) layout of M_6 [4].

negative resistance provided by the cross-coupling pair. The layout of the cross-coupled VCO is shown in Figures 7.5(c, d).

This circuit will have two resonant frequencies, ω_L and ω_H, which can be expressed as:

$$\omega_{1,2} = \sqrt{\frac{L_1C_1 + L_2C_2 \pm \sqrt{(L_1C_1 - L_2C_2) + 4k_{12}^2 L_1C_1L_2C_2}}{2L_1C_1L_2C_2\left(1 - k_{12}^2\right)}} \tag{7.3}$$

According to the analysis of the three-coupled resonator in the previous section, the main resonant loop Tank 1 oscillates at low frequency, and the resonant loop Q factor will be improved, so the phase noise will be improved, so in order to obtain better performance, the designed VCO oscillation frequency will be excited at ω_L. The best phase noise is –119.45 dBc/Hz at 1 MHz at 3.76 GHz. The output power is –8.25 dBm, and the frequency tuning range is 28.9% from 2.88 to 3.85 GHz, and during the entire tuning range, the phase noise is better than –114.7 dBc/Hz.

Figure 7.6 shows a Colpitts oscillator based on the SISL technology. The topology and the layout of the VCO used the TRS resonator [5]. The measured frequency tuning range of the proposed VCO is from 4.08 GHz to 4.25 GHz with a 4.1% bandwidth. The phase noise is better than –130.2 dBc/Hz at a 1-MHz offset frequency, and the output power is higher than 4.64 dBm.

7.2 Mixers and Multipliers

The mixer is a three-port device that performs frequency conversion in the RF front-end system. For instance, the mixer in the receiver downconverts the RF signal received by the antenna at a higher frequency into an easily managed IF signal at a lower frequency, while the mixer in the transmitter upconverts the IF signal to an RF signal, which is then transmitted through the power amplifier and the antenna.

It becomes difficult to generate the high-quality fundamental oscillation signal needed for the mixer to operate normally as the frequency of the RF front-end system increases to the millimeter band, so a frequency multiplier is required to boost the fundamental frequency's oscillation signal to the needed frequency.

Mixers and frequency multipliers are commonly used in most RF front-end circuits, such as radar, radiometer, and communication systems, especially as the system frequency rises to the submillimeter band, because of the lack of effective

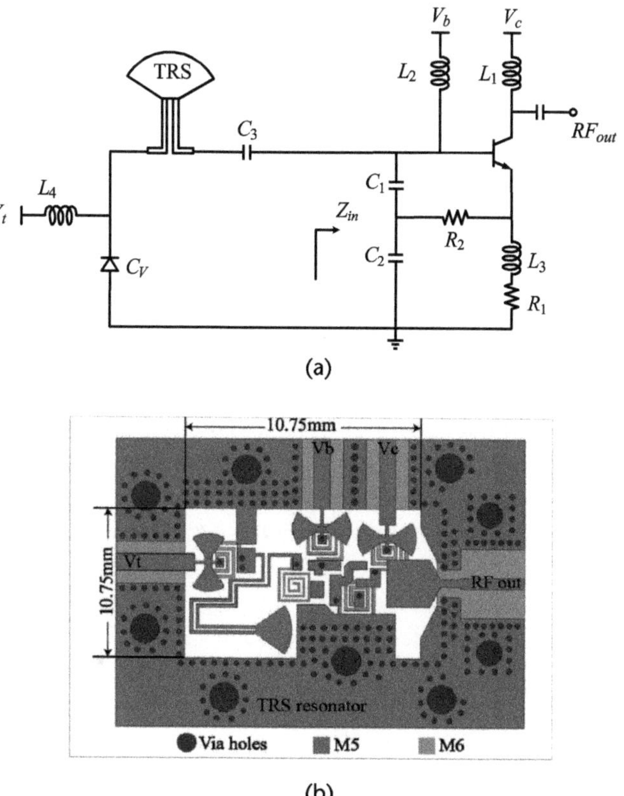

Figure 7.6 VCO using TRS resonator: (a) the topology, and (b) the layout [5].

low-noise amplification, mixers are often used in the design of RF front-end systems as the first stage of the circuit after the antenna, so often the performance of the mixer (usually measured by the conversion loss) to a large extent determines the overall system performance, while the frequency multiplier is required to provide the mixer with the fundamental oscillation power it needs to achieve optimal frequency conversion loss, and the performance of the frequency multiplier is usually measured by the output power and frequency multiplier efficiency.

As described in the previous section, the electromagnetic fields of SISL circuits can be distributed mainly in the low-loss air and rarely inside the dielectric as in the traditional suspended line, and in addition, the cavities provide space for the assembly of discrete devices such as MMIC chips, diodes, transistors, or capacitive inductors. Therefore, based on the abovementioned low-loss and high-integration features of the SISL, it provides a good platform basis for realizing high-performance millimeter-wave mixer and multiplier designs.

7.2.1 Mixer

The performance of the entire receiving system is heavily influenced by the mixer, a crucial component of the super-outlier receiver. The working frequencies of contemporary radar and guidance, electronic countermeasures, communications, radio astronomy, remote sensing, and telemetry systems have steadily been expanded to the millimeter-wave band to decrease the receiver system's size order to decrease the size of the receiver system and improve its performance. As a result, millimeter-wave mixer performance has become more critical, and research into low-cost, high-performance millimeter-wave mixers has gained significant importance [8–11].

Next, we will describe in detail two mixers based on the SISL design; the first one is a single-balanced fundamental mixer based on a 90° patch coupler with a patch phase shifter [12], and the second one is a fourth-harmonic mixer [13].

7.2.1.1 Single-Balanced Fundamental Mixer

A single-balanced fundamental mixer generally requires the use of two identical diodes; according to the relative position of the RF and LO ports and diodes, the fundamental single-balanced mixer can be divided into the LO inversion type and the RF inversion type, through the selection of a suitable coupler or phase shift device to make the RF and LO signals loaded on the two diodes to meet the specific phase relationship, in order to achieve the superposition and offset between the diode mixing components, suppress certain harmonic components and local oscillation noise, in the microwave and millimeter-wave band is widely used.

Figure 7.7 shows the schematic diagram of the fundamental LO inversion-type fundamental mixer. The ideal 180° coupler loads the RF signal into the two diodes in the same phases, while the 180° coupler loads the LO signal into the two diodes in the opposite phase. Thus, the voltages of RF and LO on the two diodes can be expressed as:

$$V_{s1} = V_{s2} = V_S \cos \omega_S t \tag{7.4}$$

$$\begin{aligned} V_{L1} &= V_L \cos \omega_L t \\ V_{L2} &= V_L \cos(\omega_L t + \pi) \end{aligned} \tag{7.5}$$

Time-varying conductance is:

$$\begin{aligned} g_1(t) &= g_0 + 2\sum_{n=1}^{\infty} g_n \cos n\omega_L t \\ g_2(t) &= g_0 + 2\sum_{n=1}^{\infty} g_n \cos n(\omega_L t + \pi) \end{aligned} \tag{7.6}$$

The RF power is usually much smaller than the LO power, the LO signal can be regarded as a large signal, and the RF signal can be regarded as a small signal; then the current of the two diodes in the single balanced working state of the mixer are:

$$i_2 = V_{s2} g_2(t) = V_s \cos \omega_s t \left[g_0 + 2\sum_{n=1}^{\infty} (\omega_L t + \pi) \right] \tag{7.7}$$

$$i_2 = V_{s2} g_2(t) = V_s \cos \omega_s t \left[g_0 + 2\sum_{n=1}^{\infty} g_n \cos n(\omega_L t + \pi) \right] \tag{7.8}$$

Further simplification leads to:

$$i_1 = V_s \cos \omega_s t \left(g_0 + 2g_1 \cos \omega_L t + 2g_2 2\cos \omega_L t + 2g_3 3\cos \omega_L t + \ldots \right) \tag{7.9}$$

$$i_1 = V_s \cos \omega_s t \left(g_0 - 2g_1 \cos \omega_L t + 2g_2 2\cos \omega_L t - 2g_3 3\cos \omega_L t + \ldots \right) \tag{7.10}$$

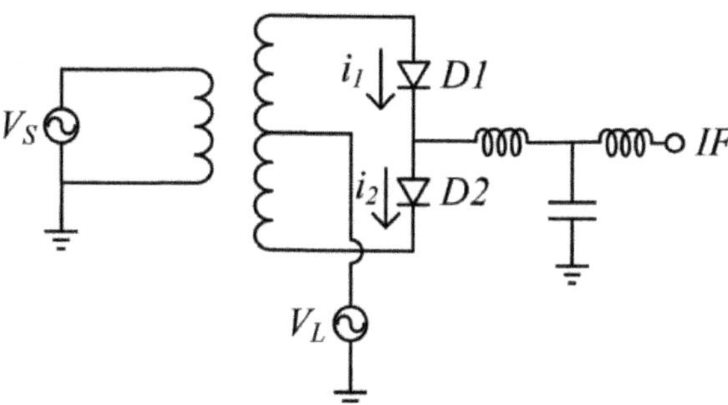

Figure 7.7 Schematic diagram of LO inversion type fundamental mixer [12].

Then the total current is:

$$\begin{aligned} i_\Sigma &= i_1 - i_2 \\ &= 4g_1V_s\cos\omega_s t\cos\omega_L t + 4g_3V_s\cos\omega_s t\cos 3\omega_L t + \ldots \\ &= 2g_1V_s\cos(\omega_s - \omega_L)t + 2g_1V_s\cos(\omega_s + \omega_L)t \\ &\quad + 2g_3V_s\cos(\omega_s - 3\omega_L)t + \ldots \end{aligned} \tag{7.11}$$

The desired IF signal is obtained after frequency selection by the IF filter:

$$i_{\text{IF}} = 2g_1V_s\cos(\omega_s - \omega_L)t \tag{7.12}$$

An SISL-based 90° patch phase shifter and a 90° patch coupler are used to form a 180° coupler and applied to design an SISL single-balanced mixer centering.

The circuit topology is shown in Figure 7.8. The RF port and the LO port are connected to the 90° coupler input port and the isolation port, respectively, and the two outputs of the coupler are connected to the reference patch and the main patch of the 90° phase shifter, respectively, and then connected to a pair of reverse cascaded Schottky diodes through a matching circuit, and finally connected to the IF output port through an LPF.

The frequency conversion loss of the mixer mainly comes from four parts: (1) circuit loss, (2) mismatch loss, (3) junction loss of the diode core, and (4) nonlinear conductance net frequency conversion loss. One of the major advantages of the SISL lies in its ability to minimize circuit loss, which generally includes dielectric loss, conductor loss, and radiation loss, and the following will introduce how the SISL reduces these three parts of the circuit loss. First, the SISL cavity is surrounded by a large number of metalized through-holes; these through-holes will constitute an approximate metal wall and, with the upper and lower metal grounding layers together, form an approximately closed electromagnetic shielding cavity, so that the internal circuit has almost no radiation and thus has a small radiation loss. Second, Substrate 2 and Substrate 4 form the upper and lower cavities by removing the dielectric, and Substrate 3 removes part of the unnecessary dielectric so that most of the electromagnetic field is distributed in the cavity, thus effectively reducing the dielectric loss. Finally, the double-layer wiring technology is used to interconnect the upper and lower layers of Substrate 3 through metalized through-holes,

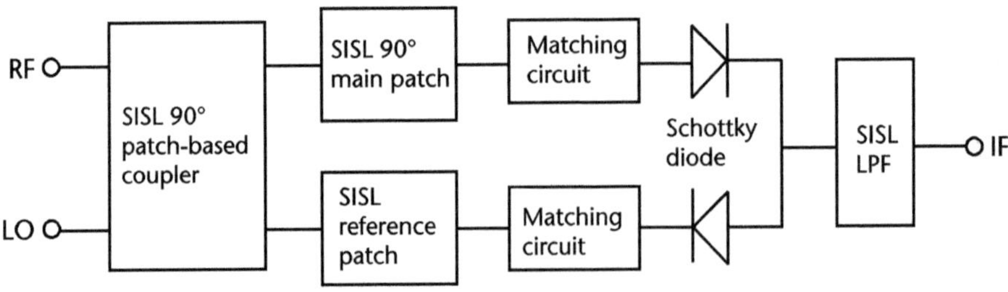

Figure 7.8 Topology of the SISL single-balanced mixer centering [12].

which increases the conductor surface area and reduces the conductor loss to a certain extent.

The mixer design of the patch coupler and the patch phase shifter both use the abovementioned SISL double-layer wiring and patterned substrate technique as a way to reduce circuit losses. Their specific design details are presented in [12].

Figure 7.9 displays the processing physical diagram for the SISL single-balanced mixer Substrate 3. In order to minimize total circuit loss, all internal modules,

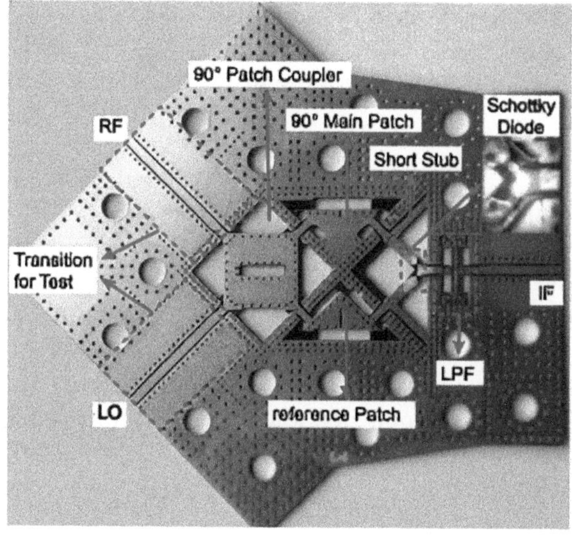

(a)

(b)

Figure 7.9 Photograph of the SISL mixer using SISL patch-based phase shifters and coupler: (a) Substrate 3, and (b) multilayer [12].

including the patch phase shifter, patch coupler, matching circuit, and LPF, are highly integrated into a single SISL platform using double-layer wiring and patterned substrate technology. While the isolation between the LO port and the RF port is primarily dependent on the isolation between the input port and the isolation port of the 180° coupler used, it is improved by the ground short branch at the matching circuit and provides a dc loop to the IF signal. An LPF at the IF port is used to suppress the RF and LO signals in the higher-frequency bands.

In addition, the mixer adopts a multicavity, honeycomb structure and uses a metalized through-hole fence for isolation between different modules, effectively reducing electromagnetic crosstalk between modules. At the same time, the modules are interconnected internally using a small-sized, ribbon line structure, which makes the overall structure compact and highly integrated. After comparison and testing, the performance parameters of the mixer are summarized as shown in Table 7.1.

7.2.1.2 Harmonic Mixer

The harmonic mixer is to use the Nth harmonic of the LO signal and RF signal mixing to get the IF signal. In the case of the same RF signal frequency, the harmonic mixer requirements of the fundamental frequency are 1/*N* times that of the fundamental mixer, which can solve the problem of difficult access to high-frequency sources. Harmonic mixing commonly used tube pair structure, that is, two consistent performances of the mixing diode in series or parallel, by suppressing the odd or even-harmonic components, and enhance the required harmonic components, thereby increasing the mixing efficiency and improving performance. This design uses an antiparallel structure as shown in Figure 7.10, and the following is a schematic analysis of it.

Table 7.1 Measured Performance Comparison of the Single-Balanced Mixer

Reference	*[14]*	*[15]*	*This Work*
Structure	SIW Magic-T + Microstrip LPF	SIW Coupler + Microstrip LPF	SISL Patch-Based Phase Shifter, SISL Patch-Based Coupler, and SISL LPF
Integration	Medium	Medium	High
Substrate	Rogers 6010LM	Rogers 6002	RF4+Rogers 5880
LO power (dBm)	13	13	3
Diodes	SMS7630-006LF	MGS901	MA4E2038
Frequency range (GHz)	7.5–11.5	20–26	24.5–26.5
Conversion loss (dB)	>7.4	>7.6	>5.8
LO-to-RF isolation (dB)	No data	20	15.3–23.8
Circuit size	~50 mm × 40 mm	~65 mm × 18 mm	21 mm × 13 mm
Self-packaging	No	No	Yes

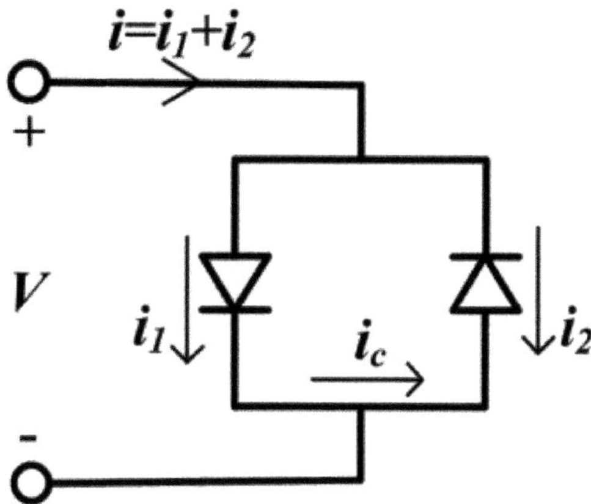

Figure 7.10 Antiparallel diode to harmonic mixing schematic [16].

The current flowing through the two diodes can be expressed as:

$$i_1 = i_s\left[e^{\alpha v} - 1\right],\ i_2 = -i_s\left[e^{\alpha v} - 1\right] \tag{7.13}$$

The total current is:

$$i = i_1 + i_2 = i_s\left(e^{\alpha v} - e^{-\alpha v}\right) = 2i_s \sinh(\alpha v) \tag{7.14}$$

The voltage of the LO signal loaded on the diode pair is:

$$v_L = v_{\mathrm{LO}} \cos\left(\omega_{\mathrm{LO}} t\right) \tag{7.15}$$

Substituting (7.15) into (7.14) and making a Fourier series expansion yields:

$$i = 4i_s\left(I_1 \alpha v_{\mathrm{LO}} \cos\omega_{\mathrm{LO}} t + I_3 \alpha v_{\mathrm{LO}} \cos 3\omega_{\mathrm{LO}} t + I_5 \alpha v_{\mathrm{LO}} \cos 5\omega_{\mathrm{LO}} t + \cdots\right) \tag{7.16}$$

Since the mixer has two input signals, the nonlinear effects generated by the LO signal and the RF signal should also be considered. The time-varying conductance of the LO of the diode pair is:

$$g = \frac{di}{dv} = 2\alpha I_s \cosh\left(\alpha V_{\mathrm{LO}} \cos\omega_{\mathrm{LO}} t\right) \tag{7.17}$$

A Fourier series expansion of the above equation yields:

$$g = 2i_s\left(I_0 \alpha v_{\mathrm{LO}} \cos\omega_{\mathrm{LO}} t + 2I_2 \alpha v_{\mathrm{LO}} \cos 2\omega_{\mathrm{LO}} t + 2I_4 \alpha v_{\mathrm{LO}} \cos 4\omega_{\mathrm{LO}} t + \ldots\right) \tag{7.18}$$

With the large signal driven by the LO, the RF small signal input to the mixer diode is expressed as:

$$v_S = v_s \cos(\omega_s t) \tag{7.19}$$

Then the total voltage of the LO signal and the RF signal loaded on the diode pair can be expressed as:

$$v = v_s \cos(\omega_s t) + v_{\mathrm{LO}} \cos(\omega_{\mathrm{LO}} t) \tag{7.20}$$

Compared with the large signal of the LO, the RF signal can be regarded as a weak small signal, so the mixing conductance of the diode pair can be approximately equivalent to that determined by the LO conductance only; then the total current can be expressed as:

$$i = gv = g\left[v_s \cos(\omega_s t) + v_{\mathrm{LO}} \cos(\omega_{\mathrm{LO}} t)\right] \tag{7.21}$$

Substituting (7.18) into (7.21), the total current after mixing is obtained as:

$$\begin{aligned} i = {} & A_1 \cos \omega_{\mathrm{LO}} t + A_2 \cos \omega_s t + A_3 \cos 3\omega_s t + A_4 \cos 5\omega_{\mathrm{LO}} t \\ & + A_5 \cos(\omega_s + 2\omega_{\mathrm{LO}})t + A_6 \cos(\omega_s - 2\omega_{\mathrm{LO}})t \\ & + A_7 \cos(\omega_s + 4\omega_{\mathrm{LO}})t + A_8 \cos(\omega_s - 4\omega_{\mathrm{LO}})t + \cdots \end{aligned} \tag{7.22}$$

As can be seen from the above equation, the combined frequency components generated by the mixing only contain the even-harmonic components of the LO signal combined with the RF signal, so for the fourth-harmonic mixer, only the $\omega_S - 4\omega_{LO}$ frequency components need to be frequency-selected output at the IF port, and the RF is about four times the LO frequency, which facilitates the design of circuits such as harmonic recycling and helps to reduce the frequency loss. The odd-harmonic components of the LO signal in the mixing process are offset in the internal loop of the diode pair, so the LO noise can be effectively reduced and the energy the LO signal can be fully utilized.

The topologies commonly used in current millimeter-wave harmonic mixers are mainly divided into two types: a 3D structure of waveguide coupler type [13], which consists of a microstrip or suspension line circuit together with a waveguide, and a planar structure of dendritic line type. In this design, the branch-and-line type structure is used to design the fourth-harmonic mixer.

Based on the traditional branch-type harmonic mixer structure, this design adds four transmission lines together with other branches to form a harmonic recovery circuit; in addition, they will also participate in the matching of diodes to both ends together with other branches to achieve the purpose of harmonic recycling and full mixing, thus reducing the conversion loss of the mixer and increasing the freedom of design.

The RF and IF ports have been placed on the same side, while the LO port is located on the other side, making it easier to construct the filter unit. The duplexer is used to create isolation between the RF and IF ports. It consists of an RF BPF and an IF LPF. The RF signal is equivalent to a short circuit, which creates a loop to ground, while the LO signal is equivalent to an open circuit, which does not affect how the LO signal feeds into the diode pair to participate in the mixing. The length of the short-circuit branch D is a quarter-wavelength of the LO frequency. Branch D will further provide to the ground loop for the IF signal as well as the dc component due to a diode imbalance because of the IF signal's lower frequency. Short-circuiting branches can be grounded directly on the same metal layer attributed to the special benefits of the SISL structure, as opposed to having to drill holes into the back as is necessary with microstrip circuits. The other branch lines mostly serve to recycle the significant harmonic components and operate on an identical basis.

The final layout is shown in Figure 7.11, where all modules are implemented and integrated on the SISL platform, using a honeycomb technique with rows of metalized vias for effective isolation between adjacent cavities of the duplexer and diode-matching circuits.

The dielectric substrate's other layers are inexpensive FR4, while the core circuit of the SISL mixer is constructed on a 0.127-mm Rogers 5880 dielectric substrate. By partially removing Substrates 3 and 5 to create the harmonic mixer's cavity and using metal layers 4 and 11 as the ground to create the harmonic mixer's electromagnetic shielding barrier, self-packaging is made possible. A short-circuit reflecting surface is provided by the metal layer 2 on the dielectric Substrate 1 when the RF signal is delivered through the WR10 waveguide from the mixer's bottom.

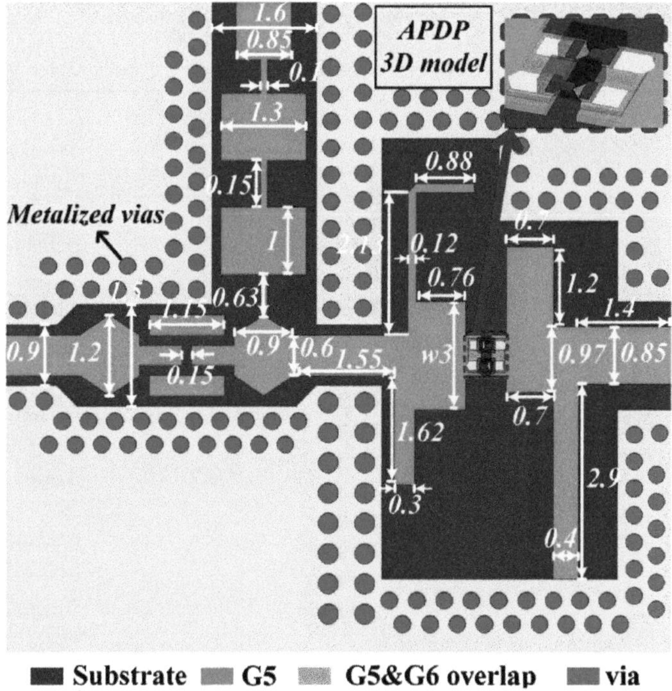

Figure 7.11 Top-side view of the Sub4 layer of the presented fourth-harmonic mixer [13].

Table 7.2 shows the comparison of the harmonic mixer in this chapter with the existing advanced level in the same frequency band. Usually, the higher the number of harmonics of the LO signal involved in effective mixing, the higher the frequency conversion loss. It can be seen that the harmonic mixer proposed in this chapter achieves low-frequency conversion loss at a low cost and can be self-packaged in the FR4 substrate, which does not require precision processing technology, and is significantly smaller than other millimeter-wave board-level mixers in terms of size and especially height.

7.2.2 Multiplier

The frequency multiplier is a very critical component in microwave and millimeter-wave transceiver systems, especially in submillimeter-wave and terahertz systems. Frequency multipliers can complete electronic circuits that multiply and transform the input signal frequency, as a way to obtain a frequency source that is not available at the fundamental frequency [9, 10].

A W-band balanced triplexer designed based on the SISL platform is described in detail next [23].

7.2.2.1 Balanced Frequency Multiplier

Operating Principle

The balanced frequency multiplier structure can suppress the odd or even harmonics generated in the diode pair because of its structural characteristics, thus greatly reducing the spurious components in the circuit, facilitating the matching and filter

Table 7.2 Measured Performance Comparison of the Harmonic Mixer

Reference	*Harmonic Times*	*Conversion Loss (dB)*	*RF (GHz)*	*IF (GHz)*	*LO Power (dBm)*	*Substrate/ Package*	*Size (mm × mm × mm)*
[17]	4th	16–17	80–103	dc–15	16	0.1-μm GaAs pHEMT	\
[18]	2nd	22.3–33	89–93	9–13	10	Rogers 5880/ Metal Block	45 × 32 × 30
[19]	6th	26.1–29.5	70–75	dc–3	13.5	Rogers 5880/ Metal Block	\
[20]	4th	18.7–20	92.7–95.5	0.1–2.9	14.2	Rogers 5880/ Metal Block	12 × 15 × 15
[21]	1st	17–21	77–81	0.7–4.7	10	Rogers 5880/ Metal Block	45 × 32 × 30
[22]	4th	15.6–17.3	76–82	dc–5	14	0.15-μm GaAs pHEMT	\
This work	4th	12.2–16.3	76–81	0.2	12	Rogers 5880, FR4 (Self-Packaging)	8.2 × 9.2 × 2.3a

design at the input and output, and making it easier to achieve broadband and high-efficiency frequency multiplication.

The schematic diagram of the odd-balanced frequency multiplication is shown in Figure 7.12, which looks in from both the input and output in antiparallel. Where the voltage of the input signal loaded on the diode pair is:

$$v = v_s \cos(\omega_s t) \tag{7.23}$$

The currents flowing through the two diodes are:

$$i_a = i_s\left[e^{\alpha v} - 1\right],\ i_b = -i_s\left[e^{-\alpha v} - 1\right] \tag{7.24}$$

The total external current is:

$$i = i_a + i_b = i_s\left(e^{\alpha v} - e^{-\alpha v}\right) = 2i_s \sinh(\alpha v) \tag{7.25}$$

Taking (7.23) into (7.25) and making a Fourier series expansion yields:

$$i = 4i_s\left(I_1 \alpha v_s \cos\omega_s t + I_3 \alpha v_s \cos 3\omega_s t + I_5 \alpha v_s \cos 5\omega_s t + \cdots\right) \tag{7.26}$$

From (7.26), it can be seen that the output component of the odd balanced multiplier circuit contains only the input fundamental signal and the odd-harmonic component generated by the diode pair, while the even harmonics are canceled out in the internal loop of the diode pair. This design is a triplexer, so only the signal output of the $3\omega_S$ frequency component needs to be selected.

The triplexer has five basic components: CPW-SISL transition, SISL-WR10 transition, input LPF, input and output matching circuit, and antiparallel diode pair.

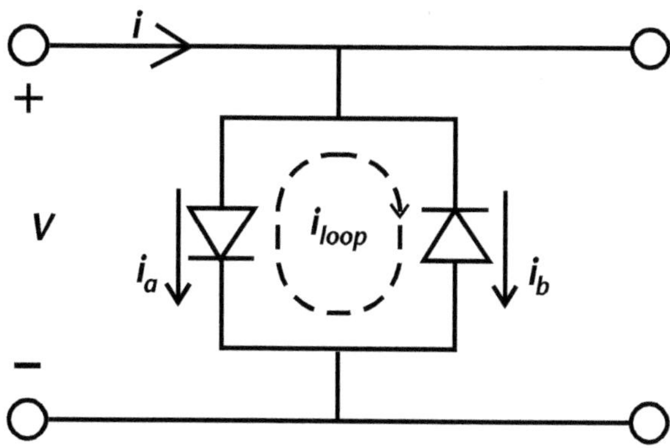

Figure 7.12 Odd times balanced frequency multiplication schematic [16].

To make the triplexer more compact, the bandpass characteristics of the SISL-WR10 transition circuit are used as a W-band BPF to frequency select the output of the third harmonic and suppress other harmonics, while avoiding the insertion loss introduced by an additional bandpass filter.

As shown in Figure 7.13, the dielectric substrate material of the triplexer designed in this chapter is FR4, and there are various thicknesses of FR4 dielectric from which to choose. Substrate 2 and Substrate 4 will be partially removed as the cavity layer of the SISL triplexer, so their thickness will directly determine the overall height of the shielding cavity, in the design of millimeter-wave board-level frequency multiplier, in order to avoid cavity resonance and high submode transmission, to ensure that the frequency multiplier works in single-mode transmission mode, the size of the shielding cavity should be small enough. However, the thickness of the dielectric substrate of the cavity layer should not be too thin; otherwise, it will lead to problems such as insufficient hardness and uneven thickness. After comprehensive consideration, we decided to select 0.3-mm FR4 as the cavity layer of the frequency multiplier. The top and bottom layers of the SISL circuit were chosen as the self-packaging cover layer for processing convenience, and the same thickness as the cavity layer was chosen. In addition, because of the waveguide transition structure required for millimeter-wave SISL circuits, an additional layer of 0.3-mm Substrate 5 was added. The thickness of this substrate was chosen with respect to the corresponding wavelength of the waveguide, and the metal layer 11 on Substrate 6 will provide a short-circuit reflection surface for the WR10 waveguide.

With the advantage of the SISL platform's extremely low radiation loss and good electromagnetic shielding characteristics, the test results show that it still has competitive output power compared to other product designs of the same type, as

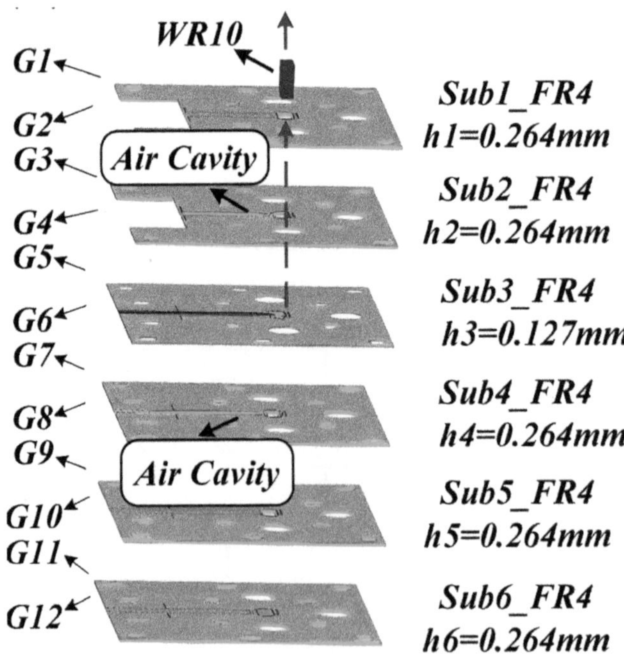

Figure 7.13 The 3D structure of the presented SISL tripler [23].

shown in Table 7.3, provided that all substrates of the multiplier are made of low-cost FR4. In addition, the SISL triplexer uses all low-cost PCB processing and is self-packaged compared to common millimeter-wave multipliers that use low-loss, high-cost dielectric substrates and require package in a metal waveguide housing.

7.2.3 Summary

This chapter introduces several millimeter-wave mixers and multipliers designed based on the SISL platform, which fully exploits the advantages of the SISL's low loss and high integration by using techniques such as double-layer wiring, patterned substrate, and honeycomb structure. In addition, all SISL-based designs feature low cost and self-packaging, which makes SISL one of the most advantageous platforms for a millimeter-wave circuit design. With the gradual rise of future application frequency and the progress of the PCB process level, it is believed that the SISL can continue to dominate in the submillimeter-wave and terahertz bands.

7.3 24-GHz FMCW Radar

With the rapid growth of people's traveling demands, vehicle safety issues are becoming increasingly prominent. The Advanced Driving Assistance System (ADAS) can utilize the sensing, decision-making, and execution devices on the vehicle to monitor the driving environment and assist the drivers. Millimeter-wave radar has become the main sensor for ADAS due to its advantages such as long detection range, moderate cost, and all-weather operation [29]. Short-range radar can be used for rear collision warning, blind spot detection, intelligent parking assistance, and other aspects in ADAS, primarily operating in the 24-GHz band.

The current mainstream solution for millimeter-wave radar systems is to use MMIC to implement the transceiver, the signal processing module is implemented

Table 7.3 Measured Performance Comparison of the Tripler

Reference	*Diodes*	*Substrate/ Package*	*Size (mm × mm × mm)*	*Input Power (dBm)*	*Min/Max Output Power (dBm)*	*Min/Max Conversion Loss (dB)*	*Output Frequency (GHz)*
[24]	BES Schottky diode	GaAs	\	19	−3.7/2	17/22.7	75–110
[25]	GaAs pHEMT diode	GaAs	\	18	−4/−6	22/24	87–102
[26]	DBES105a	Al2O3/Metal Block	\	12	−15/−3	15/27	60–110
[27]	MGS901	Neltec/Metal Block	35 × 40 × 15	15	−13/−2	17/28	40–60
[28]	MA4E1310	Rogers 5880/ Metal Block	30 × 38 × 24	21.5	−3/4	17.5/24.5	75–110
This work	MA4E2038	FR4 (Self-Packaging)	27 × 27 × 2	22	−4/4.46	17.54/26	75–110

by a baseband digital signal processing chip, and the antenna is implemented by a high-frequency PCB. However, intricate transitions are required between the MMIC chip and the board-level circuit [30–33], which can have a negative impact on radar performance. In addition, the cost of the MMIC process is high and it is difficult to achieve circuit packaging.

Many automotive radars implemented on the SIW have been reported [34–36]. However, the active devices and chips in the above systems cannot be directly integrated with SIW.

The SISL is a new suspended line based on multilayer PCB process, which can easily integrate other active and passive circuits. An FMCW radar system on the SISL platform is proposed in this section [37].

7.3.1 FMCW Radar

FMCW is the most commonly used modulated waveform for automotive millimeter-wave radar [38]. FMCW can continuously track targets and simultaneously measure the range and speed of multiple targets. It requires low transmission power and low cost, and the difficulty and real-time performance of signal processing can meet vehicle requirements.

Sawtooth frequency waveform (chirp) is a common modulation waveform of FMCW, as shown in Figure 7.14(a). B is the sawtooth frequency waveform bandwidth, T is the sawtooth frequency waveform duration time, and f_b is the frequency of the beat signal obtained after mixing and filtering. The range between the radar and the target is:

$$R = \frac{cT}{2B} f_b \tag{7.27}$$

The theoretical range resolution ΔR is:

$$\Delta R = \frac{c}{2B} \tag{7.28}$$

where T is 500 μs, B is 2 GHz, and the center frequency f_c is 25.5 GHz.

The velocity of target v is:

$$v = \frac{\lambda \Delta\varphi}{4\pi T} \tag{7.29}$$

where $\Delta\varphi$ is the phase difference between two adjacent chirps.

The theoretical velocity resolution Δv is:

$$\Delta v = \frac{\lambda}{2N_d T} \tag{7.30}$$

where N_d is the number of chirps contained in a single frame signal.

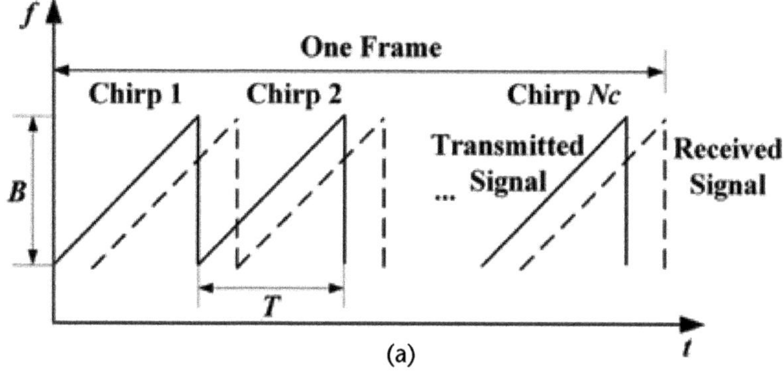

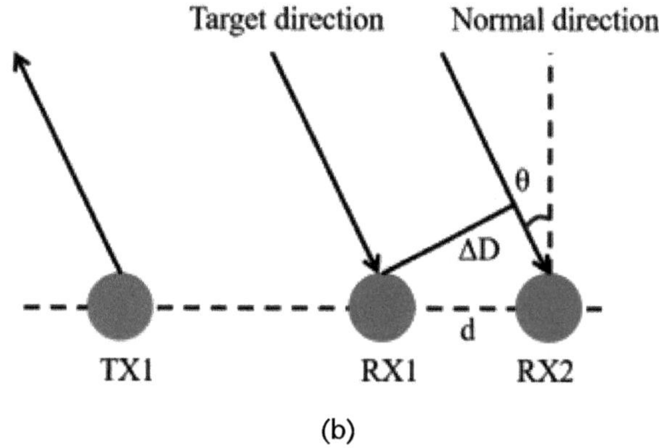

Figure 7.14 (a) FMCW waveform (sawtooth), and (b) phase angle measurement [37].

Figure 7.14(b) shows the diagram of angle measurement using the one transmitter and two receivers phase method. θ is the angle between the normal of the receiving antenna and the reflected echo of the target, d is the distance between receiving antennas, and ω is the phase difference of the signal received by two receiving antennas. The target angle θ is:

$$\theta = \arcsin\frac{\lambda}{2\pi d}\omega \tag{7.31}$$

The theoretical angular resolution is:

$$\Delta\theta = \frac{\lambda}{N_{\mathrm{RX}} d\cos\theta} \tag{7.32}$$

where N_{RX} is the number of receiving antennas.

According to the radar equation, the received power of the antenna is:

$$P_r = \frac{P_t G_t G_r \lambda^2 \sigma}{(4\pi)^3 R^4} \tag{7.33}$$

where P_t is the transmission power, G_t is the gain of the transmitting antenna, G_r is the gain of the receiving antenna, and σ is the radar cross-section of the target.

Taking the logarithm of both sides of the equation:

$$10\lg P_r = 10\lg P_t + 10\lg G_t + 10\lg G_r + G_\sigma - 2FSL \tag{7.34}$$

where FSL is the single-path, free-space loss and G_σ is the target gain factor.

$$FSL = 20\lg f + 20\lg R + 32.45 \tag{7.35}$$

where the unit of transmission frequency f is GHz and the unit of R is m.

$$G_\sigma = 20\lg f + 10\lg \sigma + 21.46 \tag{7.36}$$

where the unit of σ is m^2.

Design specifications are proposed in Table 7.4.

7.3.2 Six-Port Receiver

The six-port junction is the core passive part of the six-port receiver. A six-port circuit usually consists of a power divider and three 90° couplers [39], where the power divider can be replaced by a combination of an 90° coupler and a phase shifter or a 180° coupler. The S-parameter matrix of the six-port junction can be expressed as:

$$S = \frac{1}{2}\begin{bmatrix} 0 & 0 & 1 & -j & j & -1 \\ 0 & 0 & j & -1 & j & 1 \\ 1 & j & 0 & 0 & 0 & 0 \\ -j & -1 & 0 & 0 & 0 & 0 \\ j & j & 0 & 0 & 0 & 0 \\ -1 & 1 & 0 & 0 & 0 & 0 \end{bmatrix} \tag{7.37}$$

Table 7.4 Design Specifications of 24-GHz SISL Radar System

Specifications	*Values*
Center frequency f_c	25.5 GHz
Bandwidth B	2 GHz
Chirp duration time T	500 μs
Number of RX antennas N_{RX}	4
Number of chirps per frame N_c	65
Sampling frequency f_s	2 MHz
Theoretical range resolution ΔR	7.5 cm
Theoretical velocity resolution Δv	0.18 m/s
Theoretical angular resolution $\Delta\theta$	28.7°

A six-port network can convert RF signals into baseband signals, and after certain algorithmic calculations by the signal processing unit, the phase difference between two input signals can be obtained.

Based on the consideration of symmetry, the proposed six-port junction consists of one 180° coupler and three 90° couplers as Figure 7.15(a) [40] shows, according to the honeycomb structure, they are arranged in different cavities and connected to each other using strip wires.

According to the topology of the SISL six-port junction circuit shown in Figure 7.15(b), the four couplers are integrated on the same SISL platform according to the standard of honeycomb concept. The vias arranged around the cavity isolate the cavity. The coupler in different cavities will not affect each other, so there is basically no need for further debugging and optimization work. In addition, the proposed six-port junction uses double-metal layer with patterned substrate to reduce the conductor and dielectric losses of the circuit. Moreover, due to the electromagnetic shielding characteristics of the SISL, the radiation losses of the circuit are greatly reduced.

The photograph of the six-port junction is shown in Figure 7.15(c).

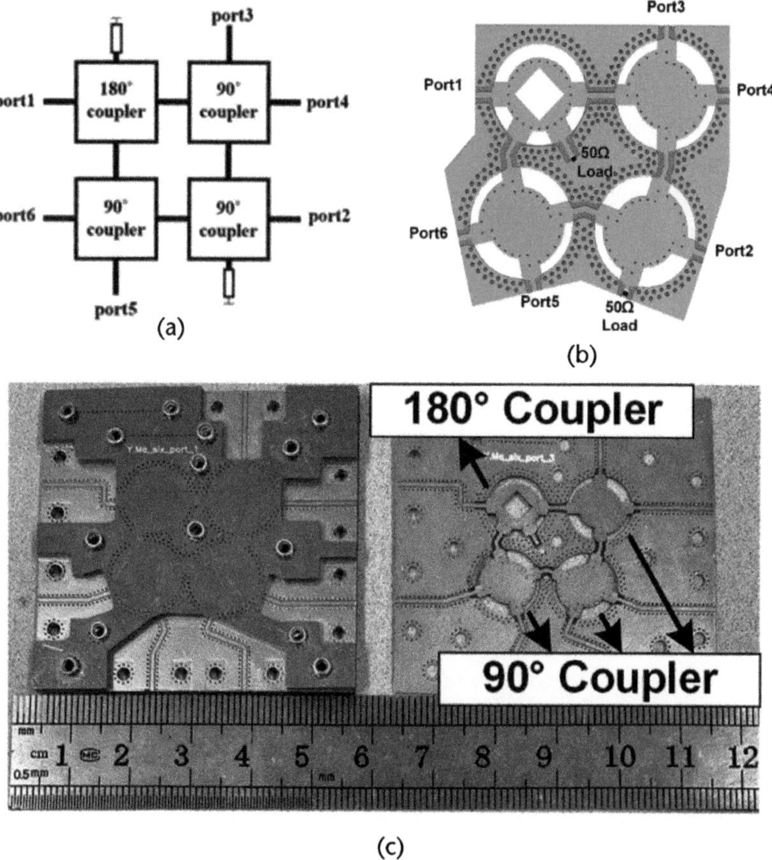

Figure 7.15 (a) Six port junction circuit [40], (b) top view of core circuit layer of six-port junction, and (c) photographs of core circuit layer of the proposed six-port junction [37].

The simulated and measured S-parameters of the six-port junction circuit are shown in Figure 7.16(a, b). When Port 1 is excited, the measured return loss is better than 15 dB from 22 GHz to 26 GHz, with a maximum value exceeding 35 dB. When Port 2 is excited, the measured return loss is better than 10 dB from 22 GHz to 26 GHz, with a maximum value exceeding 30 dB. The isolation between the two input ports is better than 16 dB, with a maximum value exceeding 40 dB. At the center frequency of 24 GHz, the measured S_{31}, S_{41}, S_{51}, and S_{61} are all within the range of −6.4 ± 0.2 dB, while the ideal theoretical value is −6 dB. Therefore, the measured insertion loss is less than 0.5 dB at the center frequency and less than 1.4 dB from 22 GHz to 26 GHz. The amplitude imbalance at the center frequency is less than 0.29 dB.

The phase relationship measured results of the six-port junction circuit are shown in Figure 7.16(c). Ideally, when Port 1 and Port 2 are, respectively, excited, the phase difference between the four output signals is 90° or a multiple of 90°. The phase error measured is less than 7° within the entire frequency range.

In practical circuits, the square rate characteristic of Schottky barrier diode (SBD) is usually used to detect power. Millimeter-wave signals generate dc, fundamental, and higher-order signals through detection diodes. For the application of FMCW radar in a six-port network, a low-frequency IF signal needs to be retained, and the fundamental and higher-order harmonic signals need to be suppressed.

The power detector based on the SISL is shown in Figure 7.17(a), composed of diodes, dc circuit, input matching circuit, and LPF. The SMS7630-061 Schottky zero bias detector diode from the Skyworks Company is selected. This diode belongs to the silicon low barrier Schottky diode and has the characteristics of ultra small 0201 occupation area, low barrier height, and low parasitic impedance. The dc loop adopts the method of loading multiple quarter-wavelength short-circuited stubs, with the length and width of the quarter-wavelength short-circuited stubs being 1.975 mm and 0.35 mm. The input matching circuit is achieved by using a quarter-wavelength impedance converter and a transmission line. The LPF is implemented by loading one high-impedance line and two open stubs.

Measured S-parameters are shown in Figure 7.17(b). The measured return loss is better than 5 dB from 24.5 to 26.5 GHz, and the return loss at the center frequency of 25.5 GHz is better than −9 dB. The relationship between the input power and output voltage of the diode power detector is shown in Figure 7.17(c).

When the Schottky diode power detector operates in the square law region, its output voltage is at the millivolt level and cannot be directly sampled by ADC devices or read by an oscilloscope. To obtain orthogonal I/Q signals, the four signals output by the six-port junction circuit need to be subtracted in pairs, so a differential amplifier structure is chosen. In order to amplify the weak signal output by the power detector to meet the input voltage range of the ADC device or oscilloscope, the amplification factor of the differential amplification link is set to about 100 times. A two-stage amplification structure is adopted, with a magnification of 10 times for each stage. The amplifier chips selected are two types of amplifiers, ADA4817 and AD8009. The first stage of the high gain operational amplifier is composed of two chips, ADA4817-2 and ADA4817-1. ADA4817-2 integrates two differential amplifiers into one chip, and it is used to achieve differential function. Then the signal is input into the ADA4817-1 chip and amplified. The second-stage

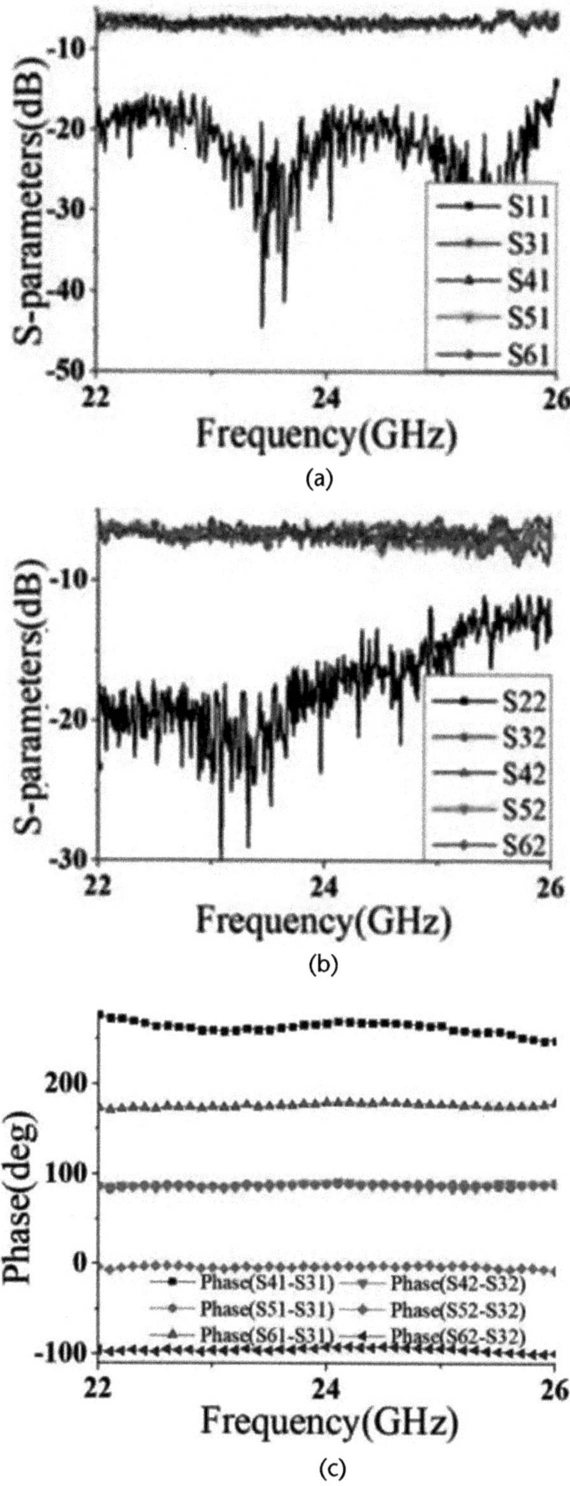

Figure 7.16 (a) Measured S-parameters of the six-port junction when Port 1 is excited, (b) measured S-parameters of the six-port junction when Port 2 is excited, and (c) measured phase relationship results of the six-port junction [41].

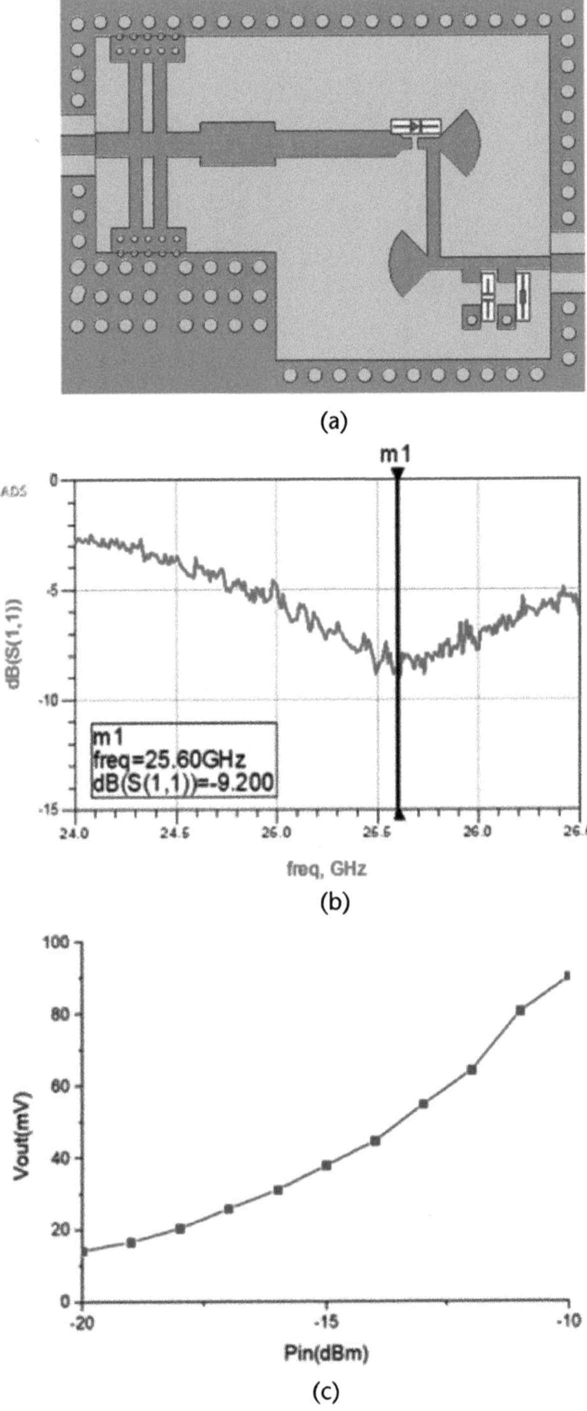

Figure 7.17 SISL power detector: (a) planar view of the SISL power detector, (b) measured S-parameters of the power detector, and (c) measured relationship between the input power and output voltage [41].

amplification circuit is a single-ended amplifier. The high gain operational amplifier module is also manufactured on the platform of the SISL.

The six-port receiver is composed of a six-port junction, two sets of differential operational amplifiers and four detectors. In the six-port receiver, the two input ports are RF signal and LO signal, and the four output ports of the six-port junction are the coupling signals of RF signal and LO signal, with a phase difference of 90°. The four detectors convert the power level proportionally to the baseband voltage. Finally, the four output signals are amplified in pairs through two sets of OPAs to obtain orthogonal IQ signals. The photograph of the six-port receiver is shown in Figure 7.18.

The measured gain of the six-port receiver versus local oscillator power is shown in Figure 7.19(a). The RF is 25.5005 GHz, the RF power is −20 dBm, and the LO frequency is 25.5 GHz. The LO power is changed in steps of 0.1 dBm from −6.5 dBm to −4.5 dBm. As the LO power increases, the gain of the six-port receiver first increases and then decreases, resulting in an optimal LO power of approximately −5.5 dBm. The measured gain of the six-port receiver versus LO frequency is shown in Figure 7.19(b). The RF power is −20 dBm, the LO power is −5.5 dBm, and the IF is 500 kHz. Change the LO frequency in steps of 0.1 GHz from 24.5 GHz to 26.5 GHz, and change the RF in steps of 0.1 GHz from 24.5005 to 26.5005 GHz. The gain of a six-port receiver at 25.5 GHz is about 12 dB.

The calibration principle of IQ demodulation in a six-port network is described in [42]. The two output signals of a six-port network are the real part (X) and imaginary part (Y) of the complex signal, expressed as I and Q signals. The phase difference between them can be obtained through:

$$\Delta\phi = \arctan\frac{Y}{X} \tag{7.38}$$

The measured demodulated I/Q signal is shown in Figure 7.19(c). Drawing curves with X and Y as the horizontal and vertical axes, it can be seen that the original measured curve is an ellipse, while the ideal curve is a standard unit circle. The reasons for the deviation include the amplitude and phase imbalance of the coupler in the six-port junction, the deviation of the characteristics of the four detection diodes, the deviation of the gain of the two differential operational amplifiers, and the deviation of processing and welding.

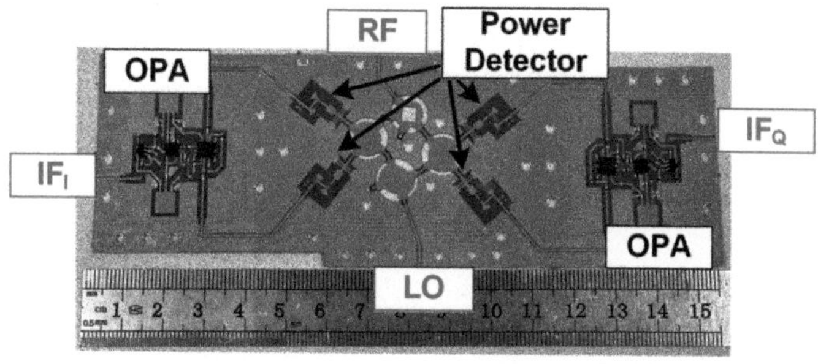

Figure 7.18 Photograph of core circuit layer of the proposed six-port receiver [37].

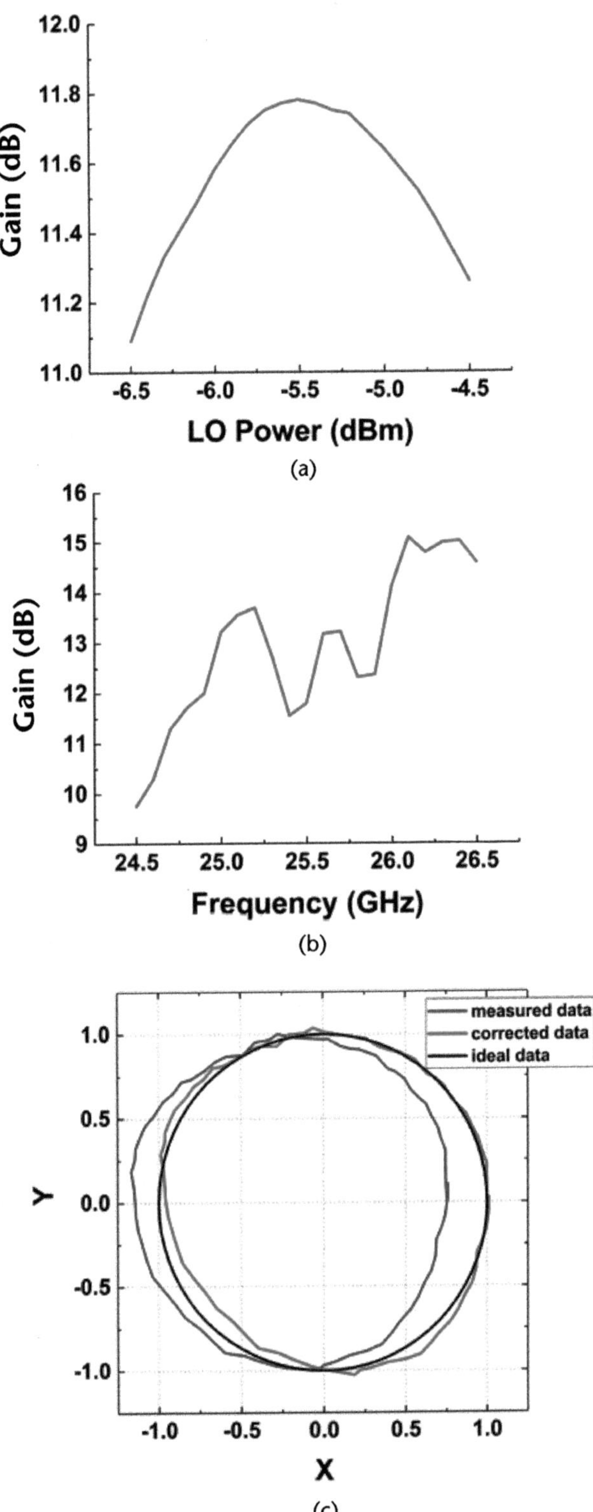

Figure 7.19 (a) Measured gain of the six-port receiver versus LO power, (b) measured gain of the six-port receiver versus LO frequency with fixed f_{IF} = 500 kHz, and (c) measured demodulated I/Q signal with fixed f_{IF} = 500 kHz [37].

There is a deviation between the original measurement data and the ideal curve, so the IQ measurement curve is calibrated. X_0 is the offset voltage of X, Y_0 is the offset voltage of Y, ρ is the phase imbalance between X and Y, and K_Y is the gain error of Y. The values of the four calibration constants are shown in Table 7.5. The data before and after calibration is shown in Figure 7.19(c). It can be seen that the calibrated data is closer to the ideal data than the original data.

7.3.3 24-GHz SISL Radar Receiver

An SISL-based 4 × 4 Butler matrix has been proposed in [43]. The eight element U-slot antenna array based on SISL was proposed in [44]. After the Butler matrix is integrated with the eight-element, U-slot antenna array, the measured gain at 25.5 GHz is greater than 12 dBi, the scanning coverage is ±47°, and the four beam-string angles are ±11° and ±35°. Combining the Butler matrix with the SP4T switch can achieve the function of beam scanning.

The structural diagram and photograph of the radar receiver are shown in Figure 7.20. In addition to the six-port receiver, Butler matrix, and U-slot antenna array mentioned above, the LNA in the radar receiver is AMMP-6233. The SP4T switch in the radar receiver is HMC1084LC4. The FMCW signal is transmitted by the transmitting antenna, and after being reflected by the target, the echo signal is received by the receiving antenna. After passing through the Butler matrix and SP4T switch, it is amplified by LNA and mixed with the LO signal by a six-port receiver, finally obtaining the IQ IF signal.

7.3.4 24-GHz SISL Radar System

The connection relationship of the range measurement instrument is shown in Figure 7.21(a, b). Any waveform generator (AWG) is implemented using M8195A. The spectrum analyzer is FSW85. The oscilloscope is implemented using DSA-X93204A. The PA is implemented using N4985A-S50. The TX antenna with a 1 × 4 cavity-backed end-fired dipole based on the SISL platform is described in detail in [16]. The power divider is implemented using PD-0165. DSP and PC are used for signal processing. The TX antenna port, LO port, and IF port use three 2.92-mm end launch connectors. The LO power of the system is fixed at the optimal LO power of the six-port receiver −5.5 dBm.

A metal plate with RCS of approximately 0.1 m^2 was used as the target. The position relationship between the radar and the target is shown in Figure 7.22(a). The target was moved in steps of 0.5m from 0.5m to 6m, preserving the IF spectrum at different ranges. The radar system link calculation obtained from the formula is shown Table 7.6. The range measurement results are shown in Figure 7.22(b), where the IF is converted to distance according to the equation, and it can be seen that

Table 7.5 Calibration Constants of Six-Port Receiver

f_0	X_0	Y_0	K_Y	ρ
25.5 GHz	229.7 mV	242.8 mV	0.974	5°

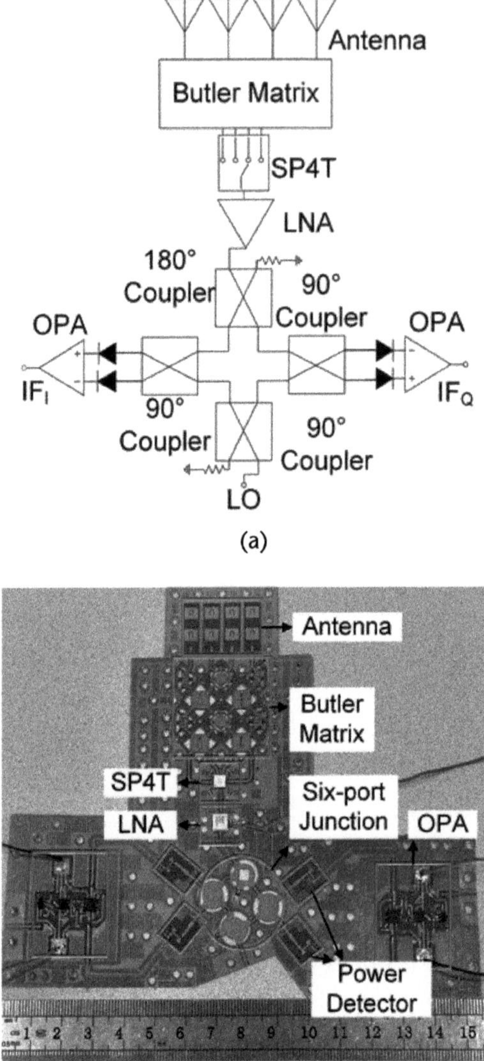

Figure 7.20 (a) Architecture and (b) photograph of core circuit layer of the proposed SISL radar receiver [37].

the frequency value of the IF peak corresponds to the distance value one by one. The power values of IF spikes above 2m follow Table 7.6. For closer distances, the measured values are less than the theoretical values, which is caused by the inaccurate alignment between the target and the radar.

In the range deviation measurement, the target is placed at 5m, and the frequency values of 1,000 IF spikes are continuously recorded, converted into distance values, and subtracted from 5m to obtain the range deviation value. The measured results of range deviation are shown in Figure 7.23(a), and it can be seen that the measured range deviation is ±1.6 cm at 5m. The deviation can be further reduced by averaging the measurement values every 10 times. Figure 7.23(b) is the corresponding histogram, which conforms to normal distribution, proving that the result is credible.

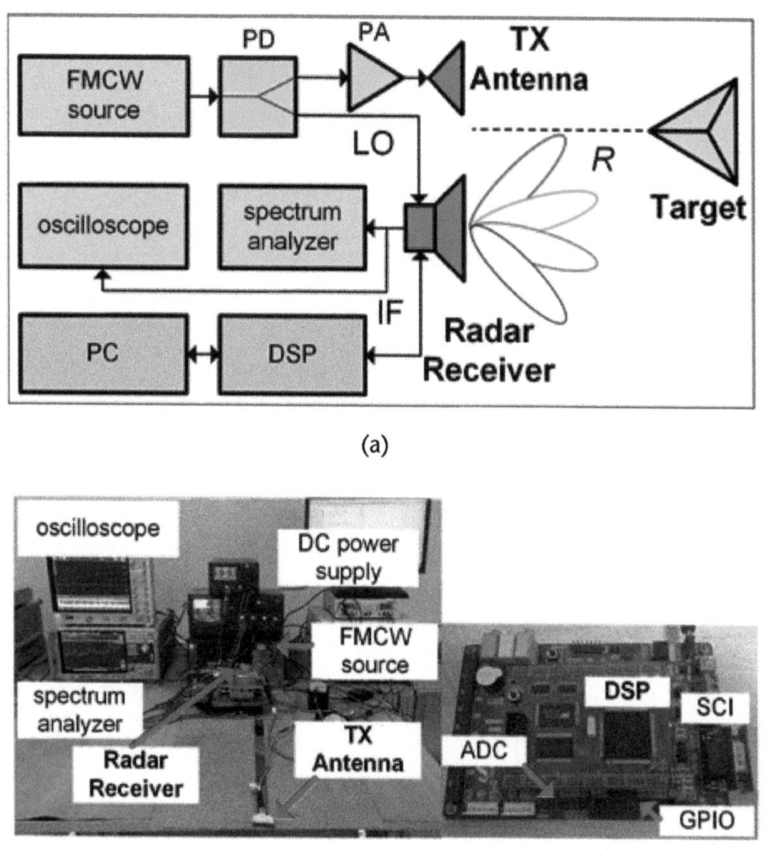

Figure 7.21 (a) Block diagram and (b) photograph of measurement setup for the proposed 24-GHz SISL radar system [37].

When the two targets are placed at 5m and 5.076m, respectively, the range resolution measurement result is shown in Figure 7.23(c), which shows two distinctive peaks, indicating a range resolution of 7.6 cm, which is 1.33% different from the theoretical value.

In angle estimation measurement, the target is placed at the same range and different angles, with a fixed range value of 1m. The target is moved in steps of 2° from −50° to 50°. The settings of the spectrograph are the same as those in the range measurement. The four conduction states of the SP4T switch are switched, corresponding to the beams in the four directions, the power values of the IF spikes are recorded. The power has been normalized for the convenience of display. The angle estimation measurement results are shown in Figure 7.23(d). The four beam angles of the Butler matrix antenna array are −35°, −11°, 11°, and 35°. For each curve, if the target is located at the beam angle, the IF power is the largest. The angle estimation measurement verified the beam scanning ability of the radar.

Two proposals suitable for embedded and PC applications are proposed: the DSP proposal and the MATLAB proposal.

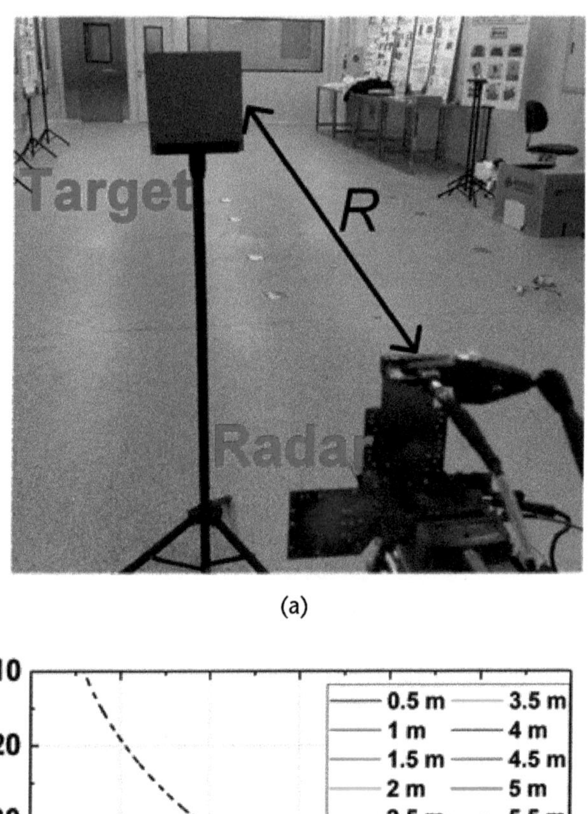

(a)

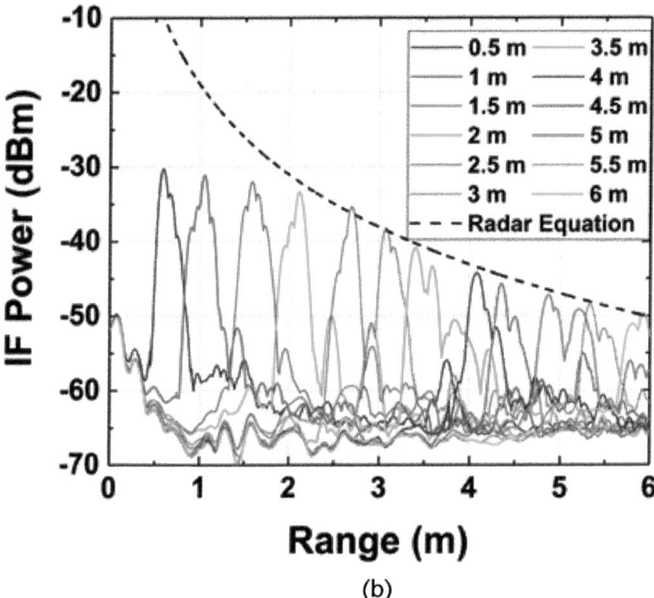

(b)

Figure 7.22 (a) Metal plate with RCS of 0.1 m^2 in an empty room, and (b) range measurement compared with radar equation [37].

In the verification of the DSP proposal, the two target locations are shown in Figure 7.24(a), placed at ranges of 4.7m and 8.2m with the angles of 8° and −8°, and a wall is located about 15m away. The interface of GUI is shown in Figure 7.24(c). By adjusting the threshold value of CA-CFAR, the two targets can be distinguished clearly through the amplitude line of RFFT and the threshold line of CA-CFAR as Figure 7.24(b) shows. The ranges of the two targets are calculated to be 4.687m and 8.203m, and the angles are 6.066° and −7.885°. The target location map was

Table 7.6 Link Budget Analysis of 24-GHz SISL Radar System

Parameters	*Values*
Transmitting power	4.5 dBm
Transmitting antenna gain	14.1 dBi
Path loss	109.1 to 152.3 dB
RCS gain	39.6 dB
Receiving antenna gain	14 dBi
Butler matrix loss	1.1 dB
SP4T switch loss	3 dB
LNA gain	22 dB
Six-port gain	12 dB
IF power	−7 to −50.2 dBm

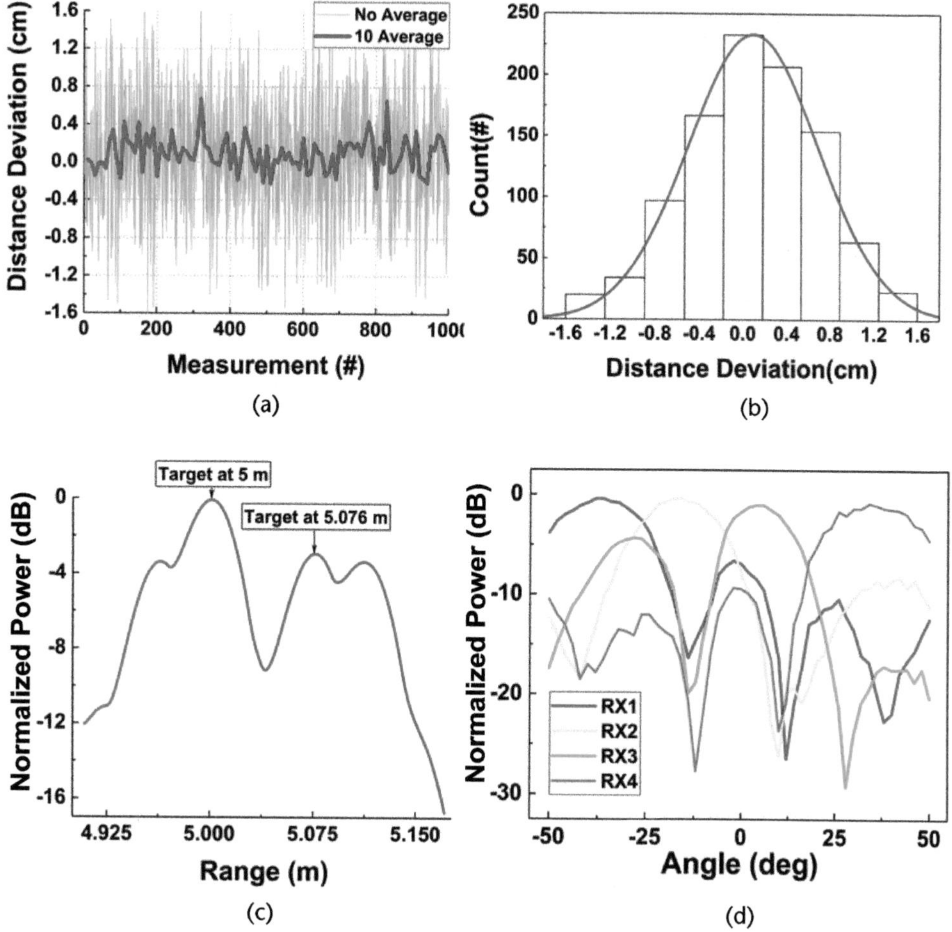

Figure 7.23 (a) Repeatability of the system demonstrated by 1,000 sets of measurements to the same target at 5m, (b) histogram for the distance, (c) range resolution measurement result, and (d) angular estimation measurement [37].

obtained; thus, the verification of DSP proposal ranging and angle measurement is completed.

In the verification of the MATLAB proposal, the values of the four beam angles corresponding to the actual SISL Butler matrix: −35°, −11°, 11°, and 35° are input into the MVDR beamformer. The positions of the two targets are the same as the DSP proposal. The range-angle map shown in Figure 7.25(a) is obtained, indicating that the radar can correctly detect the positions of the two targets and the walls located at 15m. Due to the small number of receiving antennas and low angular resolution, the display of the target is an arc. In velocity measurement, a person holding a corner reflector runs towards the radar. The range-Doppler map is shown in Figure 7.25(b). The velocity of the person is about 3 m/s, and a negative value indicates that the target is approaching the radar. The validation of the MATLAB proposal for range measurement, velocity measurement, and angle measurement is completed.

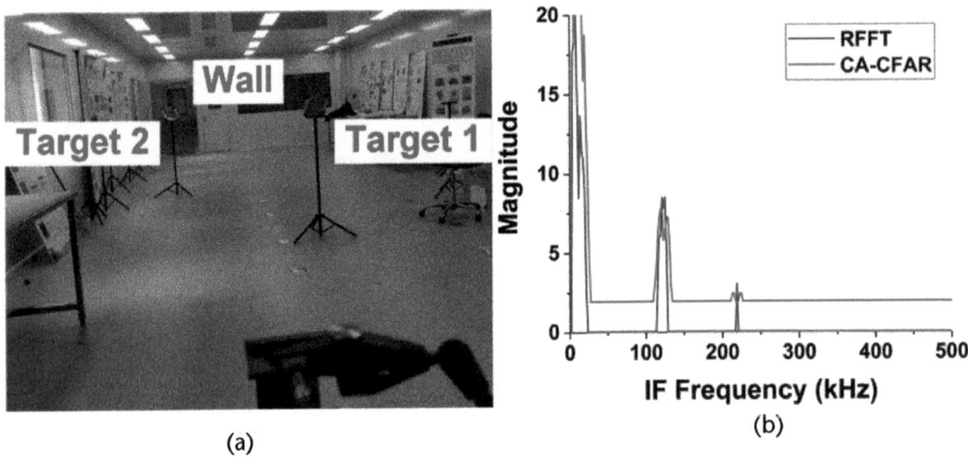

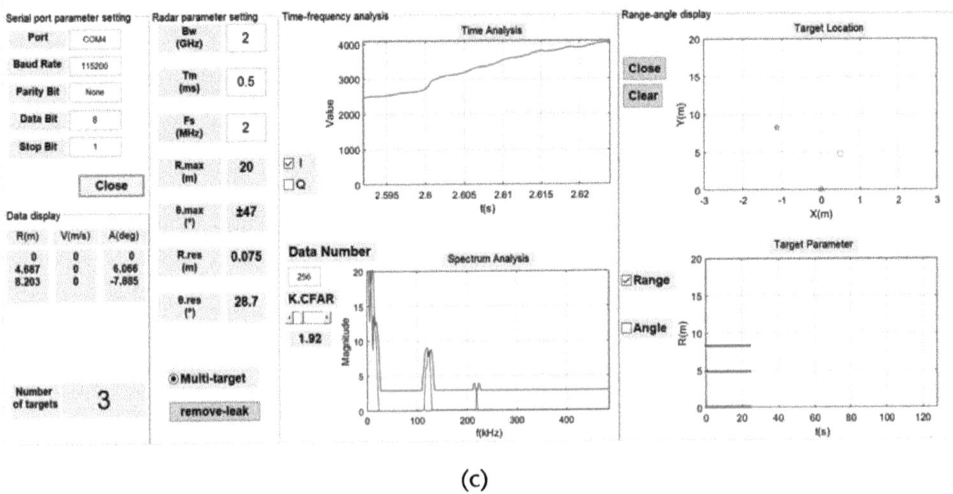

Figure 7.24 (a) Range and angle measurement scenario, and (b) magnitude of RFFT and threshold of CA-CFAR, and (c) interface of GUI [37].

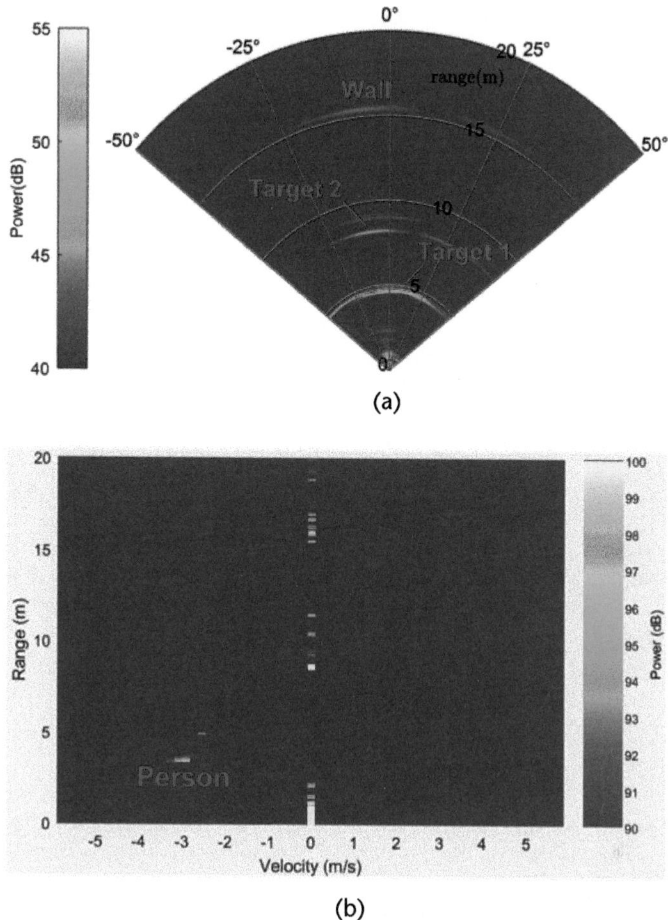

Figure 7.25 (a) Range-angle map, and (b) range-Doppler map [37].

Acknowledgments

We would like to thank Jingwen Han, Yuhao Hu, Dongqi Gu, JiXuan Ye, and Haozhen Huang for their valuable support to this chapter.

References

[1] Leeson, D. B., "A Simple Model of Feedback Oscillator Noise Spectrum," *Proc. IEEE*, Vol. 54, No. 2, 1966, pp. 329–330.

[2] Li, M., et al., "Design and Fabrication of Low Phase Noise Oscillator Using Q Enhancement of the SISL Cavity Resonator," *IEEE Transactions on Microwave Theory and Techniques*, Vol. 67, No. 10, October 2019, pp. 4260–4268.

[3] Li, L., K. Ma, and S. Mou, "Modeling of New Spiral Inductor Based on Substrate Integrated Suspended Line Technology," *IEEE Transactions on Microwave Theory and Techniques*, Vol. 65, No. 8, August 2017, pp. 2672–2680.

[4] Fu, H., et al., "A FR4-Based Compact VCO with Wide Tuning Range Using SISL Transformed Triple-Tanks," *Electron. Lett.*, Vol. 57, No. 20, September 2021, pp. 764–766.

[5] Zou, P., K. Ma, and S. Mou, "A Low Phase Noise VCO Based on Substrate-Integrated Suspended Line Technology," *IEEE Microwave Wireless Component Letters*, Vol. 27, No. 8, August 2017, pp. 727–729.

[6] Ma, K. X., et al., Integrated Circuit Architecture with Strongly Coupled LC Tanks, U.S. Patent No. 9331659B2, May 2016.

[7] Han, J., K. Ma, and N. Yan, "A Low Phase Noise Oscillator Employing Weakly Coupled Cavities Using SISL Technology," *IEEE Transactions on Circuits Systems Regul. Pap.*, 2023, pp. 1–14.

[8] Ren, Q., et al., "Novel Integration Techniques for Gap Waveguides and MMICs Suitable for Multilayer Waveguide Applications," *IEEE Transactions on Microwave Theory and Techniques*, Vol. 70, No. 9, September 2022, pp. 4120–4128.

[9] Dang, Z., et al., "A High-Efficiency W-Band Power Combiner Based on the TM_{01} Mode in a Circular Waveguide," *IEEE Transactions on Microwave Theory and Techniques*, Vol. 70, No. 4, April 2022, pp. 2077–2086.

[10] Cheng, H., et al., "Compact 31-W 96-GHz Amplifier Module in GaN-MEMS for Wireless Communications," *IEEE Transactions on Microwave Theory and Techniques*, Vol. 70, No. 2, February 2022, pp. 1233–1241.

[11] Roev, A., et al., "A Wideband mm-Wave Watt-Level Spatial Power-Combined Power Amplifier with 26% PAE in SiGe BiCMOS Technology," *IEEE Transactions on Microwave Theory and Techniques*, Vol. 70, No. 10, October 2022, pp. 4436–4448.

[12] Wang, Y., and K. Ma, "Low-Loss SISL Patch-Based Phase Shifters and the Mixer Application," *IEEE Transactions on Microwave Theory and Techniques*, Vol. 67, No. 6, June 2019, pp. 2302–2312.

[13] Zhang, N., et al., "A 77-GHz Low-Conversion-Loss Fourth-Harmonic Mixer on SISL Platform," *IEEE Microwave Wireless Component Letters*, Vol. 32, No. 6, June 2022, pp. 567–570.

[14] He, F. F., et al., "A Planar Magic-T Structure Using Substrate Integrated Circuits Concept and Its Mixer Applications," *IEEE Transactions on Microwave Theory and Techniques*, Vol. 59, No. 1, January 2011, pp. 72–79.

[15] Zhang, Z. -Y., Y. R. Wei, and K. Wu, "Broadband Millimeter-Wave Single Balanced Mixer and Its Applications to Substrate Integrated Wireless Systems," *IEEE Transactions on Microwave Theory and Techniques*, Vol. 60, No. 3, March 2012, pp. 660–669.

[16] Gong, Y., "Research on Millimeter Wave Frequency Doubling and Mixing Technology," Chengdu, University of Electronic Science and Technology, 2019.

[17] Kang, M., et al., "80–100 GHz Sub-Harmonically and Fourth-Harmonically Pumped Diode Mixers Using 0.1 μm GaAs pHEMT Process," *2012 37th IEEE International Conference on Infrared, Millimeter, and Terahertz Waves*, Wollongong, NSW, Australia, September 2012, pp. 1–2.

[18] Fujiwara, K., and T. Kobayashi, "Low-Cost W-Band Frequency Converter with Broad-Band Waveguide-to-Microstrip Transducer," *Glob. Symp. Millim. Waves*, 2016.

[19] Yao, C., and J. Xu, "An Improved Architecture of Sixth Subharmonic Mixers in E-Band," *Int. J. Infrared Millim. Waves*, Vol. 29, No. 4, April 2008, pp. 353–359.

[20] Xiang, B., et al., "Research on Fourth Harmonic Mixer at W Band in the Imaging System," *J. Electromagn. Eng. Sci.*, Vol. 10, No. 4, December 2010, pp. 316–321.

[21] Fujiwara, K., et al., "Simple-Structure and Cost-Effective FMCW Radar Test System Using a PLL-Gunn Oscillator and Fundamental Mixer in the E-Band," *IET Radar Sonar Navig.*, Vol. 13, No. 9, September 2019, pp. 1428–1436.

[22] Wang, C., et al., "W-Band Fourth-Harmonically Pumped Mixer Using 0.15 μm GaAs pHEMT Process," *2016 IEEE International Symposium on Radio-Frequency Integration Technology (RFIT)*, Taipei, Taiwan, August 2016, pp. 1–3.

[23] Zhang, N., et al., "A W-Band Broadband Self-Packaged Compact SISL Tripler Using FR4 Substrate," *Microw. Opt. Technol. Lett.*, Vol. 64, No. 6, June 2022, pp. 1023–1028.

[24] Morgan, M., and S. Weinreb, "A Full Waveguide Band MMIC Tripler for 75–110 GHz," *2001 IEEE MTT-S International Microwave Symposium Digest (Cat. No.01CH37157)*, Phoenix, AZ, 2001, pp. 103–106.

[25] Lin, K. -Y., et al., "A W-band GCPW MMIC Diode Tripler," *32nd IEEE European Microwave Conference, 2002*, Milan, Italy, October 2002, pp. 1–4.

[26] Hrobak, M., et al., "Design and Fabrication of Broadband Hybrid GaAs Schottky Diode Frequency Multipliers," *IEEE Transactions on Microwave Theory and Techniques*, Vol. 61, No. 12, December 2013, pp. 4442–4460.

[27] Jha, P., A. Basu, and S. K. Koul, "Broadband Frequency Tripler Design at 40–60 GHz," *2016 IEEE Asia-Pacific Microwave Conference (APMC)*, New Delhi, India, December 2016, pp. 1–4.

[28] Dou, J., and J. Xu, "Simple and Accurate Design of GaAs Schottky Diode Model," *Electron. Lett.*, Vol. 53, No. 13, June 2017, pp. 881–883.

[29] Waldschmidt, C., J. Hasch, and W. Menzel, "Automotive Radar—From First Efforts to Future Systems," *IEEE J. Microw.*, Vol. 1, No. 1, January 2021, pp. 135–148.

[30] Bhutani, A., et al., "Packaging Solution Based on Low-Temperature Cofired Ceramic Technology for Frequencies Beyond 100 GHz," *IEEE Transactions on Components, Packaging and Manufacturing Technology*, Vol. 9, No. 5, May 2019, pp. 945–954.

[31] Yue, C. P., and S. S. Wong, "Physical Modeling of Spiral Inductors on Silicon," *IEEE Transactions on Electron Devices*, Vol. 47, No. 3, March 2000, pp. 560–568.

[32] Chang, K. F., et al., "77-GHz Automotive Radar Sensor System with Antenna Integrated Package," *IEEE Transactions on Components, Packaging and Manufacturing Technology*, Vol. 4, No. 2, February 2014, pp. 352–359.

[33] Goettel, B., et al., "Packaging Solution for a Millimeter-Wave System-on-Chip Radar," *IEEE Transactions on Components, Packaging and Manufacturing Technology*, Vol. 8, No. 1, January 2018, pp. 73–81.

[34] Kueppers, S., et al., "A Compact 24 × 24 Channel MIMO FMCW Radar System Using a Substrate Integrated Waveguide-Based Reference Distribution Backplane," *IEEE Transactions on Microwave Theory and Techniques*, Vol. 68, No. 6, June 2020, pp. 2124–2133.

[35] Dong, H., et al., "A Low-Loss Fan-Out Wafer-Level Package with a Novel Redistribution Layer Pattern and Its Measurement Methodology for Millimeter-Wave Application," *IEEE Transactions on Components, Packaging and Manufacturing Technology*, Vol. 10, No. 7, July 2020, pp. 1073–1078.

[36] Han, L., and K. Wu, "24-GHz Integrated Radio and Radar System Capable of Time-Agile Wireless Communication and Sensing," *IEEE Transactions on Microwave Theory and Techniques*, Vol. 60, No. 3, March 2012, pp. 619–631.

[37] Liu, B., et al., "An SISL-Based 24-GHz FMCW Radar with Self-Packaged Six-Port Butler Matrix Receiver," *IEEE Transactions on Components, Packaging and Manufacturing Technology*, Vol. 12, No. 10, October 2022, pp. 1661–1672.

[38] Patole, S. M., et al., "Automotive Radars: A Review of Signal Processing Techniques," *IEEE Signal Process. Mag.*, Vol. 34, No. 2, March 2017, pp. 22–35.

[39] Gerardi, L., et al., "A New Six-Port Circuit Architecture Using Only Power Dividers/Combiners," *2007 IEEE/MTT-S International Microwave Symposium*, Honolulu, HI, June 2007, pp. 41–44.

[40] Ma, Y., M. Kaixue, and W. Yongqiang, "A Six Port Network Based on Dielectric Integrated Suspension Line," *The Chinese Institute of Electronics. Proceedings of the 2019 National Microwave and Millimeter Wave Conference*, Vol. 2, 2019, p. 3.

[41] Ma, Y., "Research on SISL Based Six Port Network Technology," Chengdu, University of Electronic Science and Technology, 2020.

[42] Laemmle, B., et al., "A 77-GHz SiGe Integrated Six-Port Receiver Front-End for

Angle-of-Arrival Detection," *IEEE J. Solid-State Circuits*, Vol. 47, No. 9, September 2012, pp. 1966–1973.

[43] Wang, Y., K. Ma, and Z. Jian, "A Low-Loss Butler Matrix Using Patch Element and Honeycomb Concept on SISL Platform," *IEEE Transactions on Microwave Theory and Techniques*, Vol. 66, No. 8, August 2018, pp. 3622–3631.

[44] Jian, Z., et al., "A Wideband SISL U-Slot Antenna Array," *2018 IEEE International Applied Computational Electromagnetics Society Symposium—China (ACES)*, Beijing, China, July 2018, pp. 1–2.

About the Authors

Kaixue Ma received BE and ME degrees from Northwestern Polytechnical University (NWPU), Xi'an, China, and a PhD degree from Nanyang Technological University (NTU), Singapore. From 1997 to 2002, he worked in Chinese Academy of Space Technology (Xi'an) as a group leader. From 2005 to 2007, he was with MEDs Technologies as an R&D manager. From 2007 to 2010, he was with ST Electronics as an R&D manager, project leader, Technique Management Committee, and technique consultant in 2011.

From 2010 to 2013, he was with NTU as a senior research fellow and millimeter-wave radio frequency integrated circuit (RFIC) team leader for 60-GHz Flagship Chipset project. From 2013 to 2018, he was a full professor with the University of Electronic Science and Technology of China (UESTC), Chengdu, China. Since February 2018, he has been the dean and distinguished professor at the School of Microelectronics of Tianjin University, PI of National IC Innovation & Entrepreneurship Platform of Tianjin, the director of Tianjin Key Laboratory of Imaging and Sensing Microelectronics Technology and the chairperson of Tianjin IC Association. Dr. Ma proposed a variety of RF and microwave integrated circuits based on advanced complementary metal-oxide-semiconductor (CMOS), silicon germanium bipolar complementary metal-oxide-semiconductor (SiGe BiCMOS), gallium arsenide (GaAs), and silicon-on-insulator (SOI) technologies, and microwave circuit and system design technology patented with "quasi-planar circuits with embedded air cavity," named as SISL in publication. The preliminary research and development results were partially transformed into the company's flagship products sold to Europe and the United States or used in China's on-orbit satellites.

Dr. Ma was responsible for designing the first low-power reconfigurable 60-GHz SiGe millimeter-wave transceiver SOC, packaging, and system testing and completed a high-speed dual-chip wireless communication system. He is currently working on silicon-based and GaAs RF millimeter-wave integrated circuits and systems. He has filed over 50 patents and has published three books in English, over 200 *SCI International Journal* papers (over 180 IEEE journal articles), and 200 international conference papers. Dr. Ma is a Fellow of the Chinese Institute of Electronics and an awardee of the Chinese National Science Fund for Distinguished Young Scholars. He received 10 technique awards, including the best paper award. He became an IEEE Fellow in 2024 for his contributions to low-loss substrate integrated suspended line technology and reconfigurable millimeter-wave front-end integrated circuits.

He was the associate editor for the *IEEE Transactions on Microwave Theory and Techniques* and the guest editor of the *IEEE Microwave Magazine* and is a current member of and organizes international conferences. He was the Coordinator IEEE

MTT-S R10 for China and Singapore from 2018 to 2022 and a current member of the Speakers Bureau of MTT-S TC-4.

Yongqiang Wang received BS and PhD degrees from the University of Electronic Science and Technology of China (UESTC), Chengdu, China, in 2014 and 2019, respectively. He is also a member of the IEEE.

From 2019 to 2020, he was a research associate with The Chinese University of Hong Kong, Hong Kong. He is currently a full-time associate professor with the School of Microelectronics, Tianjin University, Tianjin, China. His current research interest includes microwave passive circuit design and its applications. Dr. Wang has authored or coauthored over 90 journals and conference articles and holds 20 granted patents. Dr. Wang was the finalist in the Student Innovation Competition of IEEE International Workshop on Electromagnetics (iWEM) in 2014, the Advanced Practice Paper Competition (APPC) of the IEEE Microwave Theory and Technology Society (MTTS) International Microwave Symposium (IMS) in 2017, and International Wireless Symposium (IWS) in 2019. He was a recipient of the Best Student Paper of UK-Europe-China Workshop on Millimetre Waves and Terahertz Technologies (UCMMT) 2018, National Conference on Microwave and Millimeter Wave (NCMMW) 2017, and NCMMW 2019, where he twice received the Excellent Paper Award supported by the Education Development Foundation of Lin Weigan.

Dr. Wang is a reviewer for the *IEEE Transactions on Microwave Theory and Techniques*, *IEEE Transactions on Industrial Electronics*, *IEEE Transactions on Circuits and Systems I: Regular Papers*, *IEEE Transactions on Antennas and Propagation*, *IEEE Transactions on Components, Packaging and Manufacturing Technology*, *IEEE Transactions on Electron Devices*, *IEEE Microwave and Wireless Components Letters*, *IET Microwaves, Antennas and Propagation*, and *Electronics Letters*.

Index

Artech House Microwave Library

Behavioral Modeling and Linearization of RF Power Amplifiers, John Wood

Chipless RFID Reader Architecture, Nemai Chandra Karmakar, Prasanna Kalansuriya, Randika Koswatta, and Rubayet E-Azim

Chipless RFID Systems Using Advanced Artificial Intelligence, Larry M. Arjomandi and Nemai Chandra Karmakar

Control Components Using Si, GaAs, and GaN Technologies, Inder J. Bahl

Design of Linear RF Outphasing Power Amplifiers, Xuejun Zhang, Lawrence E. Larson, and Peter M. Asbeck

Design Methodology for RF CMOS Phase Locked Loops, Carlos Quemada, Guillermo Bistué, and Iñigo Adin

Design of CMOS Operational Amplifiers, Rasoul Dehghani

Design of RF and Microwave Amplifiers and Oscillators, Second Edition, Pieter L. D. Abrie

Digital Filter Design Solutions, Jolyon M. De Freitas

Discrete Oscillator Design Linear, Nonlinear, Transient, and Noise Domains, Randall W. Rhea

Distortion in RF Power Amplifiers, Joel Vuolevi and Timo Rahkonen

Distributed Power Amplifiers for RF and Microwave Communications, Narendra Kumar and Andrei Grebennikov

Electric Circuits: A Primer, J. C. Olivier

Electronics for Microwave Backhaul, Vittorio Camarchia, Roberto Quaglia, and Marco Pirola, editors

An Engineer's Guide to Automated Testing of High-Speed Interfaces, Second Edition, José Moreira and Hubert Werkmann

Envelope Tracking Power Amplifiers for Wireless Communications, Zhancang Wang

Essentials of RF and Microwave Grounding, Eric Holzman

Frequency Measurement Technology, Ignacio Llamas-Garro, Marcos Tavares de Melo, and Jung-Mu Kim

FAST: Fast Amplifier Synthesis Tool—Software and User's Guide, Dale D. Henkes

Feedforward Linear Power Amplifiers, Nick Pothecary

Filter Synthesis Using Genesys S/Filter, Randall W. Rhea

Foundations of Oscillator Circuit Design, Guillermo Gonzalez

Frequency Synthesizers: Concept to Product, Alexander Chenakin

Fundamentals of Nonlinear Behavioral Modeling for RF and Microwave Design, John Wood and David E. Root, editors

Generalized Filter Design by Computer Optimization, Djuradj Budimir

Handbook of Dielectric and Thermal Properties of Materials at Microwave Frequencies, Vyacheslav V. Komarov

Handbook of RF, Microwave, and Millimeter-Wave Components, Leonid A. Belov, Sergey M. Smolskiy, and Victor N. Kochemasov

High-Efficiency Load Modulation Power Amplifiers for Wireless Communications, Zhancang Wang

High-Linearity RF Amplifier Design, Peter B. Kenington

High-Speed Circuit Board Signal Integrity, Second Edition, Stephen C. Thierauf

Integrated Microwave Front-Ends with Avionics Applications, Leo G. Maloratsky

Intermodulation Distortion in Microwave and Wireless Circuits, José Carlos Pedro and Nuno Borges Carvalho

Introduction to Modeling HBTs, Matthias Rudolph

An Introduction to Packet Microwave Systems and Technologies, Paolo Volpato

Introduction to RF Design Using EM Simulators, Hiroaki Kogure, Yoshie Kogure, and James C. Rautio

Introduction to RF and Microwave Passive Components, Richard Wallace and Krister Andreasson

Klystrons, Traveling Wave Tubes, Magnetrons, Crossed-Field Amplifiers, and Gyrotrons, A. S. Gilmour, Jr.

Lumped Element Quadrature Hybrids, David Andrews

Lumped Elements for RF and Microwave Circuits, Second Edition, Inder J. Bahl

Microstrip Lines and Slotlines, Third Edition, Ramesh Garg, Inder Bahl, and Maurizio Bozzi

Microwave Component Mechanics, Harri Eskelinen and Pekka Eskelinen

Microwave Differential Circuit Design Using Mixed-Mode S-Parameters, William R. Eisenstadt, Robert Stengel, and Bruce M. Thompson

Microwave Engineers' Handbook, Two Volumes, Theodore Saad, editor

Microwave Filters, Impedance-Matching Networks, and Coupling Structures, George L. Matthaei, Leo Young, and E. M. T. Jones

Microwave Imaging Methods and Applications, Matteo Pastorino and Andrea Randazzo

Microwave Material Applications: Device Miniaturization and Integration, David B. Cruickshank

Microwave Materials and Fabrication Techniques, Second Edition, Thomas S. Laverghetta

Microwave Materials for Wireless Applications, David B. Cruickshank

Microwave Mixer Technology and Applications, Bert Henderson and Edmar Camargo

Microwave Mixers, Second Edition, Stephen A. Maas

Microwave Network Design Using the Scattering Matrix, Janusz A. Dobrowolski

Microwave Power Amplifier Design with MMIC Modules, Howard Hausman

Microwave Radio Transmission Design Guide, Second Edition, Trevor Manning

Microwave and RF Semiconductor Control Device Modeling, Robert H. Caverly

Microwave Transmission Line Circuits, William T. Joines, W. Devereux Palmer, and Jennifer T. Bernhard

Microwave Techniques in Superconducting Quantum Computers, Alan Salari

Microwaves and Wireless Simplified, Third Edition, Thomas S. Laverghetta

Millimeter-Wave GaN Power Amplifier Design, Edmar Camargo

Modern Microwave Circuits, Noyan Kinayman and M. I. Aksun

Modern Microwave Measurements and Techniques, Second Edition, Thomas S. Laverghetta

Modern RF and Microwave Filter Design, Protap Pramanick and Prakash Bhartia

Neural Networks for RF and Microwave Design, Q. J. Zhang and K. C. Gupta

Noise in Linear and Nonlinear Circuits, Stephen A. Maas

Nonlinear Design: FETs and HEMTs, Peter H. Ladbrooke

Nonlinear Microwave and RF Circuits, Second Edition, Stephen A. Maas

On-Wafer Microwave Measurements and De-Embedding, Errikos Lourandakis

Parameter Extraction and Complex Nonlinear Transistor Models, Günter Kompa

Passive RF Component Technology: Materials, Techniques, and Applications, Guoan Wang and Bo Pan, editors

PCB Design Guide to Via and Trace Currents and Temperatures, Douglas Brooks with Johannes Adam

Practical Analog and Digital Filter Design, Les Thede

Practical Microstrip Design and Applications, Günter Kompa

Practical Microwave Circuits, Stephen Maas

Practical RF Circuit Design for Modern Wireless Systems, Volume I: Passive Circuits and Systems, Les Besser and Rowan Gilmore

Practical RF Circuit Design for Modern Wireless Systems, Volume II: Active Circuits and Systems, Rowan Gilmore and Les Besser

Principles of RF and Microwave Design, Matthew A. Morgan

Production Testing of RF and System-on-a-Chip Devices for Wireless Communications, Keith B. Schaub and Joe Kelly

Q Factor Measurements Using MATLAB, Darko Kajfez

Radio Frequency Integrated Circuit Design, Second Edition, John W. M. Rogers and Calvin Plett

Reflectionless Filters, Matthew A. Morgan

RF Bulk Acoustic Wave Filters for Communications, Ken-ya Hashimoto

RF Circuits and Applications for Practicing Engineers, Mouqun Dong

RF Design Guide: Systems, Circuits, and Equations, Peter Vizmuller

RF Linear Accelerators for Medical and Industrial Applications, Samy Hanna

RF Measurements of Die and Packages, Scott A. Wartenberg

The RF and Microwave Circuit Design Handbook, Stephen A. Maas

RF and Microwave Coupled-Line Circuits, Rajesh Mongia, Inder Bahl, and Prakash Bhartia

RF and Microwave Oscillator Design, Michal Odyniec, editor

RF Power Amplifiers for Wireless Communications, Second Edition, Steve C. Cripps

RF Systems, Components, and Circuits Handbook, Ferril A. Losee

Scattering Parameters in RF and Microwave Circuit Analysis and Design, Janusz A. Dobrowolski

The Six-Port Technique with Microwave and Wireless Applications, Fadhel M. Ghannouchi and Abbas Mohammadi

Solid-State Microwave High-Power Amplifiers, Franco Sechi and Marina Bujatti

Stability Analysis of Nonlinear Microwave Circuits, Almudena Suárez and Raymond Quéré

Substrate Integrated Suspended Line Circuits and Systems, Kaixue Ma and Yongqiang Wang

Substrate Noise Coupling in Analog/RF Circuits, Stephane Bronckers, Geert Van der Plas, Gerd Vandersteen, and Yves Rolain

System-in-Package RF Design and Applications, Michael P. Gaynor

Technologies for RF Systems, Terry Edwards

Terahertz Metrology, Mira Naftaly, editor

Understanding Quartz Crystals and Oscillators, Ramón M. Cerda

Vertical GaN and SiC Power Devices, Kazuhiro Mochizuki

The VNA Applications Handbook, Gregory Bonaguide and Neil Jarvis

Wideband Microwave Materials Characterization, John W. Schultz

Wired and Wireless Seamless Access Systems for Public Infrastructure, Tetsuya Kawanishi

For further information on these and other Artech House titles, including previously considered out-of-print books now available through our In-Print-Forever® (IPF®) program, contact:

Artech House
685 Canton Street
Norwood, MA 02062
Phone: 781-769-9750
Fax: 781-769-6334
e-mail: artech@artechhouse.com

Artech House
16 Sussex Street
London SW1V 4RW UK
Phone: +44 (0)20 7596 8750
Fax: +44 (0)20 7630 0166
e-mail: artech-uk@artechhouse.com

Find us on the World Wide Web at: www.artechhouse.com